T0211763

Colonial Transformations

Colonial Transformations: The Cultural Production of the New Atlantic World, 1580-1640

Rebecca Ann Bach

palgrave

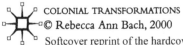 COLONIAL TRANSFORMATIONS
© Rebecca Ann Bach, 2000
Softcover reprint of the hardcover 1st edition 2000 978-0-312-23099-9

First published 2000 by
PALGRAVE
175 Fifth Avenue, New York, N.Y.10010 and
Houndmills, Basingstoke, Hampshire RG21 6XS.
Companies and representatives throughout the world

PALGRAVE is the new global publishing imprint of St. Martin 's Press LLC Scholarly and Reference Division and Palgrave Publishers Ltd (formerly Macmillan Press Ltd).

ISBN 978-1-349-62825-4   ISBN 978-1-137-08099-8 (eBook)
DOI 10.1007/978-1-137-08099-8

Library of Congress Cataloging-in-Publication Data

Bach, Rebecca Ann.
    Colonial transformations : the cultural production of the New Atlantic World, 1580-1640 / Rebecca Ann Bach.
        p. cm.
    Includes bibliographical references and index.
    1. English literature—American influences.   2. English literature—Early modern, 1500-1700—History and criticism.   3. National characteristics, American, in literature.   4. Atlantic Ocean Region—In literature.
    5. Western Hemisphere—In literature.   6. Imperialism in literature.
    7. America—In literature.   8. Colonies in literature.   I. Title.

PR129.A4 B33 2000
820.9'003—dc21   00-040513

A catalogue record for this book is available from the British Library.

Design by Westchester Book Composition

First edition: December, 2000
10 9 8 7 6 5 4 3 2 1

For my father, Sheldon Bach, my beloved and glorious example in life and work.

Contents

List of Illustrations   ix

Acknowledgments   xi

Introduction: Colonial Transformations   1

Chapter 1   Colonial Poetics in Spenser's *Amoretti*   37

Chapter 2   Bermuda's Ireland: Naming the Colonial World   67

Chapter 3   The New Atlantic World Transformed on the
               London Stage   113

Chapter 4   Colonial Transformations in Court and City
               Entertainments   149

Chapter 5   "A Virginia Maske"   191

Epilogue: Late-Twentieth-Century Transformations:
Pocahontas and Captain John Smith in
Late-Twentieth-Century Jamestown   221

Notes   233

Bibliography   259

Index   281

List of Illustrations

Figure 2.1—Robert Tindall's "Draught of Virginia" 88

Figure 2.2—*Nova Virginiae Tabvla* 90

Figure 2.3—Engraved version of Richard Norwood's map of Bermuda 100

Figure 3.1—*A Declaration for the certaine time of drawing the great standing Lottery* 139

Figure 3.2—Title page. Patrick Copland, *Virginia's God be Thanked* 142

Figure 4.1—Sir Thomas Phillips, *A Gennerall Plat of the lands Belonginge to the Cittie of London* 154

Figure 5.1—The internal title page. Ben Jonson, *Masques at Court* in Ben Jonson. *The Workes of Beniamin Jonson* 194

Acknowledgments

I feel very fortunate to have so many institutions, friends, and colleagues to thank in these pages. I wrote two chapters of this book at the John Carter Brown Library, supported by an Andrew W. Mellon Foundation long-term postdoctoral fellowship. Another chapter was completed at the Folger Shakespeare Library, which supported me with a short-term fellowship. My own institution, the University of Alabama at Birmingham, made that research and writing and other work on the book possible. The Graduate School at UAB gave me both a Faculty Research Grant and a Faculty Development Grant for two summers of research at the Furness Library at the University of Pennsylvania. In addition, the Graduate School gave me a grant to cover production costs. I could never have completed this book without the generosity and resources of these institutions.

However, institutions are really people; and though I am grateful for the anonymous support by the committees that approved my grant applications, I feel even more honored to have spent time with the people at these institutions. From the moment I walked into the John Carter Brown Library, I was welcomed. The staff were always ready to answer my questions and help me find what I needed. In addition the community of scholars at the JCB with me was inspiring; many JCB fellows shared references and helped the project in other ways. I must mention particularly Meaghan Duff, Jorge Cañizares Esguerra, David Murray, and Walter Woodward. The JCB's resources are remarkable for a project such as this, and the general atmosphere there was such that reading and writing (and eating) there each day was pure pleasure. The Folger Shakespeare Library also offered incomparable resources as well as the most fantastic reference librarian ever. Without Georgianna Ziegler's assistance with research at the library and with illustrations from a distance, this book would be much poorer. Although I would like to keep the Furness Library at Penn a secret, I must thank the staff of that library who have assisted me year after year and summer after summer. Finding materials at Furness and working in the amazing collection are always great fun, thanks in a large part to the prodigious efforts of John Pollack and Dan Traister and the rest of the library's staff. Thank you all again. Of course, if it were not for my dean and chairmen at UAB, I could never have accepted the fellowships that took me to those great libraries. Ted Benditt, former Dean of the School of Arts and Humanities, gave me leave and financial support so that I could take the JCB and Folger fellowships. My former chairman, John Haggerty, and my present chairman, Lee Person, gave me time off from teaching to research the book and to complete the revisions.

My six years at UAB, minus the eight months I spent at the JCB and the Folger, have been the most productive and delightful time I could imagine at a job. My colleagues and students have given me more than I can ever hope to return. I owe many professional and personal debts to them, and any acknowledgment here can go only a very short way toward paying them. The faculty of the English department as a whole deserves my thanks for their support and encouragement. It is very hard to single out colleagues in such a uniformly excellent faculty, but I must thank Kyle Grimes for intelligent conversations and rides, Marilyn Kurata for her inspiring example, her decency, and her listening skills, Bob Collins for many basketball conversations, and Elaine Whitaker my consulting medievalist and such a good friend in need. Jim Mersmann has been a great hall neighbor and buddy. I feel very lucky to have had Cassandra Ellis as both a colleague and friend; I can't see how I would have gotten through the final revisions without her. Our administrative associate, Juanita Sizemore, has facilitated my work every step of the way. Finally on the English department front, much of my work is made both possible and enjoyable by Amy Elias. Thank you for walks and talks and your unfailing interest in my research. Thanks are also due to the students in my Winter 1999 honors seminar, "From Columbus to Pocahontas: Colonial Texts and Postcolonial Theories," especially Molly Hurley and Jillian Van Ells.

Friends and colleagues at other universities generously gave their time and efforts to make this book much better than it would otherwise have been. Any infelicities that remain may be attributed wholly to me. However, many of the book's virtues are the virtues of its astute readers. The following people read sections of the manuscript at one time or another and prompted me to make them better: Margreta De Grazia, Myra Jehlen, Beth McGowan, Phyllis Rackin, and Peter Stallybrass. In addition to reading chapters, Phyllis Rackin offered my family her beautiful house for a summer of work at Furness. The book benefited immeasurably from two scholars who read it from cover to cover. Jean Howard, the reader for St. Martin's and one of the best readers in the world, gave me invaluable advice. Alison Chapman read the book chapter by chapter and draft by draft. This book is much easier to read because of her reading and writing skills; in addition Alison cheered me on completely unselfishly.

In the last stages of copyediting the book, I was fortunate enough to be a participant in the National Endowment for the Humanities summer institute, "Texts of Imagination and Empire: The Founding of Jamestown in its Atlantic Context." I learned an enormous amount in the institute, especially from my erudite and generous co-participants. The institute both challenged and confirmed my conclusions in the book, and the book is surely the better for it.

My editor at Palgrave, Kristi Long, has been patient and encouraging. She is a pleasure to work with. Copy editor Enid Stubin's emendations were timely and very useful, and the production editor, Donna Cherry, has answered all of my many questions graciously. Suzanne Daly did a brilliant job on the index. I am very fortunate to have had such professional, essential help with the manuscript.

I am fortunate, as well, to have a group of friends (some academics, some not) who gave me the love and support I needed to keep writing and whose influence can be felt in this book. I am very grateful to Marni Bonnin, Louise Cecil, Anne Cubilié, Suzanne Daly, Deborah Feingold, Judy Filc, David Golumbia, Elaine Jacobs, Vance Lehmkuhl, Mary Janell Metzger, Carol Jones Neuman, Nancy Sokolove, Marc Stein, and Cynthia Way for being there for me. Each of you is aware, I hope, that no mere mention can express my obligations and my joy in your friendship. I am grateful, also, for the scholarly example of Kim Hall and Jeffrey Masten. Beth McGowan listens to me forever and shares all of my sorrow and happiness. Deborah, Vance, and Beth's feminism and passion for social justice continually inspire me.

Nothing that I have accomplished in my scholarly or personal life could have been done without my family. My parents, Sheldon Bach and Phyllis Beren, have supported me every step of the way; I know that my brother Matthew will always be there for me. My enormous, joyous debt to my father is reflected in the dedication. I am not sure that I could ever express verbally what I owe to my partner in life, Brendan Helmuth. His love covers everything. Finally, my other companion over the last six years has been our wonderful daughter, Julia. I could not imagine a better family. I am very blessed.

Earlier versions of sections of chapters 3 and 4 appeared as "Ben Jonson's Civil Savages," *Studies in English Literature, 1500–1900* 37 (1997): 277-293, reprinted by permission of the Johns Hopkins University Press; and "'Ty good shubshects': The Jacobean Masque as Colonial Discourse," *Medieval and Renaissance Drama in England* 7 (1995): 206-223, reprinted by permission of Fairleigh Dickinson University Press.

Introduction
Colonial Transformations

> *The* Golden Age *was first* . . .
> *To visit other Worlds no wounded Pine*
> *Did yet from Hills to faith les Seas decline*
> *The un-ambitious Mortals knew no more*
> *But their own Country's Nature-bounded Shore.*
> (Ovid's Metamorphosis Englished by George Sandys *l.48,*
> *94-97)*

In Shakespeare's *Two Gentlemen of Verona*, probably written to be performed in the early 1590s, a Veronese nobleman, Antonio, discusses his son Protheus's education with his servant Panthino. Panthino informs his master that Antonio's brother has

> wondred that your Lordship
> Would suffer [Protheus], to spend his youth at home,
> While other men, of slender reputation,
> Put forth their Sonnes, to seeke preferment out.
> Some to the warres, to try their fortune there;
> Some, to discouer Islands farre away:
> Some, to the studious Vniuersities;
> For any, or for all these exercises,
> He said, that *Protheus* your son was meet. (TLN 306-14, 1.3.4-12)[1]

Antonio responds that he himself has been deliberating how best to educate his son: "I haue consider'd well, his losse of time,/ And how he cannot be a perfect man,/ Not being tryed, and tutord in the world" (l. 321-3,

1.3.19-21). Also in the early 1590s, between 1590 and 1596, Edmund Spenser published his mammoth though incomplete Protestant epic *The Faerie Queene,* the "generall end" of which, the poet writes, "is to fashion a gentleman or noble person in vertuous and gentle discipline" ("A Letter of the Authors" 15). Book five of Spenser's epic indicates that this proto-typical perfected gentleman or nobleman must learn, under the aegis of "Justice," to subdue native populations, if necessary with brutal force, in order to protect the knightly code to which *The Faerie Queene* subscribes.[2]

Both Shakespeare's and Spenser's texts worry the problem of gentility: what constitutes a gentleman. Both also collapse gentility and nobility, pre-sumably at least partially in the interest of their respective authors' social ambitions. The two texts treat that problem differently, and the play has a much more limited view of it; however, both texts, *Two Gentlemen* in Anto-nio's conversation with Panthino and *The Faerie Queene* in books five, six, two and elsewhere, indicate one of this book's central concerns. As early as 1590, at least one prominent, even definitional, English identity—"the gentleman"—was being formed in relation to England's colonial experi-ences.[3] If we take the problem of the Veronese gentleman that Shake-speare's text stages as a convenient screen for the problem of the English gentleman,[4] it becomes clear that, in addition to a university education and to experience as a soldier, the discovery and exploration of the "New World" was becoming a vital part of the English gentleman's education. In fact, Sir Philip Sidney, the prototypical nobleman, had "New World" dis-covery ambitions, as well as a university education and soldierly experience. Panthino's image of perfection describes that failed courtier well.[5] The "perfect man" must experience at least one of these world-broadening activities, and preferably all: he must investigate the knowledge produced in the "old world" (and the classical world), must advance that world's mili-tary causes, and must also extend its boundaries.

Unlike Shakespeare's play, Spenser's epic makes explicit the parameters of that discovery and exploration. In its episodes at Acrasia's bower, at Munera's castle, in the Amazons' land, and other locations, *The Faerie Queene* exposes what Panthino's instruction screens: that discovery requires encounter, in fact that "discover[ing] lands" and going "to the wars" may be one and the same world-expanding activity. While Panthino's "far away" islands are sanitized locations for the gentleman's exploits, existing to serve as his stage, Spenser's mounted knights face the inhabitants of the lands they enter and may even acknowledge, at least temporarily, their foes' power and alternative imaginations. But Spenser's text also makes the same point as Shakespeare's about the English gentleman: that the "perfect man" should be prepared to extend England's boundaries, that the English gentle- or nobleman is a discoverer, even an imperialist. The 1590s may seem too early to discuss an imperial England. Indeed, at least one important recent book contends that late-sixteenth-century England conceived of itself as an isolated island. That book, Jeffrey Knapp's *An Empire Nowhere*, emphasizes England's very real ineptitude in Virginia and argues that "by making the case for empire within the very terms of isolationist rhetoric, otherworldly imperialism proved especially adept at idealizing the colonial failures that kept England insular" (12).[6] In contrast, I argue that in each of England's extensions outward, as early as its 1536 annexation of Wales, it became involved in the process I call colonial transformation, a process that redefined the territory and people the English encountered, but also importantly refigured the territory and people of the metropolitan center. *The Faerie Queene's* constitutive fantasy is that its knights can encounter difference and emerge as purified versions of their intrinsic virtues; yet were Shakespeare's Protheus a real English gentleman who followed the course Antonio's brother recommends, he would emerge irrevocably transformed by his encounters, as his name might indicate. As this introduction will show, England's colonial enterprises helped to define an Englishness that was always a constructed category. English gentility and commonness were defined sometimes together against Irish and Indian otherness and sometimes against each other in the face of colonial transformation; and Englishness and savagery were both defined in the colonial crucible of the new Atlantic world.

The *Oxford English Dictionary* offers two related early modern usages of the verb "transform," both of which bear on this book's vision of England's colonial process: definition 1, a usage the dictionary finds as early as 1340, "to change the form of: to change into another shape or form; to metamorphose"; and 1b, from as early as 1556, "to change in character or condition; to alter in function or nature." In its focus on colonial transformations, this book observes and comments on colonial changes in shape and form and colonial changes in function and nature and, of necessity, it shows the interrelations of those species of change. England's interactions with Ireland, Virginia and Bermuda between 1580 and 1640 changed all four locations' shapes and forms, most basically by defining and redefining their boundaries and human settlements. And those interactions also changed those locations' characters and conditions, as human settlements and as ecological environments, as this book will show. For example, when the English government of Elizabeth I seized the Earl of Desmond's lands in 1583, it remapped southwestern Ireland as the Munster plantation;

it also attempted (unsuccessfully) to change that part of the island's character from an Old English seignory populated by Gaelic and old English farmers to a New English plantation with only English settlers, and in that attempt confronted a colonial settlement's inevitable infiltrations by its escheated land's native community. In colonial interactions, English, Irish and Native American people were transformed in both form and character. As Antonio and Panthino's discussion indicates, it was in the crucible of war, discovery, and education—all interrelated in early modern England—that the gentleman was formed. The *OED*'s equivocal verbs also record the essential instability of colonial processes, for "to change," "to metamorphose," and "to alter" can indicate both what an agent does to an object and what happens to a subject. Ovid's *Metamorphoses*, with its stories of the gods' agency and their subjects' fates, makes this clear, when, for example, Jove's transformation of Calisto into his rape victim entails his own transformation into her goddess Diana.

Of course, each of these categories of people—English, Irish, and Native American—and each of these locations—England, Ireland, Bermuda, and Virginia—signify real people and places and are simultaneously constructed and imaginary; it is the colonial process that helped to transform them into the terms late-twentieth-century readers so easily recognize. As Benedict Anderson argues, "the nativeness of natives is always unmoored, its real significance hybrid and oxymoronic . . . Nationalism's purities (and thus also cleansings) are set to emerge from exactly this hybridity . . . [and] it was . . . through print moving back and forth across the ocean that the unstable, imagined worlds of Englishness and Spanishness were created" (316). Thus in Ireland in 1580, there were Gaelic people, old English settlers, and new English settlers; each of these groups was made up of children, women, and men; each was divided by rank and status; and each of these groups had its own ideas about who was Irish and who was English. Virginia was home to many different native groups including the Chickahominy, Nansemond, Pamunkey, Monacan, Powhatan;[7] these groups were also made up of women, men, and children, and had their own status divisions. The English[8] who landed in Virginia and in the uninhabited Bermuda islands were likewise deeply divided by rank, gender, religion, age, riches, and place of origin, as were the "adventurers" and "undertakers" who sent them over the ocean. Colonial transformation was a process that produced these divided groups as "savages" and "English," and it was also a process to be feared and resisted when it threatened those groups' sanctity.

While colonizers attempted to and did transform the environments and people they encountered, they were themselves transformed by those encounters. However, as my preliminary discussion of social divisions indicates, the term "colonizer," as well as signifying a real and devastatingly powerful group, is as much a constructed and imaginary category as the term "Englishman."[9] Ann Laura Stoler cautions that in the anthropology of colonialism, for the most part, "the makers of metropole policy become conflated with its local practitioners. Company executives and their clerks appear as a seamless community of class and colonial interests whose internal discrepancies are seen as relatively inconsequential, whose divisions are blurred" (1989, 35). And she also notes that "anthropologists have taken the politically constructed dichotomy of colonizer and colonized as a given, rather than as a historically shifting pair of social categories that needs to be explained" (136).[10] Of course these theoretical blindnesses are not those of anthropologists alone; we literary critics and cultural studies practitioners pay attention to the hybridity of the native subject but too often see the colonizer as a monolithic representative of governmental or technological power, rather than as a constructed category painfully maintained in order to preserve English power in the face of appealing or divisive difference.

As England's experiences in Virginia reveal, the colonized are also a constructed category. In 1609, the year the Virginia Company made its first public stock offer and the year Sir Thomas Gates, Sir George Sommers, William Strachey and others set out for the colony and accidentally landed in the Bermuda islands, Richard Hakluyt, the younger, translated and published a mid-sixteenth-century description of Fernando de Soto's expedition into the parts of the United States we now call Florida, Georgia, the Carolinas, Tennessee, Mississippi, and Texas. Hakluyt called his translation *Virginia richly valued, By the description of the maine land of Florida, her neighbor,* and he dedicated the text to the Virginia plantation's counselors and adventurers, writing, "This worke . . . doth yeeld much light to our enterprise now on foot: whether you desire to know the present and future commodities of our countrie; or the qualities and conditions of the Inhabitants, or what course is best to be taken with them" (A2). Reading Hakluyt's translation now, one might find it difficult to see how the many Indian[11] groups de Soto and his men encountered in their travels could be generalizable into one group of "inhabitants," from the encounter with which Virginia adventurers could determine any singular course of action. De Soto's group confronted "warlike" and peace-loving Indians, taller and shorter groups, Indians governed by women and Indians governed by men, and the Spanish employed groups of Indians against each other (O2v). But Hakluyt's dedication shows that this prominent colonial mover and a considerable number of Virginia adventurers could easily make this and a similarly destructive mental leap: somehow in the face of massive evidence to the contrary, they could see the native people encountered and to be encountered in Virginia as one undifferentiated group, essentially the same as all other Indian groups everywhere.

Since the only texts we have from the Virginia encounter were authored by English men, these foundational assumptions make the scholarly work based upon them repeat their terms, collapsing the at least thirty-six native groups in seventeenth-century Virginia into one generic group of "Indians."[12] As Frantz Fanon observes, "colonialism is not satisfied merely with holding a people in its grip and emptying the native's brain of all form and content. By a kind of perverted logic, it turns to the past of the oppressed people, and distorts, disfigures, and destroys it" (154). This book maintains and details differences between Gaelic Irish, Old English and new English inhabitants of Ireland; but it also exposes both their hybridity and English fears that these groups have collapsed or will collapse into one savage Irishness. It also exposes divisions between English metropolitan people and English settlers, and divisions within those groups, as it details English fears of internal transformation, that are concurrent with some adventurers' and settlers' struggles to construct a coherent English identity. This book can successfully analyze those differences because to a greater or lesser extent it has sources written by members of each of these groups. But *Colonial Transformations* is necessarily and unhappily less successful at documenting Native American group identities stripped away in the process of colonial transformation, often reduced at best to a name—and, as we will see in chapter 2, often an inaccurate name. While the Virginia Company Records and (more often) texts published by colonists distinguish Indian groups, especially when one native polity resists English encroachment more than another or when one group may be enlisted against another, many of the texts refer to generic Indians, as do most late-twentieth-century popular American texts.[13] While resistance to this hegemonic understanding is imperative, the sources' biases frequently make it impossible. And this very impossibility points to the potential difficulties with theories of hybridity that obscure essential power differentials between groups.[14] Thus in this book, as well as attending to the complexity of "the colonized" and "the colonizers" as terms and as actual signified groups, I will carefully mark power differences that transcend hybridity to freeze difference in the interest of violent oppression—all-too-common instances in which the colonizer and the colonists become real and significant constructed categories.

Seventeenth-century English people as well as late-twentieth-century observers recognized the pathos and stupidity of Hakluyt's assumption that all Indians are the same. Writing in 1623, a Virginia settler, Peter Arundell, blamed his English cohorts for the violence they provoked:[15]

> *Wee our selues haue taught them how to bee trecherous by our false dealinge with the*
> *poore kinge of Patomeche that had alwayes beene faythfull to the English, whose peo-*
> *ple was killed hee and his sonne taken prisoners brought to Jeames towne, brought*
> *home agayne, ransomed, as if [he] had beene the greatest enemy they had: Spilmans*
> *death is a just reuenge,* it was done about *that part of the Country:* If wee had
> sufficient prouision, wee should not neede to seeke after the Indians: It is a
> great loss to vs for that Cap. was a grea [a grea canc.] {the best} linguist of
> the Indian Tongue of this Countrys. (*Records* IV 89)

Arundell refers here to the official English policy, after the 1622 Indian attack—what has been labeled the 1622 massacre—of using "any means necessary" to punish the Indians for the "massacre" and to force surrounding Indian groups to supply the colony with corn (*Records* III 672, 697, IV 9). During one of those attacks, Englishmen acting under the authority of their colonial government had killed a number of Patawomeke Indians and kidnapped the "king of Patomack" and his son, disregarding the Patawomeke-English alliance. Probably in response, some Patawomeke Indians executed Henry Spelman, one of the colony's translators. As Arundell noted, the English policy of not distinguishing among Indians inspired enmity on the part of former allies, and it also disabled English communication with Indian allies and enemies. This policy could be interpreted as a deliberate attempt to wipe out the Indian population surrounding the colony,[16] but the extant English records, including Arundell's fragmentary letter, make it look as much like a desperate search for provisions combined with a self-destructive inability on the part of governing settlers to recognize the differences that could have saved English lives and that their own actions have erased from history.[17] In their violence, those English settlers froze the Indians around them into the category of "the colonized," denying difference and consolidating themselves as "colonists" in that process.

As this book attempts to add a new perspective that avoids homogenizing native groups and thereby forming a single "colonized" population, it also looks at the imported metropolitan divisions that must lead us to question the category "colonist" when it is applied as a blanket term for English settlers. The governor of Sir Walter Raleigh's short-lived Roanoke colony, Ralph Lane, opens his August 12, 1585 letter to Sir Philip Sidney, "Albeyt in the myddest of infynytt busynesses, as hauyng, emungst savvages, the chardge of wylde menn of myne owene nacione, whose vnrulynes ys such as not to gyve leasure to yhe gardes to bee all most att eny tyme from them" (17). Lane's colonial government has a double mission: to civilize the "wylde menn" of his "owene nacione" as well as the "savvages" surrounding them. Thirty-nine years later, on March 30, 1624, Jane Dickenson petitioned Governor Francis Wyatt of the Virginia plantation and his council:

> The humble Petition of Jane Dickenson Widdowe
> Most humblie sheweth that whereas her late husband Ralph Dickenson Came ouer into this Cuntry fower Yeares since, obliged to Nicholas Hide deceased for the tearme of seauen yeares, hee only to haue for himselfe & your petitioner the one halfe of his labors, her said husband being slaine in the bloudy Masacre, & her selfe Caried away with the Cruell saluages, amongst them Enduring much misery for teen monthes At the Exspiration it pleased God so to dispose the hartes of the Indians, that for a small ransome your petitioner with diuers others should be released, In Consideration that Doctor Potts laid out two pound of beades for her releasement, hee alleageth your petioner is linked to his seruitude with a towefold Chaine the one for her late husbandes obligation & thother for her ransome, of both which shee hopeth that in Conscience shee ought to be discharged, of the first by her widdowhood, of the second by the law of nations, Considering shee hath already serued teen months, tow much for two pound of beades
> The promises notwithstanding Dr Pott refuseth to sett your peticioner at liberty, threatning to make him serue her [sic] the vttermost day, vnles shee procure him 150li waight of Tobacco, shee therfore most humbly desiereth, that you wilbe pleased to take what Course shalbe thought iust for her releasement fro' his seruitude, Considering that it much differeth not from her slauery with the Indians, & your peticioner shalbe bound to pray etc.
> (*Records* IV 473)

Can Lane's "wylde men" be classified with Lane as colonists? In the dispute between Dr. Potts and Jane Dickenson, who is the colonist? Or to put the question another way, can Jane Dickenson, Dr. Potts, Ralph Lane, and the "wylde menn" be thought of as one and the same English colonist, divided as they are by gender and social status?

Living in Virginia at a time when the Indians dwelling with them were not yet subjected to the English settlers, those settlers were also deeply divided within themselves, so much so that Jane Dickenson found it difficult to choose between servitude to her English social superior and "slauery with the Indians." Indeed she compares Potts's behavior, his refusal and his threats, unfavorably with the Indians' easily moved hearts, although she attributes that movement not to their natures but to God's influence. By implication, Potts—presumably John Potts, an important member of the Virginia council—responded less readily to God's will than did her original kidnappers. At the same time, Dickenson's category, "Cruell Savages," and Potts's ransom of her for two pounds of beads both indicate a colonial relationship between two groups, although the colonial relationship signified by the exchange of beads—what Knapp would term "trifles"—for people is complicated by Potts's value system. One thing that distinguishes the English from the Indians in the colonial literature is the "true" understanding of value that the colonizer holds: beads are essentially worthless but can be exchanged with the feckless "savages" for corn and other useful commodities.[18] This is the topos behind Othello's characterization of the "base *Indian*"[19] who throws away a pearl of great worth. But in Dickenson's petition, it is Potts who overvalues the beads, assessing them at "150li waight of Tobacco," a considerable sum in Virginia's tobacco economy and one clearly out of Dickenson's reach.

The "bloudy Masacre" Dickenson refers to was the concerted Indian rebellion of 1622 that prompted the policy that in Arundell's opinion got Henry Spelman killed. Peter Hulme has brilliantly shown that that rebellion enabled the English to carry out a "justified" campaign of extermination against the groups that had been feeding them and whose resources the expanding settlement was massively straining (160-68). After the Indians in the settlement areas joined together to rebel, Governor Wyatt authorized numerous raids to pillage, burn, and exterminate native crops and people indiscriminately, such as the raid in which the English captured the "king" of the Patawomeke and his son; many settlers, despite their different ranks, assented to an identity formed in total opposition to the Indians. Letters written to English people in England by lower middling settlers and their superiors use a rhetoric of innocent English against vicious savages,[20] but the Virginia Company records also indicate the instability of that newly consolidated English identity in the colony.

Writing to Mr. Farrer (John Ferrar), a prominent metropolitan adventurer, in March 1622/3, George Sandys, Sir Edwin Sandys's brother, complained of Sir William Nuce's arrival in the colony "with a very few of weake and vnseruiceable people." After Nuce's death, Sandys relates, "11 men were all that remained for the Company, whom for want of prouision I was enforced to sell . . . Fower that were left one my handes, I was fayne to send to my owne plantation; two of these a little after ran away (I am afraide to the Indians) and noe doubt the other two would haue consorted with theire companions if sicknes had not fettered them" (*Records* IV 22).

Thus although the "massacre" did provide English rhetoric, as Hulme argues, "a huge infringement of Natural Law which left its victims free to pursue any course they wanted, unregenerate savagery having forfeited all its rights, civil and natural," it did not succeed in convincing all people from England that the Indians were any more savage than their English masters (172). And some English people's desire to run to the Indians seems reasonable given their superiors' attitudes towards them. Writing in defence of his and Sir Thomas Smith's government of the colony until 1619, Alderman Robert Johnson declared that the colony had achieved prosperity "with the losse of a very fewe of his maiesties Subiectes (those alsoe beinge People for the most parte of the meanest Ranke)" (*Records* IV 4). Those mean English people, characterized by their social superiors as "weake," "unserviceable," and "ragged," might well prefer to live among the Indians than to form part of an English settler society defined against the natives.[21]

As Stoler's caution indicates, we cannot see even the social superiors—noble men and women and rich merchants—involved in England's colonial enterprises in Virginia as a homogeneous group of colonizers with one coherent colonial ideology. Interestingly, the 1622 "massacre," as well as consolidating an English identity defined against the ruthless "savages," exposed the divisions between the Virginia Company in London and the council in Virginia. Writing to that council in August of 1622, the treasurer and counsell for Virginia in London "grieved" that the colonists brought the attack upon themselves and advised them to attend "to the humble acknowledgment and perfect amendment whereof together with our selues, we seriously advise and invite you; and in particular earnestly require the speedie redresse of those two enormous exesses of apparell and drinkeing" (*Records* III 666). According to the company in London, those two moral faults along with "the neglect of the Devine worshipp" caused the Indians to prevail, more than the colonists' "weaknes" (*Records* III 666). Thus the metropolitan adventurers quickly moved to accuse the colony's governors of having insufficiently disciplined their people.

For the sumptuary complaint, the Company probably relied on one of John Pory's letters home in 1619, which had accused lower-ranked settlers of transcending their status by wearing their betters' apparel:

Nowe that your lordship may knowe, we are not the veriest beggers in the worlde, our Cowe-keeper here of Iames citty on Sundayes goes acowterd all in freshe flaming silkes and a wife of one that in England had professed the black arte not of a scholler but of a collier of Croydon, weares her rough beuer hatt with a faire perle hattband, and a silken suite thereto correspo[o canc.]{e}ndent. (*Records* III 221)

Pory was the council in Virginia's secretary. His ironic commentary heaped disdain on his inferiors: a keeper of cows who thought his profession gave him the right to "acowter" himself in the fabric (silk) and color (crimson) of a nobleman, and a blackened man, who came by his color digging coal but whose wife equated his blackness with the black arts of the scholar and decorated her "rough" animal hat with a noblewoman's jewels. In the social system Pory supported, this man and woman belonged with the animals and dirt they tended, but in the "free" space of the colonies they paraded their equality with their betters, many of whom were starving and dying as quickly as they were.

Responding to potential transgressions of the status system by commoners adopting fabrics and jewels reserved in England for the nobility, the Virginia council passed its own sumptuary law forbidding anyone "but the council and heads of hundreds [plantations] to wear gold in the cloaths, or to wear silk till they make it themselves" (quoted in Brown 89, from William Waller Hening, ed., *The Statutes at Large; Being a Collection of All the Laws of Virginia* 13 vols., 1823; reprint Charlottesville 1969, July 24, 1621, I , 114). This law reveals the settlement's internal divisions and the highly ranked colonists' fears that their servants, apprentices, and workmen might dissolve the differences between them and their masters; but it also reveals the ruling class's inability to get their inferiors to produce goods for them, in this case the silk that the company at home and their ultimate master, King James, desired as a colonial commodity. The law, somewhat despairingly, holds out the hope of sumptuary advancement for any cowkeeper industrious enough actually to produce the silk he might legally only admire on his betters' backs. Commoners should not attempt to transform themselves into their betters unless they can transform England's colonial economy to benefit their betters. The colony's 1619 sumptuary law employed the same status carrot as the Shakespearean King Henry V, who in the eponymous play promises his troops that fighting the French against overwhelming odds will make his troops into gentlemen and the king's brothers.

The Virginia Company's members used the settlement's internal division to shift the blame for the "massacre" off themselves, since they anticipated that the settlers and the English public, informed of the colony's poor health and lack of supplies by disgruntled returning settlers and disaffected colonists' letters home, would accuse them of poorly protecting and badly supplying their English abroad—the situation the Company obliquely refers to as their "weakness." Writing from London in 1612, one of the company's most vehement promoters, Alderman Johnson, had attempted to cut off such accusations:

> And as for those wicked Impes that put themselues a shipboard, not knowing otherwise how to liue in England; or those vngratious sons that dailie vexed their fathers hearts at home, and were therefore thrust vpon the voyage, which either writing thence, or being returned back, to couer their owne leaudnes doe fill mens eares with false reports of their miserable and perilous life in *Virginea*, let the imputation of miserie be to their idlenes, and the blood that was spilt vpon their owne heads that caused it. (C2)[22]

The Company's officials assumed that ten years later those "wicked impes" and "vngratious sons" along with newly minted, angry, former, and current settlers, would blame the London Company for the colony's misfortune, and they again used the strategy of attack against their internal detractors. Depicting the colonists as parading around Virginia drunk, godless, and dressed in their betters' clothing, they shifted the blame from their metropolitan masters. In response, the council in Virginia wrote that English people at home supplied the English in Virginia with alcohol, and that instead of being overdressed, the English in Virginia lived in "povertie and nakedness," a gibe at the Londoners for sending unfurnished colonists and poorly supplying the colony (*Records* IV 11). The council in Virginia also accused the Virginia Company's colonial "instructions" of causing the attack, for the company had formerly told them, they argued, to let the Indians "Cohabitt with us." According to the council, the company was at fault for turning the Indians into "secrett Enemies that live promiscouslie amongst [them], and are harbored in [their] bosomes" (*Records* IV 10). That policy had long been a matter of dispute within the settlement itself. In the wake of the 1622 attack, then, fissures in both colonial policy toward the Indians and relations between the enterprisers in England and in the settlement threatened any solidified English colonial identity.

In the Bermudas, unlike in Virginia, English settlers did not have to contend with a native human population, but divisions between colonizers were no less apparent. A prominent metropolitan backer of the Bermuda (Sommer Islands) colony, Sir Nathaniel Rich, in his notes on the tobacco contract of 1622, records his fears that those settlers might revolt from the company and crown's rule altogether:

> And for Virginia Tobago when it comes it may please his maiestie to take what course he please. But it were fitt that this parcell which is the Whole

cropp of the Sommer Ilandes should be first sould least through want and
discontent they should reuolt and then extreame Inconveniences would
ensue. For His Maiestie should not only loose the strongest fort of Chris-
tendom, and place of singular importance to fasten his alliance with the K.
of Spaine in respect of their Neighbourhood to the W. Indies, but if they
should reuolt to the Pyrattes they would as well infest his Maiestie as his
Allyes: strength ["strength" canc.] aduance the strength of the Pyrattes to and
make them more fe [from "to" to "fe" canc.] to the destruction of {merchant
&} Merchandize . . . And lastly euen to the plantacion of Virginia the reuolt
of these Ilandes would giue a deadlie and a fatall blow. (*Records* IV 29)

Perhaps founded in his fear of losing his considerable investments in
Bermuda, Rich's apocalyptic vision of the Sommer Ilands' settlers
"reuolt[ing] to the Pyrattes" feeds metropolitan fears that, distant from their
noble masters, English settlers might transform themselves into England's
enemies. Richard Norwood's 1617 map of Bermuda, which chapter 2 discusses at length, had poised the settlement equidistant between Virginia
and the West Indies. Strategically placed, then, the settlers were always on
the brink of becoming either Spanish or pirates, both identities posing
threats in every direction—to the European balance of power,[23] to English
ships and the growth of merchant enterprise, and to the struggling Virginia
colony. In Rich's perhaps self-serving vision, the Bermuda settlers are only
tentatively *English* colonists, firmly English only if the crown would firmly
and generously support their agricultural efforts. His notes show that Rich
was prepared to warn the king that he would have to bribe Englishmen to
remain his subjects.

Even in the more heavily supervised Virginia colony, governors could
not insure that English men would remain firmly English. Writing in
response to accusations that Sir Thomas Dale's early administration of the
colony was too harsh, Ralph Hamor commented that Dale needed to suppress brutally the evolving network of English traitors and conspirators in
the colony in 1609:

I see not how the vtter subuersion and ruine of the Colony should haue bin
preuented, witnesse Webbes and Prises designe the first yeere, since that
Abbots and others more daungerous then the former, and euen this summer,
Coles and Kitchins Plot, with three more, bending their course towards the
Southward, to a *Spanish Plantation* reported to be there, who had trauelled (it
being now a time of peace) some fiue daies iorney to *Ocanahoen*, there cut

off by certaine Indians, hired by vs to hunt them home to receiue their deserts. (27)

Hamor's discourse again complicates our terms "colonist" and "colonized." In which position can we put the English rebels Coles and Kitchins? In which can we put the Indians that the English governors employed to catch them? In addition, Hamor's rebuttal to Dale's enemies illustrates the Virginia government's difficulty with securing the colony members' loyalty to their colonial government and the English crown. Even defection to England's long-term European enemy Spain might be preferable to life in the colony for those not in power there.

The specter of rebellion in a distant land lived always in metropolitan adventurers' thoughts. Writing a promotional tract for the Virginia plantation in 1609, Johnson assured his readers that the colony would recruit men "of euery trade and profession . . . which will be glad to goe, and plant themselues so happily and their children after them, to hold and keepe conformity, with the lawes, language and religion of England for euer" (C4v). Johnson's hyperconfident tone is belied by his insistence on "conformity," which betrays the terror of settlers' transformation into lawless, Algonquian or Spanish-speaking atheists or Catholics, a group Johnson wanted carefully excluded from the plantation.[24] Assuring potential adventurers that not only these original settlers but also the settlers' children would remain law-abiding, English-speaking Anglicans, Johnson warded off the frightening possibility that colonial transformation might be particularly potent in the first and subsequent Creole generations, a fear that I discuss in chapter 1 in relation to Spenser's progeny.

Even the Company in London, of which Rich was a prominent member, was deeply divided as to policy in Virginia and leadership at home. The Company was divided into factions—Alderman Johnson, and Sir Thomas Smith's (Johnson's father-in-law and the Company's first treasurer) group; Sir Nathaniel Rich and his brother-in-law Robert, Earl of Warwick's group and the Ferrar family; Sir Edwin Sandys (the Company's second treasurer), and Southampton's (Shakespeare's patron and treasurer of the Company from 1620 to 1624) group in London. And those factional struggles finally caused James I to dissolve the Virginia Company in 1624 and take the colony into royal hands.[25]

Surveying Rich's fears, George Sandy's complaints, and Dickenson's petition for a theoretical understanding of how we must complicate the terms "colonizer" and "colonist" would seem to have taken this introduction far from its beginnings in Shakespeare's *Two Gentlemen of Verona*, but all of these texts signal both early modern England's social divisions and how those social divisions became exacerbated in that culture's colonial expansion. In Rich's and Sandy's worries, we see highly ranked English men complaining that the middling and lower orders might ally themselves with Indians and Pirates and so, rather than consolidating an imperial England, dissolve it from its margins. Dickenson's petition demonstrates one reason why such an escape and rebellion might be both sought by an oppressed English subject and feared by her social superior. It also indicates how gender might influence activity on the margins of this expanding culture, another pressing concern of this book. Shakespeare's play points us to divisions in that culture's upper ranks, as Panthino's discourse envisions nebulously ranked "other men, of slender reputation" whose sons might advance beyond Proteus by "discover[ing] islands" while Antonio's son languishes at home. Those other slenderly reputable men's dreams of social advancement in the colonial field[26] were the dreams of such colonists as Edmund Spenser and Captain John Smith, a Virginia colonist whose fantasies are the subject of chapter 5. Like Spenser, Smith was a man who dreamed both of his own transformation at England's margins and of the complete transformation of the native cultures he confronted.

As the complication of the term "colonizer" indicates, however, Smith's changing colonial policies toward the native population were only one of a series of competitive transformative visions. In May 1609, for example, the Virginia Company issued its "Instruccions Orders and Constitucions to Sir Thomas Gates knight Governor of Virginia." Among many other directions in a lengthy document, the company instructed Gates to "endeavour the conversion of the natiues to the knowledge and worship of the true {god} of [canc.] and their redeemer Christ Jesus, as the most pious and noble end of this plantacion, which the better to effect you must procure from them some convenient nomber of their Children to be brought vp in your language, and manners." To further that pious end, the company directed Gates to remove the Indians' "Iniocasockes or Priestes" and even to kill them as necessary (*Records* III 14-15). The Company here seems to envision a Christian native population, a noble end in their terms. Without regret, the Company envisions a complete transformation of those societies—the removal of the Indians' children, their indoctrination in both the English language and English modes of behavior (clearly inseparable here as in Johnson's New Britain), and the total destruction of native religion. In this vision of transformation, Virginia's Indian cultures would be remade as English culture, as a people speaking English, behaving as the English do, and believing in and obeying the English God's dictates.

The instructions ignore the fact that the English themselves were far from unified in language, manners and, most especially and fatally, in religious beliefs, so that this transformation was fantastic on more than one level: it contained a fantasized English union as well as a fantasized English Powhatan culture. As Ashis Nandy has suggested, "colonizers . . . came from complex societies with heterogeneous cultural and ethical traditions . . . it is by underplaying some aspects of their culture and over-playing others that they built the legitimacy for colonialism" (12). The Virginia Company's instructions to Gates pretend that the people living on the English island had agreed on what the "knowledge and worship of the true God" might mean. But Johnson's *Nova Britannia*, written the same year as those instructions, belies their confidence in an English unity upon which to base colonial power. Johnson makes his own virulent anti-Catholic stance clear, apotropaically invoking a vision of England undermined by its Catholic minority. Associating all Catholics and recusants with the 1605 Gunpowder Plot, he asks, "How like you these Catholikes and this divinity? If they grow so bold and desperate in a mighty settled state, howe much more dangerous in the birth and infancy of yours?" (D2v). *Nova Britannia* makes it clear that only certain members of England's polity will be counted as English for colonization purposes, but it also reveals that the Company could define English manners, language and religion only in contrast to those of the alien native population, and then by denying England's own differences.

However insecure the Company's vision of an unified England on which it could base its policies, its fantasized transformation of Indians in Virginia into this extended "New England" was even less plausible, as Gates's other instructions reveal. While the English government might require its English subjects to attend Anglican church and swear oaths of allegiance,[27] the colony in 1609 could not even feed itself without the Indians' assistance, much less require the Indians to release their children to be educated as English men and women. Later in the same document, the company informed its governor that the native people "will never feede [the English] but for feare" (*Records* III 19). The company sharply commanded Gates to keep the Indians away from vital instruments of the very English culture into which it meant to inculcate them:

> You must constitute and declare some sharpe lawe with a penaltie thereon to restrayne the trade of any prohibited goods especially of Swordes, Pikeheads gunnes Daggers or any thinge of Iron that may be turned against you . . . haue also especially regard that no arte or trade tendinge to armes in any

wise as Smythey Carpentry of or such like be taught the Savages or vsed in their Presence as they may learne therein. (*Records* III 20-21)

The Company's information about the Indians' character, which implies that Gates must keep them frightened of the settlers, shows that on one level at least the Company did not believe that Gates could make the Indians English but rather desperately desired that he make them England's thralls. And Gates's instruction to keep the Indians away from firearms shows that the Company feared the Indians as much as it hoped the Indians would fear its settlement. The Company had learned from England's experience with the Gaelic population in Ireland: Sir John Perrot had advocated training the natives to use English weaponry, and in the 1590s that trained population had turned those weapons against the English (Highley 88). On the one hand, the Virginia Company wanted Indians with English religion, language, and manners; on the other, it wanted to keep them far away from the martial power it hoped would distinguish the two groups of people.

Of course, the colony could not keep the Indians away from English weapons, just as it was unable to keep some of its members from choosing to live with the Indians instead of their English cohorts. In 1619, the assembly in Virginia again issued a law "that no man do sell or give any Indians any piece [firearm] shott or poulder, or any other armes, offensive or defensive upon paine of being held a Traytour to the Colony, and of being hanged" (*Records* III 170-71). Writing from London about conditions in the colony after the 1622 Indian rebellion, Alderman Johnson blamed Sir Edwin Sandys's administration for "th'arming of the Sauages with weapons and teaching them the vse of gunnes" (*Records* IV 180), but under Johnson and Sir Thomas Smith's reign, the Indians had at least as much access to English guns. Despite their governors' warnings, settlers willingly traded guns for food, and the Indians living among and around them quickly adopted English technology and military skills. Writing to his father from the colony in 1623, an ordinary settler, Richard Ffrethorne, begged to be brought back to England, bemoaning the settlers' precarious position: "we heare that there is 26 of English men slayne by the Indians, and they haue taken a Pinnace of Mr Pountis, and have gotten peeces, Armour, swordes, all thinges fitt for Warre, so that they may now steale vpon vs and wee Cannot know them from English, till it is too late, that they bee vpon vs" (*Records* IV 61). Just as the company had earlier feared, the Indians' acquisition of English arms rendered them indistinguishable from the English in the most vital area: military power. But Ffrethorne's letter also makes it clear that the English have promoted this acquisition as well as legislating against it: "but now the Rogues growe verie bold, and can vse peeces, some of them, as well or better then an Englishman, ffor an Indian did shoote with Mr Charles my Masters Kindsman at a marke of white paper, and hee hit it at the first, but Mr Charles Could not hit it" (*Records* IV 61). As in Ireland, English men trained the native population to use English weaponry. Clearly, rather than being attacked and deprived of his weapon, Richard Ffrethorne's master's kinsman participated in a friendly target practice with this Indian. Ffrethorne's story illustrates the impossibility of separating English people and Indians, and therefore the inevitable failure of the Company's policy of keeping arms from the Indians.

While the Company dreamed of taking Indian children and transforming them into obedient Englishmen (whatever that might mean), the Indians around them were instead transforming themselves into Indians armed with English weapons. Although this was indeed a colonial transformation, it was hardly the one desired by either the colony's government in London or its governors in Virginia. But despite acute resistance from the native population to this English dream of little Indians growing up as good English adults, before the 1622 rebellion the dream lived on, as it clearly appealed to an English hierarchical system in which children largely existed to be indoctrinated into a set place in their economic culture.[28] On June 8, 1617, John Rolfe, later Pocahontas's English husband, declared in a letter to Sir Edwin Sandys that "the Indyans [are] very loving, and willing to parte with their childeren" (*Records* III 71). That fantasy, which many other documents contest, proved an important fundraiser for the Company. In 1619 the Company's treasurer's report noted that "*Another* vnknowne person, (together with a godly letter) hath lately sent to the *Treasurer* 550. pounds in gold, for the bringing vp of children of the *Infidels:* first in the Knowledge of God and true Religion; and next, in fit trades whereby honestly to live" (*Records* III 117). Nicholas Ferrar, the elder, dying in 1620, bequeathed "24 pounds by the yeare to be dispersed vnto three discreete and godly men in the Colonie, which shall honestly bring up three of the Infidels children in Christian Religion, and some good course to live by" (Ellyson plate [18]). These gifts expose the structure of thinking behind the fantasy of converted children: not only are all Indians the same, but their mode of living (which was producing vastly more food than the settlers' mode of living on the land) was inherently not honest work; work could only be defined as a place in England's nascent capitalist economy.

While Rolfe waxed enthusiastic, other settlers' reports reveal that the Indians surrounding and living with the English actively resisted parting with their children; in fact it was the English who could more easily conceive of disposing of their powerless members to further the plantation's goals. On June 21, 1619, the Company urged their latest colonial governor, George Yeardley, to banish the Chickahomini from the settlements' territories but to "mainteyne amity" with the rest of "the natives, soe much as may be and procure their Children in good multitude to be brought vpp and to worke amongst vs" (*Records* III 147–48). Presumably in response to this urging and because of their prior failure to carry out this design, the Virginia Generall Assembly convened at James City at the end of July 1619 issued an order: "for laying a surer foundation of the conversion of the Indians to Christian Religion, eache towne, citty, Borrough, and particular plantation [to] obtaine unto themselves by just means a certine number of the natives' children" (*Records* III 165). But Yeardley wrote privately to Sir Edwin Sandys in London that "the Spirituall vine you speake of will not so sodaynly be planted as it may be desired, the Indians being very loath vpon any tearmes to part with theire children" ([1619] *Records* III 128). Yeardley proposed instead that Powhatan's brother, (ironically the man who would lead the 1622 rebellion),

> Opachankeno would apoynt and Cuse out so many . . . families, as that in every Corporation and . . . plantation there myght be placed a howshould promising him they should have howses built in every place and ground to sett Corne and plant vpon to which he willingly condisended and promised he would apoynt the ffamilies that should remove to vs, which yf he doe we shall then . . . have the opertunity to Instruct theire Children. (*Records* III 128)

Yeardley wrote that he was at a loss how to proceed otherwise, but he would still do his "best to purchase some Children . . . acording to [the Company's] former Directions, as like wise by putting some of the Companyes boyes among them [the Indians] to learne the Language" (*Records* III 129). The Governor's official report to the Company in November 1619 recommended capturing Indian children in war "seinge those Indians are in noe sort willinge to sell or by fayer meanes to part with their Children" (*Records* III 228).

Yeardley's letter to Sandys indicates the clash of cultures in Virginia; while the Indians refused to sell their children, the English were more than willing to send "the Companyes" boys to live with the Indians so that they might obtain skilled translators. In its comic turn on Protheus's boy Launce, Shakespeare's *Two Gentlemen* dramatizes the position of mastered boys wrenched from their families according to their masters' whims:

Laun. Loose the Tide and the voyage, and the Master, and the Seruice, and the tide: why man, if the Riuer were drie, I am able to fill it with my teares: if the winde were downe, I could driue the boate with my sighes.

Panth. Come: come away man, I was sent to call thee.

Lau. Sir: call me what thou dar'st.

Pant. Wilt thou goe?

Laun. Well, I will goe. (TLN 643-51, II. iii.50-8)

While the play clearly expects a laugh out of its boy/clown Launce's overwrought and misdirected grief over his parting (Launce seems at least as concerned about his dog's lack of emotion), this set-piece might well have drawn a sigh of recognition from a mastered boy lucky enough to attend the theater. Certainly not all English boys wanted to become Company property. On January 28, 1619 (old dating), Sir Edwin Sandys reported to James's principal secretary Sir Robert Naunton that

> The Citie of London have by Act of their Common Counsell, appointed one Hundred Children out of their superfluous multitude to be transported to Virginia; there to be bound apprentices for certaine yeares . . . Now it falleth out that among those Children, sundry being ill disposed, and fitter for any remote place then for this Citie, declare their vnwillingnes to goe to Virginia: of whom the Citie is especially desirous to be disburdened; and in Virginia vnder severe Masters they may be brought to goodnes. (*Records* III 259)

While Shakespeare's culture took such disposal of boys as a matter of course, the Indians they confronted in Virginia were unwilling to participate in this system (although Powhatan did send a few older boys as envoys to the English and to England).

Instead Opechancanough's assent to an alternative plan that involved bringing entire families to live with the English shows that he envisioned an alternative transformation: rather than Indian children growing up English, the English settlements would become integrated with Indian families. Despite the English colonists' poor success at keeping their settlements absolutely separate from the natives around them, this proposal was unlikely to be regarded favorably as it violated the rules the English had painfully learned in Ireland about the dangers of integrating natives into their colony—a subject chapter 1 addresses. The 1619 Generall Assembly enjoined each settlement to let in five or six Indians "and no more . . . Provided that good guarde in the night be kept upon them generally (though some amongst many may proove good) they are a most trecherous people and quickly gone when they have done a villany. And it were fitt a house were builte for them to lodge in aparte by themselves, and lone inhabitants by no meanes to entertain them" (*Records* III 165). Powerful English men were accustomed to enculturating powerless boys, but at the prospect of admitting Indian families, vistas of treachery and possibly even apostasy by contaminated settlers opened before the settlements' eyes. While the Assembly made the fear of Indian treachery explicit in its caution, it was much more guarded about revealing the second and perhaps deeper fear: that the colony would, because of the transformation of the English therein, become not a new England but a new Virginian Indian nation, containing and transforming the English themselves.

The settlements' laws, legal records, and their recorded negotiations with the Powhatan Indians, however, speak directly to those fears. William Strachey's 1612 *Lawes* for the colony include the injunction that "No man or woman, (vpon paine of death) shall runne away from the Colonie, to Powhatan, or any sauage Weroance else whatsoeuer" (C3v). Apparently that threat didn't stop determined runaways. Sir Thomas Dale (the Virginia governor who kidnapped Pocahontas) wrote from Virginia in 1614 that he sailed with her to her father Powhatan's residence to exchange her for "all the armes, tooles, swords, and men that had runne away" (Hamor 52). Powhatan's negotiators agreed to fetch a man named Simons "who had thrice plaid the runnagate" (53), but later a messenger returned to tell Dale that "Simons was run away to Nonsomwhatcond" (54). On October 20, 1617, Governor Argall pardoned two men for preferring Indians to their fellow white settlers: "Geo. White pardoned for running away to the Indians with his arms & ammunition . . . Henry potter for Stealing a Calf & running to Indians. death the others the same crimes" (*Records* III 74).[29] In order to warn other English colonists that they must stay with the people who had brought them over, the English colonial government had to kill men (and perhaps women) who wanted to abandon the colony and run to the Indians. Thus, to the Company's caution that the Indians would feed the colonists only out of fear, the English could have added a caution that less privileged English would stay English only out of fear. Such an explicit caution, however, would have undermined all the governing Englishmen's efforts to forge a solidified English identity in the colony.

No official measures—laws, punishments, exhortations—could overcome the central problem that the colonists were living among the native population and were dependent upon them for food. They needed to communicate with the Indians, and with communication came the danger of contamination, a problem obsessively reiterated in texts by new English colonists in Ireland. In 1619 the Virginia Generall Assembly tried Captain Henry Spelman—the man Peter Arundell called "the best linguist of the Indian Tongue of this Countrys"—for speaking "very unreverently and maliciously against this present Governor [Yeardley], whereby the honour and dignity of his place and person, and so of the whole Colonie, might be brought into contempte, by which meanes what mischiefs might ensue from the Indians by disturbance of the peace or otherwise, may easily be conjectured" (*Records* III 174). Spelman had been discussing Yeardley with Opechancanough, Pocahontas' uncle. The Generall Assembly's court spared Spelman's life but stripped him of his rank and condemned him to serve the colony for seven years as the governor's interpreter. John Pory, the colony's secretary, reported that "this sentence being read to Spelman he, as one that had in him more of the Savage then of the Christian, muttered certaine wordes to himselfe neither shewing any remorse for his offenses, nor yet any thankfulness to the Assembly" (175). Spelman gained his facility with Indian languages when Captain John Smith sold him to Powhatan shortly after the boy arrived in Virginia in 1609 (Spelman 9). Spelman's *Relation of Virginia* offers a short account of his time serving Powhatan and carrying messages to the English. On one of those trips the Indians tricked Spelman into unwittingly tempting a ship full of Englishmen into a trap baited with corn (10). This experience made other English people leery of trusting the boy. Early in his period of "captivity," Spelman left Powhatan for Jamestown, but when another English boy who had been living with the Powhatans, Thomas Savage,[30] came to the settlement with venison for the Virginia council's president George Percy and wanted to bring "sum of his cuntrymen" back with him, the settlement appointed Spelman to return to Powhatan. Spelman was not reluctant to leave the English colony: "I was apoynted to goe, which I the more willinglie did, By Reason that vitals were scarce with us, cariinge with me sum copper and a hatchet which I had gotten" (10).

Spelman's Virginia career touches all the human bases of colonial transformation in Virginia. His countryman sold him to the Indians in order to facilitate communication between the two groups. After his return to the colony, he found that he preferred life (or at least would be able to eat and therefore live) with the Indians; later he left Powhatan's group for the Patowomeke where he found more favor. Choosing to live with the Indians, Spelman made no mistakes about the essential similarity of Indian groups. He knew the differences between native people. Caught in the impossible position of the translator,[31] Spelman was eventually killed by Patowemeke Indians in 1623 after the English had retaliated indiscriminately against all Indians, even their Patowemeke allies, for the 1622 rebellion. Although he died because he was an Englishman, once he could speak well enough with Indians, the English could not trust him to be securely English. His compatriots punished him for being able to talk to Opechancanough, and they eventually got him killed in revenge for their violent reaction to the situation they had created with the Indians. In that reaction they produced frozen colonial categories—the English colonists and the colonized Indians—in order to ward off the threatening hybridity Spelman signified. The translator is one powerful figure of colonial transformation. He or she can be read as representing the larger transformations that made the new Atlantic world.

Not all of those transformations involved English people's interactions with native people. The English also transformed the land they encountered in Ireland, Virginia and Bermuda. In a 1994 *New Yorker* essay, "Alien Soil," Jamaica Kincaid, an Antiguan living in Vermont, compares her American garden to the gardens she saw in Antigua as a child. Kincaid asks herself, "what did the botanical life of Antigua consist of at the time . . . Christopher Columbus first saw it?" She concludes: "to see a garden in Antigua now will not supply a clue. I made a visit to Antigua this spring, and most of the plants I saw there came from somewhere else" (212).[32] In addition to importing plants from around the world, Kincaid explains, "soon after the English settled in Antigua [in 1632], they cleared the land of its hardwood forests to make room for the growing of tobacco, sugar, and cotton and it is this that makes the island drought-ridden to this day" (213).[33] Kincaid's righteous ire at the natural depredations of a people she says "have [no] more important things to do than make a small tree large, a large tree small, or a tree whose blooms are usually yellow bear black blooms," a people she calls "spiritually feverish, restless, and fully [sic] of envy" (216), brings us to the twentieth-century consequences of the colonial transformations planned and executed while Shakespeare, Spenser, and Jonson were writing drama and poetry for their largely metropolitan audiences.

Just as the English who settled Antigua in 1632 totally transformed the island's habitat, so they saw Virginia, from their first landfalls, as a natural world to be molded and re-created in any image that would provide profit for the home country. And they found a model for this depredation in their contemporaneous devastation in Ireland: "While fully one eighth of Ireland was covered with forests in 1600, a century later the country was virtually cleared. Irish timber was exported to England, and an English-controlled iron industry was also developed" (Hechter 84). The colonial documents from Virginia document a similar transformation as well as recording its—to a postcolonial observer—blithe and unconsidered character. English industrialists and merchants had denuded England's own woods in pursuit of profit, and in that model they considered Virginia's natural forests to be both a hindrance to settlement and industry and a commodity waiting to be exploited. While Elizabeth I could pass an act to prohibit the extension of ironworks in Sussex and neighboring counties because too much timber was being consumed, her subjects and her successor's subjects were eager to consume the resources of the places they saw as their colonial margins.

A 1620 "Declaration" by the Virginia Council in London announced propagandistically that "The *Iron*, which hath so wasted our *English* Woods, that it self in short time must decay together with them, is to be had in *Virginia* (where wasting of Woods is a benefit) for all good conditions answerable to the best in the world" (4). Nine years earlier, writing in the same promotional mode after his abrupt and notorious return from Virginia, Thomas West, Lord De La Warre, lauded "a goodly River called Patomack vpon the borders whereof there are growne the goodliest Trees for Masts, that may be found elsewhere in the World" (B4v). De La Warre imagines those trees awaiting their transformation into masts that will enable further sea travel, further discovery, further production of commodities for English consumption. In these projections one can easily see the object of Kincaid's contempt: the restless, insatiable Englishman. However, even the English project of wasting woods in their various colonial territories can reveal divisions among Englishmen. As we will see in chapter 4, the London Livery Companies' harvesting of Ulster's woods came under attack by the king's representative in Ulster.

In their projected clearing of Virginia's woods, English colonial projectors envisioned a vast iron industry, but they also enabled two agricultural activities that had quicker and even more devastating ecological effects on the land they settled: planting English seeds and grazing English cattle. In a September 30, 1619 letter to his patron in England, John Pory declared, "three thinges there bee, which in fewe yeares may bring this Colony to perfection; the English plough, Vineyards, & Cattle" (*Records* III 220). Pory could not have been more accurate as long as "perfection" is defined as complete devastation of the environment the Indians had been producing and living within for centuries. English colonial "perfection" meant rapid ecological and social change and devastation for all of the cultures the English encountered. Michael Hechter notes that "the development of the London food market from the mid-sixteenth century on led some southern Welsh gentry to breed cattle for sale to the metropolis. To further this end they sought to enclose, or restrict tenant access to common fields. In lowland areas this caused much displacement of peasants and an increase in vagabondage" (82). As in Wales, supporting cattle in Virginia meant transforming land from forest to pasture and transforming land use from Indian agriculture to English animals' grazing territory. All of the Company's declarations and many of its letters and relations document the importation of this massive transformative force, a focus of English efforts from their first arrivals in Virginia. By 1619 the Virginia Company (probably optimistically) counted five hundred cattle in Virginia, "with some Horses and Goates; and infinite number of Swine, broken out into the woods" (*Records* III 118). As livestock reproduced geometrically, the English settlements needed to fence off more and more land. The settlements' expansions provoked Indian attacks, which were then recorded as unprovoked massacres that justified more expansion.[34]

Since the Indians in Virginia had no domesticated livestock before the English arrival, the land they had dwelled upon bore no signs of the pressures livestock places on ecology. Their land had to be transformed for livestock, and English livestock, driven by the English, irrevocably transformed the land. Writing about colonial New England, William Cronon notes that "livestock not only helped shift the species composition of new England forests but made a major contribution to their long-term deterioration as well" (145).[35] He argues that in New England plowing by oxen and horses "which stirred the soil much more deeply [than hoes and arms] destroyed all native plant species to create an entirely new habitat populated mainly by domesticated species" (147). A May 1621 letter from the settler Jabez Whittaker to Sir Edwin Sandys indicates both the explosive growth of livestock in Virginia and their transformative power: "Since the writing of my last letter I have receaved ten young Kine, they thrife very well and I thinke are all with calfe, I have railed in for them with a firme substantiall rale two hundred acres of ground" (*Records* III 441).

As in New England, livestock transformed Virginia into "a world of field and fences" (Cronon 156). And those transformations forced the Indians to live in an English world as powerfully as did any attack upon them perpetrated with English weapons. One of the many extant accounts of supplies brought to Virginia, here in a ship explicitly named "Supply," includes the following items: "ffor .40. weedinge howes [at] liijs iiijd; ffor .30. spades [at] xxvs; ffor .2. sithes [at] vs; ffor .10. fellinge axes[36] at xvs; ffor .6. swordes at xxiiijs vid; ffor .200. of lead shot . . . [at] xxviijs; ffor garden seeds vzt. parsnip, carret, cabbage, turnep, lettuice, onyon mustard and garlick &c at xixs viijd" (*Records* III 387-88). Cronon's historical analysis of New England and Kincaid's essay on Antiguan gardens make the relatively few shillings and pence spent on seeds and spades look at least as destructive to native Virginian lifeways as the pounds paid for shot and swords.[37]

Colonial Transformations takes as its object the material reality of cultural transformation, including its operations in texts. This book is centrally about how texts of all sorts contributed to and described the material transformations of England, Ireland, Virginia, and Bermuda into a new Atlantic world. *Colonial Transformations* reads literary productions along with a wide range of colonial documents as part of an ongoing critical effort, stimulated by Stephen Greenblatt[38] and other New Historicists, to break down the artificial separation of that species of material textual work, the literary, from other texts.[39] English culture's expansion into its colonial margins in the Atlantic world, and its constitution as an imperial center, permeated all of its cultural productions, and those productions played multiple parts in that expansion and constitution. Despite official Virginia Company sermons that preach against the calumnies of players, the stage cannot be read as simply "the enemy of the colonial plantation" (Greenblatt 1988 158). As we shall see in chapter 3, plays did criticize England's colonial efforts, but they also publicized those same efforts, contributing to a public consciousness of England as an incipient imperial center, one of the central colonial transformations this book documents. The court masques written for James I explicitly took a British empire as their subject, proleptically transforming the court—many members of which were investors in Ireland, Bermuda, and Virginia—into triumphant imperialists. In this book I read poetry as well as plays and masques for a heuristic point, as well as because poetry, along with masques, reached a more exclusive audience, the audience that was most responsible for funding colonial enterprise. England's colonial experiences permeated its textual productions, and poetry was directly involved in the transformation of England into an imperial culture.

However, in order to transform England into an imperial culture, determined noble and merchant adventurers had to mobilize all segments of the population, women as well as men, the poor as well as the rich. On March 5, 1609, Don Pedro de Zuñiga, Spain's ambassador to England, reported wrathfully to his king that the English had become reckless in their pursuit of a North American empire. Zuñiga wrote that he would send Philip a broadside "which has been issued to all officials, showing what they give them for going; and there has been gotten together in *20 days a sum* of money for this voyage which amazes one; among fourteen Counts and Barons they have given 40.000 ducats, the merchants give much more, and there is no poor, little man, nor woman, who is not willing to subscribe something for this enterprise" (Brown, *Genesis* I 245-46). In his effort to mobilize the Spanish king against English colonial efforts in America, the ambassador overestimated public support, but his estimate was a measure of the London adventurers' dreams, and the period's "literary" texts responded to those dreams and their growing reality, as did early modern England's other texts. Early modern English texts responded in a similar manner to England's increasing attempts to dominate Ireland, the island to its northeast.

The heuristic point involves our reading and teaching of early modern English literature and history. This book joins an effort, most beautifully expressed in Kim Hall's *Things of Darkness*, to show that reading the canonical and noncanonical literature of the period as if it was divorced from England's expansion outward is simply false reading. During the magnificent literary explosion of the late sixteenth and early seventeenth centuries, England was reconstituting itself as an empire globally, in Ireland, in the Americas, and in Africa and Asia.[40] In so doing it encountered difference; it transformed and was transformed by the people it encountered. Plays, poetry, and masques participated fully in those transformations. Our most canonical literary figures, poets all, wrote about and actively participated in those processes, as Hall's work on Sidney, the flourish of recent work on Spenser and Ireland and Shakespeare and Ireland, and Andrew Murphy's work on John Donne and Spenser, among other projects, so amply demonstrate.[41] To read that literary explosion completely we must see it participating in the transformed Atlantic world, just as to read that world completely we must read what we now call English literature of the period. This is why, as well as reading explicitly colonial documents like the Virginia Company records, this book reads Spenser's sonnet sequence, the *Amoretti,* a text critics have invariably seen as divorced from the Irish world in which it was written. Although this book alone cannot elucidate the implications of England's canonical and noncanonical literatures in colonial transformations, chapter 1 on Spenser's *Amoretti* points the way to that end.

In addition, this book makes the point that just as the division between literature and colonial texts is imaginary, so is the division between English and American literature in the seventeenth century. This division structures English departments and critical journals throughout the United States, but it is a division that no English person in the early seventeenth century would understand about his or her own texts. All of the letters, propaganda from the Virginia Company, and relations and narratives written about experiences in America that I examine in this book were written by people who identified themselves as English, and all of these documents were printed in England, or (in the case of letters) written to people in England by people who thought that they were writing home. Just as Thomas Scanlon's recent book does, I want this book to point to "the artificiality of the boundaries between the two seventeenth centuries—the British one and the American one" (7). *Colonial Transformations*'s focus on the new Atlantic world should help to dissolve these boundaries.

Just as this introduction reads the *Supply*'s bill of lading through Kincaid's essay on Antigua, so *Colonial Transformations* brings a postcolonial lens to bear on all of the texts it reads. Mary Fuller is certainly right to assert that "one need not deny the violence of colonialism to say that the earliest history of the English in America was hardly one of mastery or proficiency," but I am less persuaded by her important book's claim that "it was hardly possible to predict what happened later from reading the history of 1576-1624" (12). That history contains, within its documents of English ineptitude and self-destruction, the physical and ideological tools that would continue to force the Indians into English self-destructive lifeways. This is evident from an ecological understanding offered by contemporary observers such as Kincaid. Likewise, late-twentieth-century ignorance about the diversity, names, and even continued existence of Native Americans, potently criticized by Native American theorists like Ward Churchill (Keetoowah Cherokee) and Joy Harjo (Muskogee Creek), is the legacy of early seventeenth-century essentialisms such as Hakluyt's. And the United States Government's behavior today and in the past toward Native American land claims is eerily reminiscent of the behavior of English colonial officials toward native and old English land claims in Munster in 1588, one of the concerns in chapter 1. If the English could not successfully buy or steal Indian children for reeducation in 1619, the dream did not die forever, as the tragic history of Indian schools in North America, which reaches well into the twentieth century, attests. Late-twentieth-century white-Indian antipathy in North America and violence between Protestants and Catholics in the two

Irelands of the present have their historical roots in Shakespeare's world. And contemporary white-black struggles in the United States, brought vividly to mind recently in the backlash attempts to end affirmative action in this country and particularly in the academy, also have their genealogical roots in the processes of colonial transformation this book explores. From the beginnings of the Virginia Company's settlements, the practice of planters reimbursing the Company for indentured servants transported by London adventurers laid the ground for perpetual black slavery in the colonies.[42] A Dutch man of war brought the Virginia colony its first African slaves in 1619, and African slaves began arriving in Bermuda in some numbers in 1616-17, the year Shakespeare died and Ben Jonson published his folio works (*Records* III 243; *Rich Papers* 17). The early-seventeenth-century manuscript *The Historye of the Bermudaes or Summer Islands* records that sometime in 1616-17 the *Edwin* made "her returne out of the West-Indies, whether . . . she was sent for plants; of which she came furnished with diuers sortes, as plantans, suger-canes, figges, pines, and the like, all of which wer presently replanted, and are since encreased into great numbers, especially the plantans and figges, very infinitely: she brought with her also one Indian and a Negroe (the first thes Ilands euer had)" (Lefroy 84). The *Edwin's* lading contained seeds of the immense ecological transformation the English immediately performed on the Bermuda islands—further explored in chapter 2—and it also initiated black slave labor in this English colony, which soon became its economy's motor, as the manuscript's parenthetical phrase reveals: the first "Negroe" to land there becomes noteworthy because he was first of many to be exploited there. Sometime before October 1619, the *Historye* records, "a smale frigate descried vpon the coast occasioneth a present distraction from all other thoughts; and especially, after she was knowen to be a good-fellowe, manned for the most part with English, who haueinge played some slie partes in the West-Indies, and so gotten some purchase, part wherof consisted of negroes (a welcome for a most necessary commoditie for thes Ilands), she offered to leaue and giue them to the Gouernour" (144). In just two years a colony of English gentry and indentured servants became a slave-holding economy, enslaved black people being "necessary" for the colony's growth.[43] Paul Gilroy has argued the "the history of slavery . . . [is] part of the ethical and intellectual heritage of the West as a whole" (49). In addition, Gilroy writes that "the time has come for the primal history of modernity to be reconstructed from the slaves' point of view" (55). The energy of Gilroy's essential critique stands behind this book's attention to both the production of blackness and whiteness in early modern texts and the voices of resistance embedded in those texts.

Kincaid, a descendent of black slaves brought to Antigua, a later English colony, reminds us forcefully of the connections between English ecological rapacity and English oppression and exploitation of people. And Kim Hall's account of whiteness and blackness constructed in the literature and material practices of England's Renaissance reveals that this exploitation belonged to and transformed the metropolitan world that authorized it.[44] England's colonial transformations constructed the black and white world Americans live in, just as they constructed the reservation system and the two Irelands that form today's Atlantic world. In addition, each of these oppressions takes and has taken energy from the other, as we will see in chapter 3 in the discussion of Michael Drayton's satire and Ben Jonson's plays.

In 1612, William Strachey published a volume of *Lawes Divine, Morall and Martiall* for the Virginia colony with an obsequious dedicatory sonnet to the Lords of London's Virginia council:

Noblest of men, though tis the fashion now
      Noblest to mixe with basest, for their gaine:
Yet doth it fare farre otherwise withyou,
      That scorne to turne to Chaos so againe,
And follow your supreme distinction still,
      Till of most noble, you become diuine
And imitate your maker in his will,
      To haue his truth in blackest nations shine.
What had you beene, had not your Ancestors
      Begunne to you, that make their nobles good?
And where white Christians turne in maners Mores
      You wash Mores white with sacred Christian bloud
This wonder ye, that others nothing make
      Forth then(great LL.)for your Lords Sauiors sake.

In this poem Strachey mobilizes status oppression, a vision of England as an evolved nation, and racism, along with a theory of racial degeneration, in the service of English colonization of Virginia. Noblemen who consort with base men invite chaos, a chaos identified with England's ancient past, with present-day black Africa and, by means of the poem's location and purpose, with Virginia's native inhabitants. Colonial violence—the "sacred Christian bloud" shed in Virginia and identified with Christ's blood—becomes a justified cleansing process directly opposed to the defilement of Christians going native by adopting Moorish (and certainly Gaelic and Old English and Indian) manners.[45] Ben Jonson's *Masque of Blackness*, which also turns on washing blackness away, played for the court that laughed and danced at his *Irish Masque at Court*, his *Gypsies Metamorphos'd* and the submissions portrayed in *For the Honor of Wales*, all texts I will discuss in chapter 4. As we will see in that chapter, London's Lord Mayor's pageants and court masques postulated an undifferentiated Indian blackness that served to consolidate a white and therefore pure English identity. And as much recent work has shown, cartoon stereotypes of black and Irish people were drawn from the same imagery in nineteenth- and twentieth-century England and America. While the connections between these events and histories are never simple, they cannot be ignored by ethical scholars. *Colonial Transformations* traces early modern England's contributions to oppressions as well as the resistances recorded in the transformations that made a new Atlantic world.

Edmund Spenser has been seen as both a powerful spokesperson for English imperialism and a preeminent poet of English nationhood, yet his sonnet sequence *Amoretti* has been cordoned off by critics from his own colonial transformation. Chapter 1, "Colonial Poetics in Spenser's *Amoretti*" reads Edmund Spenser's sonnet sequence in its Irish milieu, arguing that the sequence depicts a male mastery that wards off the possibility of its author's colonial transformation into a feminized barbarian. The chapter demonstrates that the love depicted in the sonnets is inextricable from both Spenser's own colonial ambition and his embattled status as the occupier of escheated land. The *Amoretti* speaks both to Spenser's epic poem *The Faerie Queene*, written contemporaneously, and his *A View of the Present State of Ireland*, reworking crucial events, like Lord Grey's massacre of Spanish prisoners at Smerwick, and ensuring that an encounter with feminized alterity will leave the Colin-izer on top. Roland Greene argues that "the international appeal of Petrarchism in the sixteenth century is largely political, or to be more specific imperialist; because of its engagement with such political issues as the distribution of power among agents, the assimilation of difference, and the organization of individual desires into common structures of action and reaction, Petrarchan subjectivity becomes newly immediate in the age of Europe's discovery and administration of the New World" (1995 131). The heightened nature of the lover-beloved conflict in the *Amoretti*, which has long disturbed its critics, and which has been interpreted as a critical commentary on Petrarchism, is, I argue, an accurate reflection of the historical situation Spenser thematized and within which he found himself embroiled in Ireland.

In its gendering of Ireland, the *Amoretti* joins a long and complicated tradition of depicting Ireland as a woman. Chapter 1 addresses the sequence's place within that tradition. The "bardic" poetry that Spenser castigates in *A View* posits a powerful, female, Gaelic nation. I argue that in the *Amoretti* Spenser answers that vision with a subservient, submissive, female beloved. The chapter contends that when we place Spenser within his Irish career, as so much powerful recent work has done, we cannot exempt any of his poetry from its imbrications in its place of origin; an exemption based on "love" ignores that concept's changing valences over time. Specifically in Spenser's case, his love was about alliance, economic ambition, and domination. Thus, the chapter forcefully demonstrates the insight a new Atlantic world perspective offers about literature that has been treated as separate from the world. Finally, chapter 1 treats the *Epithalamion* and reveals how the process of colonial transformation foiled Spenser's dream that his progeny would be an unbroken line of Protestant saints, untouched by their patriarch's colonial ambitions.

Chapter 2, "Bermuda's Ireland: Naming the Colonial World," argues that naming in the seventeenth-century Atlantic world transformed that world. English literary and other colonial texts renamed all of the lands and people the English encountered. The names the English gave reveal their central concerns with hierarchy, dominion over animals, and their Christian destiny. But naming is not only a reflection; it is transformative in itself. The names given in the late sixteenth and early seventeenth centuries still determine the way we understand the Atlantic world. Place names also reveal contests for power among colonists and between settlers and native populations. The chapter looks at English naming practices and native naming in Virginia and Ireland.

Chapter 2 also delineates transformative naming in Bermuda. The chapter reads names in printed texts and on maps through postcolonial and social cartographic theories. Its comparison of names on Richard Norwood's early-seventeenth-century map of the Bermuda islands and the primary texts of the Bermuda settlement reveals the Bermuda Company's ambitions for an ordered, easily ruled colony as well as the pleasures and dangers of colonial transformation that challenged that rule. The map also exposes settlers' cavalier attitudes toward the animals without whom the islands' colony and the Virginia colony itself would never have survived. The naming of an unfortified small island as "Ireland" on the Norwood map also indicates England's ambitions for its closest and most troublesome colonial subjects.

Chapter 3, "The New Atlantic World Transformed on the London Stage," looks at how ideas of savagery and civilization honed in England's Virginian and Irish encounters were displayed on London's stages, especially in Ben Jonson's plays. I contend that Ben Jonson became early modern England's primary theatrical interpreter of England's colonial efforts and imperial ambitions. Thus Ben Jonson's plays need to be central texts for colonial reading in the period. I argue that Jonson's plays take England's imperial ambitions as a measure of its own unity and social health; the plays equate home and colony and measure each by the other.

The chapter reads Virginia, Bermuda, and Ireland staged in Jonson's plays, and it also reads the lottery broadsides, pamphlets, declarations, and sermons with which the Virginia Company was trying to reach the same metropolitan audiences. Because Jonson's texts see English identity as troubled, because they define many English people as savages, and because they are metadramatically aware of their divided audiences, they reveal the contradictions in England's colonial transformations. With the efforts to expand into and control territory in the new Atlantic world came contentious transformations of London and Englishness itself.

Chapter 4, "Colonial Transformations in Court and City Entertainments," argues that London's Lord Mayor's pageants and Jacobean and Caroline court masques collectively shaped city and court into an imperial England. Lord Mayor's pageants acknowledged the London companies' roles in England's colonial efforts in Ireland, Virginia, and Bermuda, and they were motors of those efforts. Actively mobilizing London's mayors and guild members as explorer-colonists, these once-yearly lavish entertainments satisfied their audience members' visions of themselves as world-dominators, and they produced expectations of dominance that would fuel continuing mercantile expansion. Both pageants and masques displayed what I call an "undifferentiated Indian," often in blackface, in order to establish an English imperial whiteness.

Representing both a contiguous empire in Wales and Ireland and a trans-Atlantic empire in North America, Jacobean and Caroline masques transformed their royal and court audiences into successful imperialists while they projected an English empire for England's guests and ambassadors who might be (and demonstrably were) less enthralled with that vision. Comparing the lists of masque participants with lists of English soldiers in Ireland and lists of Virginia and Bermuda adventurers, and detailing London's livery companies' investments in the Atlantic world, the chapter shows those English people's material interests in representations that magically resolved the intergroup resistances and intragroup disputes over that empire's early constitution. Placing those masques within the actual resistances and disputes already documented in the book's earlier chapters, chapter 4 examines what Richard Dyer would call their "Utopian" aspirations (1993). The masques enabled courtiers and rulers to dance out their empire, literally to trace it out with their feet and body movements.

Chapter 5, "A Virginia Maske," moves back across the ocean to argue that what Captain John Smith depicted, in his 1624 *Generall Historie of Virginia, New-England, and the Summer Isles*, as a masque by Pocahontas and a number of naked, dancing Indian women was instead a native dance misread or, in Slavov Žižek's terminology, "mis-recognized," as a Jacobean masque at Smith's imagined Virginia court. I argue that in writing that "masque" Smith fashioned himself as an author on Jonson's folio's model, in order to display his authority over the Indians he encountered in Virginia. Yet that authority was both founded on a profound misunderstanding and constituted in retrospect and defensively. By 1624, when Smith published *The Generall Historie* with its women's dance rewritten as masque, Smith was physically and socially removed from any power over England's colonial efforts in North America. Rather than offering any ethnographic "truth," the "Virginia Maske" functioned wishfully to make Smith central and authoritative, as it let his reading audience comprehend another culture in England's court's terms.

Using Ward Churchill's theory of "cultural genocide" and anthropological and theoretical understandings of Indian women's communities and status, I challenge Smith's transformation of the women's dance. By exposing what Smith's construction of it as a masque ignores about the dance and its cultural meanings, and by showing Smith's interest in defining it as he does, the chapter opens up other possibilities for reading the history of women in colonial Virginia. The chapter also places Smith's transformation within a long and brutal history of transformation in which native art and culture have been appropriated and used for enhancement of white American health and lifestyles at the same moments when the majority of American Indians today live in debilitating poverty due to the U.S. government's appropriations of their lands.

Colonial Transformations's epilogue, "Late-Twentieth-Century Transformations: Pocahontas and Captain John Smith in Late-Twentieth-Century Jamestown," reads the histories and myths offered to the American and foreign tourists who visited the colonial Jamestown recreations in the summer of 1998. In the epilogue, I examine the Jamestown Settlement Museum's indoor exhibits, its living history outdoor displays (a re-created Powhatan Village, a 1610 English fort, and facsimiles of English boats), its gift shops, and its literature. Informed also by contemporary critical perspectives on museum history and purpose, this epilogue sees today's tourist attractions as another step in the ongoing colonial transformations initiated in early modern England. I argue that, despite the evident curatorial intention to display a balanced social history, the museum assumes the white English-identified perspective that early-seventeenth-century London entertainments were working so hard to create. In many ways, the museum's celebration of a shared history represses the alternative voices and alternative histories that this book uncovers in its early modern sources. The epilogue elucidates the assumptions that the late-twentieth-century tourist attraction and Virginia company texts share. With those assumptions laid bare, perhaps we can imagine radically different histories for the future.

Chapter 1

Colonial Poetics in Spenser's *Amoretti*

lthough until recently most early modern English poetry has been read as inhabiting a world apart from poets' colonial ambitions, for a long time critics have read Edmund Spenser's *The Faerie Queene* in relation to his personal career in Ireland. Recently Spenser's colonial ambitions and his colonial texts have attracted substantial critical attention. Spenser was a prominent colonist as well as the major epic poet of the late sixteenth century, and at the end of the twentieth century, *The Faerie Queene* can hardly be read without some mention of Spenser's Irish career. However, the critical inattention to colonial issues in Spenser's other poetry demonstrates the power of the belief that English literature and England's colonial expansion are essentially unrelated. This chapter argues that the poetry that has been seen as Spenser's most personal, his little love poems, the *Amoretti,* is as deeply imbricated in his colonial career as his public epic poetry. Indeed, it is precisely the *Amoretti's* personal vision that links these poems to Spenser's colonial desires, desires which were clearly at the heart of his life and his ambitions. I argue in this chapter that the realm of Spenser's writing that has been most clearly cordoned off from Spenser the planter and colonist is implicated in a colonial fashioning of both Ireland and the planter's personae. Indeed, while Petrarch and his continental and English followers use war as a trope for love, the *Amoretti* uses love as a trope for war. This chapter will also show that the Ireland that Spenser is so careful to conquer and defend against in his poetry and prose transformed the destiny of his family in exactly the way the poet most feared. Just as the colonial transformation that Spenser effects in his poetry and prose had enormous ideological and material consequences for Ireland, the planter himself was unable to defend his family from the material consequences of the colonial encounter.

In sonnet LXXX of his 1595 sequence, the *Amoretti*, Edmund Spenser's poet-persona expresses his exhaustion after finishing six books of *The Faerie Queene*. Addressing perhaps his queen, his courtly readership, or his comrade in Ireland, Lodowick Bryskett, the poet promises that following a space to gather breath and write these little love poems he will complete his epic:

Out of my prison I will break anew:
And stoutly will that second worke assoyle,
with strong endevour and attention dew.
Till then give leave to me in pleasant mew,
to sport my muse and sing my loves sweet praise:
the contemplation of whose heavenly hew,
my spirit to an higher pitch will rayse. (LXXX. 6-12)[1]

Because the sonnets culminate in a marriage song, and because the Elizabeth that Spenser married in 1594 seems so present in the sequence, the poems have been presumed to depict a love isolated from Spenser's worldly preoccupations. Even some of the most insightful New Historicist critics see these particular poems as divorced from the public sphere. Richard Helgerson asserts that for Spenser in the *Amoretti*, "poetry serves the truant passion of love, while expository and argumentative prose does the work of the active world" (88). Likewise, Louis Adrian Montrose explains that "the success of the erotic poet in reconciling tensions within the private world of *Amoretti* is contrasted to the epic poet's inability to work such a reconciliation within the larger public world" (1979, 55). These critics are following Spenser's own lead in *Amoretti* LXXX, in which the poet depicts his composition of these sonnets as a rest from his work on the epic poem. But the material conditions of the sequence's production, the sonnets' worldly lexicon, and even the circumstances of Spenser's courtship situate these poems in precisely the public and active world that *Amoretti* LXXX rejects.

In the context of the sequence's previous sonnets, *Amoretti* LXXX's prison is overdetermined, at once the persona's worldly occupations and his captivity to the love motivating this poetry. The persona/lover's prison is also and simultaneously the place of the poems' genesis: Ireland. As Spenser's worldly occupations in Ireland—his service to his queen's colonial government as clerk of the Council of Munster and as undertaker in the Munster plantation—have imprisoned him, so has his courtship of a New English mistress, Elizabeth Boyle, kinswoman to Richard Boyle, the future Earl of Cork. Thus Ireland is at once a "prison," keeping him from his serious poetic work and his beloved England, and at the same time a "pleasant mew," his sonnet-beloved's physically beautiful home.[2]

In late-twentieth-century America, the word "mew" has been deracinated, coming to signify a cul-de-sac or even an enclosed or enfolded suburban housing development. But for Spenser's readers, "mews" refers less to housing than to cages for hawks, especially those who were "in mew," "in process of moulting." "In mew" could figuratively signify "in process of transformation." "In mew" could also mean "in hiding or confinement," with "mew" signifying "a secret place, a place of concealment or retirement" (*OED* "mew"). The English word was adapted from the French "*muer* to moult, also to shed horns," which in Old French had the "wider sense, to change" (*OED* "mew"). The *Oxford English Dictionary* offers a slew of figurative examples from the sixteenth and seventeenth centuries as well as much earlier, such as Gower's 1309 "As a brdd which were in Mue Withinne a buissh sche kepte hire clos" (*OED* "mew" 1b).

In book two of *The Faerie Queene*, Spenser realizes all of the sixteenth-century connotations of "mew" when he describes the temptress Acrasia's power:

> The vile *Acrasia*, that with vaine delightes,
> And idle pleasures in her *Bowre* of *Blisse*,
> Does charme her loures, and the feeble sprightes
> Can call out of the bodie of fraile wightes:
> Whom then she does transforme to monstrous hewes,
> And horribly misshapes with vgly sightes,
> Captiu'd eternally in yron mewes,
> And darksome dens, where *Titan* his face neuer shewes. (II.V.27)

Acrasia imprisons weak male lovers in her den, which is an "yron mewe" not only because it is dark, enclosed and impossible to escape, but also because within it "her lovers" are transformed, like animals who molt, and indeed sometimes into animals and, most appallingly, into women. In his influential analysis of the Bower of Bliss episode in canto twelve, Stephen Greenblatt connects that episode's destructive energy with Spenser's and other English colonists' fears of being transformed into the colonial other—in the "New World" and also in Ireland (1980).[3] Acrasia's "mewe," like Ireland itself, threatens to turn men into monsters, especially into monstrous not-men, into men who are like boys and women,[4] men like her lover Cymochles who, "Hauing his warlike weapons cast behind, / . . . flowes in pleasures, and vaine pleasing toyes, / Mingled emongst loose Ladies and lasciuious boyes" (II.v.28.7-9). In Acrasia's bower, Cymochles is not "*Cymochles,* oh no, but *Cymochles* shade, / In which that manly person late did fade,". . . "all his force forlorne, and all his glory donne" (II.v.35.4-5,9); he is not masculine but "womanish" and "weake" (II.v.36.2). Even after Guyon destroys Acrasia's bower and his Palmer strikes the transformed beasts with his staff to reverse her spells, her mewed victims have lost their masculine power: "Yet being men they did vnmanly looke" (II.xii.86.3). The *Amoretti's* "pleasant mew" threatens the same transformation: like Guyon's, the sonnet sequence's energies are directed toward destroying the threat of emasculation and transformation into the colonized other inherent in Spenser's affair with Ireland. In the sonnets, rather than facing the "monstrous hewes" created in Acrasia's bower, the poet can contemplate the "heavenly hew" of his ultimately submissive beloved.

Although earlier centuries dismissed the *Amoretti* as inferior poetry, in recent years the poems have attracted a flurry of critical attention. At least eight dissertations written since 1980 either take the sequence as their sole subject or treat it extensively; and during the last twenty years, critics have produced two books (Gibbs and Johnson) and a number of articles devoted to the poems, ranging in their approaches from exploring the poems' numerological and calendrical significance (Dunlop 1969, Fukuda, Kaske, and Lowenstein), to seeing them as documents of the poet's personal growth or sexual anxieties (Dunlop 1980 and Marchand), to considering the sonnets in relation to Petrarchan conventions (Dasenbrook, Klein, Neely, and Quitslund).[5] Robin Headlam Wells and Catherine Bates have also broadened the context of reading to consider the poems as they may have functioned within Spenser's attempts to secure patronage, especially from Queen Elizabeth. Wells sees the queen as the primary addressee of the sonnets; Bates sees the conflicted portrait of love as reflecting the poet's position vis-à-vis the queen whom sonnet LXXIIII explicitly names and sonnet LXXX refers to as "the Faery Queene."

At the same time that critical attention has reenvisioned the *Amoretti,* there has been a resurgence of work on Spenser in Ireland, concentrating on *The Faerie Queene* as well as on his polemical *A View of the Present State of Ireland,* and initiated by Greenblatt's groundbreaking *Renaissance Self-Fashioning* and by historians' reconsideration of the literary text as supplying crucial evidence about this planter's political views and colonial aspirations (Baker 1997, Bradshaw 1987, Bradshaw et. al., Canny 1983, Carroll, Cavanagh, Coughlan, Hadfield 1997b, Highley 1997, Maley 1997b, McLeod, Murphy 1999, and Stillman).[6] But despite Patricia Coughlan's astute caution that "The work of Greenblatt and Norbrook has shown clearly that th[e] myth" of an apolitical Spenser—"the comfortable notion of a dreamy creature of the imagination, discussable in a sphere set apart from the iron insistencies of the *View*—cannot be sustained" (48), this important work on Spenser and Ireland has not led critics to reconsider the *Amoretti* as implicated in his colonial career.[7] Recently, Montrose has revised his claim about Spenser's lyric poetry, arguing that the *Epithalamion* celebrates the founding of Spenser's home in Ireland (1996, 109). But even this careful reading of Spenser's shorter poetry fails to situate the *Amoretti* within Spenser's Irish career and ambitions. Neither of the two books devoted to the *Amoretti* since 1980 mentions Spenser's Irish experience. Indeed, most critics of the sonnets see the poems as retreating from Spenser's public career to the realm of the personal, the domain of love.[8]

However, Kim Hall has brilliantly demonstrated that "the supposedly insulated language of the [sixteenth and seventeenth century] love lyric is shot through with reference to foreign difference and foreign wealth" (65). As Hall indicates, "New World" colonial interests underlie and determine Sidney's lexicon in *Astrophel and Stella*. Likewise, Spenser's *Amoretti* shares a vocabulary of conquest, rebellion, and domination with Spenser's colonial life and writings. In a way that Spenser's actual experience in Ireland did not, the love lyric enabled the poet to effect a conquest in the face of rebellion; and the lyric representation of conquest was predicated on a feminization of Ireland. Clare Carroll notes that Spenser's *A View of the Present State of Ireland* "represents the Irish as a feminized, culturally barbaric, and economically intractable society" (163). As the *Amoretti* represents an "ideal" love relationship, it provides a space for the poet to play out the dynamics of male control over a woman that the colonist so deeply desired over a feminized Ireland.

The sequence thus participates in an English personification of Ireland as a subservient woman, a vision that we will see in chapter 4 in Ben Jonson's *Irish Masque at Court*.[9] Indeed, the image of submissive, feminized, colonial territory is ubiquitous in English colonial discourse of the period. In a prefatory poem to "A Relation of the Second Voyage to Guiana," George Chapman pictures a feminized Guiana paying homage to Queen Elizabeth: "on her tip-toes at fair England looking, / Kissing her hand, bowing her mighty breast, / And every sign of all submission making" (39).[10] Chapman's embodied Guiana attests to England's superiority, affording her all the manifold physical gestures of deference characteristic of

Elizabethan society. Thomas Morton in his *New English Canaan* (ca. 1635) envisioned that territory

> Like a faire virgin, longing to be sped,
> And meete her lover in a Nuptiall bed,
> Deck'd in rich ornaments t'advaunce her state
> And excellence, being most fortunate,
> When most enjoy'd, so would our Canaan be
> If well employ'd by art and industry
> Whose offspring, now shewes that her fruitfull wombe
> Not being enjoy'd is like a glorious tombe. (10)[11]

Here a barren but potentially fecund New England desires nothing more than a colonial embrace; only through colonization can she "advance."

In Richard Hakluyt's long dedicatory epistle to Sir Walter Raleigh, prefaced to a 1587 edition of Peter Martyr's *Decades of the New World,* Hakluyt exhorts Raleigh to pursue his Virginia, who awaits his "embraces" to "bring forth new and most abundant offspring," as "no one has yet probed the depths of her hidden [but absolutely available] resources and wealth" (*Original Writings* II 367; Taylor trans.). While Hakluyt's virgin land, like Morton's new Canaan, awaits sexualized exploration to prove her fertility, other colonial promoters used the virgin image to project purity and/or violation, but all in the service of furthering plantation. In a 1609 sermon for the Virginia Company, William Symonds prayed, "*Lord finish this good worke thou hast begun;* and marry this land, a pure Virgine to thy kingly sonne Christ Iesus; so shall thy name bee magnified: and we shall haue a Virgin or Maiden Britaine, a comfortable addition to our Great Britaine" (A3v). By 1609, England's Virgin Queen was dead, and the virgin land named for her could be envisioned as a submissive sister to James's embodied greatness. In his report on the colony, William Strachey presented Bermuda and the Virginia colonial project to a noblewoman patron:

> once more this famous businesse, as recreated, and dipped a new into life and spirit hath raysed it (I hope) from infamy, and shall redeeme the staines and losses vnder which she hath suffered, since her first Conception: your Graces still accompany the least appearance of her, and vouchsafe her to bee limmed out, with the beautie which we will begge, and borrow from the faire lips: nor feare you, that she will returne blushes to your cheekes for praysing her, since (more then most excellent Ladie) like your selfe (were all tongues dumbe and enuious) shee will prayse her selfe in her most silence: may shee once bee but

seene, or but her shadow liuely by a skilfull Workman set out indeed.
(*A true reportory* 1756)

Here the relative colonial success in Bermuda beautifies a stained Virgin, but it is the noble woman's patronage that will lend grace to the bedraggled, feminized "businesse." Virginia is a silent woman, taking her beauty and character from the English enterprises involving her. These texts do not offer a single feminized colonial other, but whether a lady-in-waiting, a chaste or an eager virgin, she submits happily and completely to England and its virile male representatives. Of course, the colonial other as subservient woman is not an image confined to English discourses; as Kadiatu Kanneh notes in her discussion of Fanon's work on images of Algeria, "the feminizing of colonized territory is, of course, a trope in colonial thought" (346). Spenser's sonnet sequence is an early modern English poetic example of the construction of that trope.

The Epistle Dedicatory published with the sonnets in 1595 dedicates the sequence to Sir Robert Needham, Knight, in order to congratulate him on his "safe return from Ireland." This letter reminded the *Amoretti*'s original public that the poems were crafted at the site of England's primary colonial encounter, and that they and their maker were, therefore, potentially as endangered as their dedicatee. The letter also explicitly informs its readers that Spenser remains in Ireland: his "sweete conceited Sonets" must be published "in his absence" from the England that has long awaited more poetry from "this gentle Muse." In light of their accompanying epistle, the poems can be seen as a species of news from the battlefront. And Spenser's readers and the poet himself would have been well aware that Ireland was a battlefront.

The *Amoretti* was published in 1595, the second year of Hugh O'Neill, the Earl of Tyrone's, Ulster rebellion that was already spreading to the south and west. The year 1595 was disastrous for English forces in Ireland: Enskillen fell to the Irish; Tyrone's brother burnt Blackwater fort, and Tyrone defeated Marshal Bagenal's relief army at Clontibret. Even in 1594, the news from the North of Ireland was frightening. Enskillen was under siege by the earls O'Donnell and Maguire, and on August 7 Maguire defeated the troops sent to relieve the royal army (Ellis). In 1621, Philip O'Sullivan Beare described O'Donnell's 1595 invasion of Connacht:

> In his raids extending far and wide he destroyed the English colonists and settlers, put them to flight, and slew them, sparing no male between fifteen and sixty years old who was unable to speak Irish ... After this invasion of Con-

nacht, not a single farmer, settler or Englishman remained, except those who were defended by the walls of castles and fortified towns, for those who had not been destroyed by fire and sword, despoiled of their goods, left for England, heaping curses upon those who had brought them into Ireland. (211)

Beare's account may well be wishful, written as it is by an Irishman after the flight of the earls, but, if we cannot read it as history, it certainly reflects Gaelic Irish sentiment about the English planters.

In 1595 the Munster plantation, of which Spenser's Kilcolman estate and lands formed a small part, had been established for only about six years and was soon to be destroyed by a rebellion of the Irish population joined by many of the Old English settlers (Quinn 1966, 21; MacCarthy-Morrogh 107).[12] David Beers Quinn describes the plantation's provenance: "every piece of land in Munster was held by some old English or Irish landholder and the planters were being inserted only into certain fragments seized into the hands of the crown as a result of the Desmond rising of 1579-83" (28). The previous owners of every piece of seized land petitioned the English government to get their property back, but those petitions were dismissed summarily by the Englishmen planted on the stolen land: "albeit their evidences be fair and very law-like without exception, yet because fraud is secret and seldom found for her Majesty by jury, we have put the undertakers, for the most part, in possession, who, dwelling but half a year upon the lands, shall have better intelligences to discover the false practices than the commissioners can possibly learn out" (*Irish History* 245). Elizabeth's solicitor-general blatantly stated, "we have directly refused for this time to take proofs by witnesses, for that admitted, her Majesty should have little land left" (246).[13] After the 1588 commission on land dispute judgments, "one of the dismissed O'Mahony claimants moved in and burnt Castlemahon, now the undertaker's residence, also removing £200 of possessions." Other undertakers were robbed, assaulted, had property burned, and Spenser himself was boycotted (MacCarthy-Morrogh 131). In 1594, rebels actually started killing English settlers (MacCarthy-Morrogh 132).

As one of the Munster undertakers living in an escheated castle, Spenser was surrounded by Irish and Old English resentment. It is hardly surprising that these earlier landowners "could see no valid reason whatever" for the new plantation (Quinn 1966, 28). On October 19, 1599, four years after Spenser published the *Amoretti* and a year after Kilcolman was burned in the rebellion, the Chief Justice of Munster, William Saxey, wrote to the Earl of Essex that "the inhabitants of the province are grown into such a hatred of the English Government, that no service can be done by any of

her Majesty's forces, unless they be able to fight as well against the pre-
tended subjects as the open rebels; for in that action, against the English,
either they shrink from her Majesty's forces, and are lookers on, or [they]
join with the rebel" (191). Saxey describes a situation grounded in a long-
standing animosity that was the atmosphere Spenser breathed in Cork. The
poet described his position in Ireland through negation in *Colin Clouts
Come Home Againe:* England, unlike Munster, is a land where

> No wayling there nor wretchednesse is heard,
> No bloodie issues nor no leprosies,
> No griesly famine, nor no raging sweard,
> No nightly bodrags, nor no hue and cries;
> The shepheards there abroad may safely lie,
> On hills and downes, withouten dread or daunger:
> No ravenous wolves the good mans hope destroy,
> Nor outlawes fell affray the forest raunger. (312-19)

Although *Colin Clouts* was published the same year as *Amoretti* and *Epithalamion,* Spenser dated its dedication 1591 and the poem narrates his
1591 trip to England with Raleigh. By 1595 *Colin's* "bloodie issues," "rag-
ing sweard[s]," and "outlawes" had manifested in the rebellion threatening
Spenser's escheated castle: "my house of Kilcolman," as it is named in the
poem's dedication to Raleigh (526). The *Amoretti* was written in a state of
potential siege, and the sonnets' account of a gendered conquest must have
been much more reassuring than the news of actual conditions in Ireland.

Even if we were to see Spenser's courtship of Elizabeth Boyle as the
poem's sole topical reference, that courtship could only have signified in its
Irish context. As Kanneh notes, "black and female identities are not simply
figurative or superficial sites of play and metaphor, but occupy very real
political spaces of diaspora, dispossession and resistance" (348). While we
cannot read Elizabeth Boyle as a figure of colonial resistance, we can see
her as a woman brought to Ireland by her father's first cousin's colonial
interests, and she was overtly courted within a colonial context. We know
little about Spenser's second wife, but her most important kinsman for
Spenser was surely Richard Boyle, one of the most successful New English planters who became Earl of Cork in 1620 and Lord High Treasurer
of Ireland in 1631. Boyle came to Ireland to make his fortune in June of
1588, and presumably Elizabeth followed him there; certainly they were
still connected significantly after Spenser's death, as Elizabeth named
Richard Boyle godfather to her first child by her second husband.[14] In

1590, during Spenser's lifetime, the English crown made Boyle deputy escheator to John Crofton, escheator general of Ireland. Boyle was thus responsible for confiscating land for the crown from Irish "rebels," estates like the one Spenser occupied in Cork. In 1595 Boyle married Joan Ansley, a Limerick heiress; in 1602 he acquired a seignory of twelve thousand acres in Cork, Waterford, and Tipperary, including Sir Walter Raleigh's Munster estates, and by 1622 Boyle held seven seignories in Ireland (*DNB, Repertory,* Canny *1982*).

Had Spenser lived, his alliance through marriage with Boyle might have granted him the extensive economic interests the poet had hoped for in his Irish career. Throughout Boyle's lucrative Irish career he took pains to appoint his relatives to prominent positions and to connect himself by marriage to Irish noble families, and the "relatives of [England's king's] most prominent courtiers" (Canny 1982, 32). In order to raise and cement his own social position, the earl ceaselessly promoted his relatives' interests (Canny 1982, 44-45), and he was particularly interested in establishing himself and his family within English noble society. His marital negotiations with English nobility and his loans of great sums to Charles I were serious efforts to make the Boyles into a great English and Irish noble family (Canny 1982, 34). Elizabeth Boyle, therefore, has to be seen not only as a love interest for Spenser, whose great dream was to achieve the most noble patronage, but also as an object of the poet's ambition. Spenser and Elizabeth married in Cork, and her own future and her subsequent marriages took place wholly in Ireland. Because of the marriage, presumably, all of Spenser's descendants lived and died in the colonial stronghold (Coleman 1895). Thus Spenser's triumph in love as depicted in the sonnets and the accompanying *Epithalamion* ensured his family's future in Ireland.

The *Amoretti* directly mentions only three historical personages besides Elizabeth Boyle: his mother Elizabeth, Queen Elizabeth, and Spenser's friend, Lodowick Bryskett. Except for the poet's mother, each of these figures was intimately connected to Spenser's Irish experience. As Spenser's hopes for increased fortune in Ireland rested at least in part on his wife, Elizabeth, his quest to return in glory from his exile there rested solely on the queen.[15] *Colin Clouts Come Home Againe* compliments Queen Elizabeth as the refuge for the poet who describes his Irish plight: "That banisht had my selfe, like wight forlore, / Into that waste, where I was quite forgot" (182-83). Spenser's compliments to Elizabeth in that text and the sonnet sequence are part of his attempt to gain English patronage and eventually English residence from the queen. Had those attempts succeeded, Spenser would no longer have had to defend himself from the menacing transformative experience of Ireland. Bryskett, the friend to whom Spenser addresses sonnet XXXIII, shared that transformative experience. Both men served in Lord Grey de Wilton's colonial administration of Ireland; both served on the council of Munster, Bryskett as secretary in 1582 and Spenser, we believe, as clerk in 1584 (*DNB,* Jenkins 1932). His direct address to Bryskett, like the poems' involvement in Elizabeth Boyle's courtship, situates them in their colonial milieu.

In sonnet XXXIII, Spenser's persona complains to "lodwick" about *The Faerie Queene*'s unfinished state. The sonnet laments that the poet's "wit" is "tost with troublous fit / of a proud love, that doth [his] spirite spoyle" (11-12). The love relationship depicted in *Amoretti* has disabled his work and even undermined his confidence in its feasibility. To Bryskett's straw-man importunities, the persona pleads his distraction and tortured "spirite." This reference to the beloved's "troublous" pride is only one of the *Amoretti*'s numerous complaints about its beloved's central flaw. Despite assessments like Arthur Marotti's, in his brilliant and seminal essay on Elizabethan sonnet sequences, that "unlike the other sonnet sequences of the 1590's, the *Amoretti* celebrates a relationship of amorous mutuality," the sequence's portrait of the beloved as a cruel rebel whose pride injures her lover has long disturbed its readers and critics (416). In 1956, J. W. Lever suggested separating eighteen sonnets from the sequence based on their brutal depiction of the beloved: "The lady is represented not only as proud and disdainful—familiar attributes of the Petrarchan heroine—but also as sly, licentious, savage, and guileful . . . Precedents may be found for one or another in French or Italian sonnets of the time, but their cumulative effect is quite distinctive" (102). Donna Gibbs agrees that "it is conventional for a mistress to be accused of having a hard heart or behaving tyrannical or cruelly but Spenser's lover goes beyond this" (38). Bates also calls attention to the "persisting images of constraint and imprisonment in the *Amoretti*" (82).

In an effort to deal with the distinctive anger in these sonnets, Louis Martz calls them "close to mock heroic" and suggests that they are forms of extreme tribute (156). Alternatively, Dasenbrook sees them as metasonnets that are "almost parodically Petrarchan" (45). At least two critics compare the *Amoretti*'s beloved to *The Faerie Queene*'s warrior heroine Britomart (Bieman, Villeponteaux). Mary A. Villeponteaux notes that "the lady as combative warrior is a common trope, almost a necessary part of sonnet convention, but this trope receives particular emphasis in *Amoretti*" (31). But the *Amoretti's* beloved is far more cruel than the epic poem's female warrior; *The Faerie Queene* never describes Britomart bathing "her cruell hands" in the blood of wretched "thralls" (*Amoretti* XXXI, 10-11). When we place the love depicted in the sonnets in the context of the colonial front, however, rather than requiring explanation, their language sounds familiar. Spenser borrows the *Amoretti's* vocabulary as much from his colonial experiences and writings as from the sonnet tradition that he aims to join and extend.

A *View of the Present State of Ireland* takes as its central problem the lack of proper submission on the part of the Irish and Old English. The text presents the Irish people as justly·conquered but refusing that status, as Spenser laments that the Irish have "quite shaken of theire yooke and broken the bandes of theire obedyence" (6-7). Many Irish, he decries, have never acknowledged their "subiection" (8), and many of those who did submit "so soone as they were out of sight by them selues shooke of theire brydles" (9). The Irish, Eudoxus declares, stand "stiflye against all rule and gouerment" (19). The Anglo-Normans who settled Ireland and owed their allegiance to the English crown equally stubbornly refused their proper submissive relationship to their queen. A *View* explores both the cause of this stiffness and the means by which the English may bridle the Irish and Old English. Irenius plans, after his projected military campaigns, to ask all outlaws—by which he means rebels—to "freelie come in and submite them selues to her maiesties mercie." "I nothinge doubte," he tells Eudoxus and his audience, that after they are properly punished, "they will all most readylie and vpon theire knees submitte them selues" (159).

Like *A View of the Present State of Ireland*, the *Amoretti* worries the problem of submission. But unlike many other Petrarchan English sonnet sequences that also concern themselves with an unwilling beloved, the *Amoretti's* lexicon for the problem derives from an actual battlefield.[16] The sequence's preoccupation with the problematics of submission and its bellicose vocabulary both appear in its first sonnet, which figures the poems themselves as "lyke captives trembling at the victors sight" (1.4). That first sonnet uses conventional sonnet imagery: the sonnet mistress's neglect will kill the beloved; the lover is a "dying spright"; the beloved has "lily hands," "lamping eyes," a "blessed looke"; she is an angel who is both the lover's muse and his "heavens bliss." But the image of the lover's poetic leaves shaking like captured soldiers in her hands is considerably less conventional. In this first poem, the love that ties the *Amoretti's* lover to his beloved is intimately related to military conquest, for those shaking leaves are held in "loves soft bands." Many sonneteers of the period complain of the beloved's tyranny when she will not accept their lovers' pleas, but Spenser's lover here represents a joyous relation based in captivity—the leaves are "[h]appy" to be held by the beloved and yet they are "lyke captives trembling at the victors sight."

This is exactly the happy relationship that Spenser's Irenius recommends and projects for the recalcitrant Irish. The Irish will never be happy except in complete submission, when they kneel trembling before the victorious English. But while this relation has the proper power dynamic, the opening sonnet improperly genders the captivity scenario. Rather than the beloved submitting to the lover and thereby constituting the orthodox hierarchical dyad of early modern English marital relations, the first sonnet's opening lines picture a miniature world turned upside down in which the lover metonymically trembles at the beloved's hands. Not only does the lover beg for favors—the customary position of an early modern English sonneteer—but he also is reduced to leaves that happily tremble at the hands of the beloved as "victor." The sonnet sequence will continue to meditate on captivity, but it will also labor to gender this scenario properly.

As Eudoxus and Irenius discuss English laws against Irish apparel in *A View*, Eudoxus explains that men's "condycions are oftentymes governed by theire garmentes" (90). He attributes to Aristotle the following story:

> Then when *Cyrus* had overcome the *Lydians* that were a warlike nacion, [and] devised to bringe them to a more peaceable lyfe, he changed theire apparrell and musicke, and insteade of theire shorte warlike Coates, clothed them in longe garmentes like wyves, and in steade of theire warlike musicke, appointed to them certaine lascivyous layes and loose gigges, by which in shorte space theire myndes were [soe] mollyfied, and abated that they forgott theire former fiercenes and became most tender, and effeminate. (90-91)

Eudoxus's story illustrates the conventional early modern links between conquest and gender.[17] Victory in war is a triumph of masculinity, and the victor has the right to emasculate the conquered soldiers in order to signify their submission and ensure its longevity. In Eudoxus's story, the Lydians become women, and as women should, they submit to their conquerors whose military victory both confirms and constructs the conquerors' masculinity.

Spenser's poetic texts also obsessively ponder the problem of who should submit—who will become "like a wife" in a war scenario. In *The Faerie Queene*'s book five, Spenser's knight of Justice, Artegall, encounters a group of Amazons led by Radigund, who proposes to settle the war between Artegall's forces and her own in a single combat. Artegall is winning the battle, until unlacing her helmet, he relents, pitying her beauty:

> For though that he first victorie obtayned,
> Yet after by abandoning his sword,
> He wilfull lost, that he before attayned.
> No fayrer conquest, then that with goodwill is gayned.
> (V.v.17)

Emasculated in his submission ("abandoning his sword"), Artegall is immediately further feminized when, in a much-discussed episode, Radigund dresses him "in womans weedes," hands him a distaff, and sets him to spinning among other similarly attired captured knights. Here in the midst of the narrative of colonial conquest—framed as a reclamation of one's own—that is book five's mission, defeat at the hands of a colonial other can only come through self-abjection, and that defeat leads immediately to a gendered disempowerment.

The Amazons' central fault—that which defines them as Amazon in book five—is their prideful unwillingness to submit. When Artegall rescues Sir Terpin from their hands, he asks, "have ye yeelded you to proude oppression / Of women's powre, that boasts of mens subjection" (V.4.2). In response, Terpin explains his misfortune by his knightly desire to confront the "proud Amazon [who] did late defy / All the brave knights that hold of Maidenhead" (5.4.29). Just as "proude oppression"—subjection to women's power—shames a knight completely, so adherence to the code of knighthood requires one to confront that power. And that power, that refusal to submit, threatens not just the knight in the encounter but all knights who serve under the shield of Maidenhead. Throughout the episode of the Amazons, the narrator constantly reiterates their pridefulness as well as that gendered sin's identification with women's power and male subjection. In book five's world, women's pride always implies women's power, which in turn always implies male subjection. When Artegall, Sir Terpin, and Talus enter the Amazons' city, the narrator warns that they confront "Vnknowen perill of bold womens pride" (5.4.38). The Amazons pose a threat not because of their military power, not as alien troops, but as the sin itself. Women's pride is the alien; the alien is women's pride.

The Faerie Queene always describes its exemplary Amazon, Radigund, in terms of her pride. The text introduces her to the reader as a "Princesse of great powre, and greater pride" (5.4.33). Her aspect is prideful and even her apparel demonstrates her sin; when she enters the single combat she wears "an embroidered belt of mickell pride," and she bears "stately port and proud magnificence" (5.5.3,4). And just as her appearance denotes her dangerous character, her action in the poem stems from that sin: when Artegall is winning the single combat, she reacts out of her "vengeful wrath & sdeignfull pride half mad" (5.4.43). Spenser often used the form "sdeign" where we now—and many of Spenser's predecessors and contemporaries— would write "disdain." He uses various manifestations of the word at least five times in *The Faerie Queene* and also at least twice in his minor poetry (*OED* "sdeign"). His employment of the word in the *Amoretti* indicates its resonances for Spenser; he compliments the prideful beloved: "For in those lofty lookes is close implide, scorn of base things, and sdeigne of foul dishonor" (V.5-6). What characterizes Radigund's pride as it appears in the combat is its haughtiness, its refusal to bow to any other power, its thinking that Artegall is "(a thing) unworthy of (something)"—that is, Radigund's submission (*OED,* "disdain").

One of the *Oxford English Dictionary's* other definitions for "disdain," derived from its usages in the infinitive or gerundive, is "To think it beneath one." This definition precisely captures the danger of Radigund and the Amazons as well as any alien group such as the Irish and rebellious Anglo-Irish who will not submit to the dominant group or order—that that group will think itself above and may, therefore, rise above the dominant group. It is not accidental that Milton, one of Spenser's most committed and astute readers, has his Satan complain: "lifted up so high / I sdeind subjection, and thought one step higher would set me highest" (4.49-51). Satan's refusal to submit to God mirrors Radigund's Amazonian rejection of properly gendered subjection. As Clare Carroll suggests, "the depiction of Radigund as a rebellious female, who inverts the subjection of woman to man decreed by natural law, makes her analogous to the Irish, whose 'barbarity' demands they be 'reduced to civility' by the English" (182).

Especially in the sonnets that have so disturbed critics, the *Amoretti* repeatedly accuses its beloved of the Amazonian sin that in *The Faerie Queene* imprisons women in Lucifera's dungeon: "Proud wemen, vaine, forgetfull of their yoke" (1.5.50). In Lucifera's dungeon and throughout the epic, a proud woman by definition will not be yoked, will refuse her place under a man; and in the sonnet sequence, the beloved's pride is what enables her to refuse her properly yoked position. While pride is indeed, as Lever notes, a familiar Petrarchan accusation, the *Amoretti* concentrates its lover's anxieties on that alien sin. According to Herbert S. Donow's concordance to five Elizabethan sonnet sequences, the word "pride" occurs nineteen times in Spenser's relatively short *Amoretti* (eighty-nine poems). In contrast, Shakespeare's 154 poem sequence uses the word eleven times, Sidney's *Astrophel and Stella* six times, Samuel Daniel's *Delia* six times, and Michael Drayton's *Idea* only twice. The *Amoretti's* lover addresses his beloved as "that fayrest proud" and "Faire proud" (II.9, XXVII.1). Her "rebellious pride" prevents her from yielding to him (VI.2), and she will not submit to Cupid himself: she "proudly disobayes" love's "precept" (XIX.2). In sonnet X, the lover asks Cupid to punish "her proud hart" and force the beloved to bring low her "high look, with which she doth comptroll / all this worlds pride" (9, 10-11).

In *A View of the Present State of Ireland*, Spenser complains that Irish bards celebrate those criminals who are "most dangerous and desperate in all partes of disobedyence, and rebellious disposition" (95). Likewise, in the *Amoretti* the beloved's pride defines her as "a rebell" (XIX.14), who endangers her lover, causing him to plead, "better were attonce to let me die, / and shew the last ensample of your pride: / then to torment me thus with cruelty, / to prove your powre, which I to wel have tride" (XXV.5-8). Sonnet XXXVIII also accuses her of killing the lover because of her pride: "But in her pride she dooth persever still, / all carelesse how my life for her decayse" (9-10). And in a later sonnet, the beloved fishes for men with her smiles and then "kills [them] with cruell pryde" (XLVII.7). Frequently the poems complain that her pride obscures her good qualities, particularly those which would denote her submission to the lover: a "cloud of pryde" hides "her goodly light" (LXXXI.7-8); her pride "displace[s]" her "meeknesse" and "mild pleasance" (XXI.3,5) and it "depraves each other better part" (XXXI.3). Like Radigund's, her pride is willful, and it gives her the power to resist: sonnet XXXII compares the beloved to a piece of iron made so hard in its "wilfull pride" (10) that it resists all the efforts of the lover as smith and finally turns to stone. Like Ireland, the beloved remains "altogeather stubborne and vntamed" (*A View*, 6).

Unlike *The Faerie Queene*, which carefully cordons off pride from its exalted women, not all of the *Amoretti's* sonnets reject the beloved's pride.[18] But the few sonnets that seemingly praise what the epic rejects as sin divide into two groups: one group of poems, some of which seem directed to Queen Elizabeth, commends the pride of an exalted beloved as fitting her station; the other group values the beloved's pride for the piquancy it adds to submission. Sonnet V praises the beloved's "portly pride," under attack by the world, as a shield against what is base and dishonorable. Like sonnet V, sonnet LXI commends the lover's "Idoll"'s rejection of what is base. Sonnet LXI seems unmistakably to be addressed to the queen as its rhetoric resembles the language of the cult of Elizabeth: the addressee is the image of God's beauty (1), she is "divinely wrought" (5), she is addressed as "soverayne" (2), the lover is tied to her by "bounds of dewtie" (3), she is of the same stock as the angels and "adorne[d]" with all the gifts of the "Saynts" (6-8). The sonnet characterizes her in the same terms that Spenser uses in his epic to praise Belphoebe, an image of Elizabeth who is the "glorious mirrhour of celestial grace" and is also addressed as "soveraine" (*FQ* 2.3.25.6,7). The lover shares his position of worship toward this object with other "men of meane degree" (14).

Two sonnets play on the theme that proud obduracy will make victory sweeter. Sonnets VI and LI both posit a more valuable, enduring, and "stedfast" love as the result of difficult achievement, and VI specifies "rebellious pride" as the obstacle the lover must clear. And unlike the Irish of *A View,* whom the text can only desire to see prostrated before their conquerors, the *Amoretti's* beloved does submit within the sequence. Where Artegall bows before the embodiment of alien female pride, the *Amoretti's* prideful beloved properly abandons her resistance, and her female pride sweetens the victor's pot. The sonnet's "proud fayre" yields in much the same way that Artegall willingly submits to Radigund; but in the beloved's case that submission completes a proper conquest and will not need to be brutally reversed. Book five of *The Faerie Queene's* scenario and the *Amoretti's* beloved's submission show the opposite sides of a colonialist coin. In book five, the conqueror shamefully yields and enacts a self-feminization. In the *Amoretti,* the pursued beloved yields on the same terms and produces a properly gendered disempowerment. As Gary Waller notes, the *Amoretti* "espous[es] a view of the relationship between poet/lover and his beloved that unambiguously advocates a Christian hierarchy of male control and female subordination" (66). In so doing, the sequence adjusts the scenario it adduces in its first poem in order to correctly gender the colonial scene.

Sonnet LXVII, perhaps the sequence's most commented upon poem, reprises Petrarch's *Rime* 190, to show the beloved as an escaped deer turning towards her hunter until he "in hand her yet half trembling tooke, / and with her owne good will hir firmely tyde" (11-12).[19] The lover wonders at the "strange" sight, "to see a beast so wyld, / so goodly wonne with her own will beguyled" (13-14). Now, instead of seeing the lover as we do in sonnet I, troped as his poetic leaves, trembling in the beloved's powerful hands, we see the lover's hands holding the beloved's/deer's shaking body. He stands victorious before her. Rather than being held in love's bands, he will tie his beloved firmly so she cannot escape. In both *The Faerie Queene* 5.5.17 and *Amoretti* LXVII's conquest scenes, the victory gains its value from the willingness, the assent of the victim. And in each case colonial control and gendered submission are inextricable. Radigund gains complete control of Artegall because he submits himself to her: "no fairer conquest, then that with goodwill is gayned" (5.5.17). It is this "willful" submission—this lack of the will to resist—that enables the Amazon queen to imprison and emasculate him completely. Likewise in sonnet LXVII, the beloved's willful submission enables her to be tamed by the hunter who has been exhausted by the chase. Anne Lake Prescott notes that

> many deer legends . . . have a deep and archaic connection with the foundation of cities, churches, empires, and dynasties . . . Stories of Caesar's long-lived and collared deer, the cruciferous deer of saints like Eustace, the deerchase leading to sovereignty in Celtic legend, the deer associated with the beginnings of cities, even the deer who guide Charlemagne or Roland to places of crossing or safety—all tie the establishment or attaining of something new (religious or political) to a creature whose pursuit nonetheless normally involves pain and exhaustion. (36)

In LXVII's "trembling" deer, "firmely tyde," and willingly "beguyled," Spenser imagines a new colonial settlement, the end of proud, rebellious resistance, an acquiescent Ireland who will happily obey her master's summons and commands. The deer's submission converts a disempowering chase to a "firm" establishment of masculine power.[20]

As a result of this joyous denouement, the sonnet lover imagines the beloved employing the hands that once held his written work captive to embroider a scene of potential captivity, embellished "with woodbynd flowers and fragrant Eglantine" (LXXI.10). Although a number of critics have been charmed by sonnet LXXI's spider and bee vignette, the story realizes LXVII's ominous implications for the "beguyled" deer. Where once the beloved's hands had the power to hold the lover's trembling poetic leaves captive, now they are engaged in copying his work—that work depicting and glorifying her captivity—in the proper female medium: she will embroider a vision of herself

> . . . caught in cunning snare
> of a deare foe, and thralled to his love:
> in whose streight bands ye now captived are
> so firmely, that ye never may remove. (LXXI.5-8)

In *Colin Clouts Come Home Againe*, as in this sequence, Spenser imagines his Ireland as a beautiful prison; in the *Amoretti*, the lover imagines that she who submits will live in his prison and embellish her cage. This poem has enchanted readers; however, in the context of the sequence and its colonial concerns, sonnet LXXI reveals its stakes: its fantasy of peaceful and willing submission may seem heavenly, but in the real world, spiders trap and then consume "the gentle bee."

As Prescott suggests, as well as transforming many of Spenser's sources, sonnet LXVII brings together several of the *Amoretti*'s themes: Christian sacrifice, erotic love, and what Prescott sees as mutual constraint. That sonnet also marks the happy end of the sequence's extensive meditation on captivity and submission. That meditation, which could look conventional in a genre that often plays on the trope of being captured by love, is instead in the *Amoretti* another significant reason that many of its sonnets have disturbed critics like Lever and Gibbs. For the *Amoretti* phrases its meditation on captivity in graphic terms from a real world of kidnapping and military conquest. Although the sequence has its share of conventional love imagery that uses similar language—the beloved's golden hair captures the lover's eyes (XXXVII); his heart is trapped in her body (LXXIII)—many of its tropes and scenarios of captivity are genuinely frightening, envisioning a bloody world in which submission transcends metaphor: the world in which a royal proclamation against Tyrone declared that "he has since taken by force two others of the said Shane O'Neill's sons, holding them captives in places unknown. Aspiring to live like a tyrant over a great number of good subjects" ("Royal Proclamation" 175). That same proclamation offers the Ulster rebels, "upon their submission," "pardon of their lives and lands" if they come in to the Lord Deputy when Elizabeth's army reaches Ulster (175). Elizabeth's offering Tyrone a complete pardon in 1595 indicates the English army's precarious position in the North, and Spenser might well have anticipated that the situation in the South could easily become as unstable. By 1596, a Limerick rebel, Rory McSheehy kidnapped an Englishman and wrote to Sir Thomas Norris, "I will burn, spoil, kill and hang as many of your countrymen as I can catch" (quoted in MacCarthy-Morrogh 133). Spenser's 1595 sequence's preoccupation with captivity echoes his daily world's threatening possibilities.

The *Amoretti*'s sonnet X accuses the beloved of "massacr[ing]" "captives" with her eyes—this "whiles she lordeth in licentious blisse / of her freewill" (6,7,3-4). Each trope standing alone could easily take its place in any Elizabethan sonnet, but together in this poem they shock readers. However, their language would be familiar in the embattled colonial front the poet inhabited, and the vocabulary of sonnets X, XI, and XII emerges from that front. The beloved is a "warriour" who "addresse[s]" herself "to battell" against a lover reduced to offering "hostages" "for peace" (XI.1-4); she suspends him between life and death "with torment and turmoyle" (11) in a state of endless war. Sonnet XII continues this scenario, describing a day on which the lover sought "a truce and termes" but was "ambush[ed]" by a force in the beloved's eyes that forced him to commit treason against himself and to "yeeld," and remain captive in her "cruell bands" (2,6,4,10,12). Although the complaint in the sonnet's final couplet—"So Ladie now to you I doo complaine, / against your eies that justice I may gaine"—seems to change the poem's register, indicating that this is only love poetry after all, the sonnet's diction embeds it in the real world of war that English contemporary texts on Ireland describe.

In *A View of the Present State of Ireland,* Irenius explains to Eudoxus that the English should offer terms to the meaner sort of rebels but must always beware the possibility of treason. Generally, he says, the Irish are only concerned with the "mauntenance of theire owne lewde libertie" (96). They are not lawful enemies but rebels and traitors. Like the "arch-rebel" Tyrone described in *A View,* the sonnet's beloved is ruthless:

> But she more cruell and more salvage wylde,
> then either Lyon or the Lyonesse:
> shames not to be with gultlesse blood defylde,
> but taketh glory in her cruelnesse. (XX.9-12)

Like the English in Ireland according to Irenius, the lover has sought vainly for a peaceful union but has encountered only fierce and bloody defiance:

> In vaine I seeke and sew to her for grace,
> and doe myne humbled hart before her poure:
> and whiles her foot she in my necke doth place,
> and tread my life downe in the lowly floure. (XX.1-4)

And not only is the beloved a recalcitrant and pitiless rebel, the lover finds himself fighting his own desires, which also act like the bloody Tyrone: "Mongst whome the more I seeke to settle peace, / the more I fynd their malice to increace" (XLIIII.13-14). The *Amoretti'*s lover finds himself in the planter's predicament: surrounded by hostile natives who refuse subjection and battling the settler's impulse to join the culture he has to live with and depend upon.[21] Both externally and internally, planters like Spenser feared betrayal, and regardless of their actual depredations as landlords and invaders, they represented themselves as innocent men seeking peace.[22]

Neither the *Amoretti* nor *A View* acknowledges that the peace their protagonists seek might come with terms quite rationally unacceptable to their objects—that from the subaltern position, the beloved's/colonial subjects' resistance could be eminently justifiable.

While *A View* speaks from a panoptically secure position, the *Amoretti* explores all the permutations of captivity, temptation, treason and rebellion. Irenius is meant to guide plantation policy from his absolute knowledge of the Irish situation—the text persuades Eudoxus and, ideally, the reader that Irenius speaks from the authority of experience. And indeed, after the text was finally published, it became almost a bible for the planters:

> successive generations of English settlers in Ireland, at least until the end of the seventeenth century . . . had resort to Spenser's ideas (and they even referred to and imitated his *View*) with such frequency that we can accept the ideas enunciated by him as having provided them with an identity and sense of moral purpose which sustained them throughout the travails of the seventeenth century. (Canny 1983, 2)

On the other hand, like *The Faerie Queene*, the *Amoretti* offered the planter Spenser a way to explore the implications of his own colonial situation while assuring the ideal final outcome: a feminized Ireland willingly submitting to the victorious English. Thus in sonnets like XX and XXIX and others, the lover experiments with captivity, viewing its terrors and even its delights. In XXIX, the beloved accounts him "her captive quite forlorne," and he accepts that position in order to sing her praises as her "faithfull [poet] thrall" (4,10); in sonnet LXXIII his heart becomes imprisoned in her bosom. Even bloody, cruel captivity based in treason is safe in poetry with the knowledge that the beloved will, like the fantasy Ireland Irenius projects, willingly and happily submit.

The singular English war crime that *A View* seeks to excuse was perpetrated by Spenser's mentor in Irish affairs, the commander that he first served in Ireland, Lord Grey de Wilton. *A View* acknowledges that Grey had often been accused of massacring Spanish captives at Smerwick on November 10, 1580. The story *A View* counters says that the captives had surrendered and been promised mercy but that upon their submission, Grey slaughtered them. Irenius explains that Grey never guaranteed their safety, as the Spaniards were "noe better then roges and Runnagates . . . so as yt should be dishonorable for him in the name of his Quene to Condycion or make tearmes with such raskalls," and he asserts in addition that sparing them would have left Grey's own troops endangered and would have emboldened the Irish (*View* 140). Another Grey defender, William Camden, wrote that the Spaniards "presently surrendered, and were most of them put to the sword, which in the then situation of the kingdom, and the danger from rebels on every side, was thought the wisest and safest measure" (258).[23] But the explanation from expediency was not enough for Spenser, who argued that the military situation demanded Grey's measures, which were absolutely morally justified. The overdetermined nature of Irenius's excuse for Grey's actions—actions that history has generally deemed inexcusable—[24]indicates how much this incident challenged Spenser's sense of his commander's probity and of his own loyalty to Grey.

That paradigm of captives surrendering and being massacred haunts the *Amoretti*'s sonnet X, in which the beloved massacres her captives, and XX, the final couplet of which cautions, "Fayrer then fayrest let none ever say, / that ye were blooded in a yeelded pray." Sonnet XLVII similarly warns lovers not to trust

> . . . the treason of those smyling lookes,
> until ye have thyr guylefull traynes well tryde:
> for they are lyke but unto golden hookes,
> that from the foolish fish theyr bayts doe hyde:
> So she with flattring smyles weake harts doth guyde
> unto her love and tempte to thyr decay,
> whome being caught she kills with cruell pryde,
> and feeds at pleasure on the wretched pray. (1-8)

In this sonnet the beloved finishes with bloody hands—"yet even whylst her bloody hands them slay, / her eyes looke lovely and upon them smyle" (13-14)—as she also does in sonnet XXXI in which she has tempted her victims to worship her only in order to watch them die: "But my proud one doth worke the greater scath, / through sweet allurement of her lovely hew: / that she the better may in bloody bath / of such poore thralls her cruell hands embrew" (9-12). By ultimately metamorphosing into the sequence's subsequent contented love images, these massacre-and-bloody-hands sonnets condense and finally excuse Grey's crime. The slaughter of captives tempted to surrender becomes a little love poem, and the perpetrator a beautiful beloved who will finally submit herself to a properly dominated position. The fictional love setting enables the poet to explore a colonial crime, while it reduces that crime's potency and finally erases it. Thus the sonnets expiate Lord Grey's potential guilt as well as does *A View*'s overdetermined erasure of it, and they are finally more satisfying, as they leave the persona in absolute control of the colonial scenario.

Sonnet LXVII's "halfe trembling and "fyrmely tyde" deer tropes that control, and the sonnets that follow represent—though sometimes ambiguously—that control's satisfactions. Interestingly, in those succeeding sonnets, the poems' lexicon returns to book two of *The Faerie Queene's* primal scene of colonial conquest. Written perhaps three or four years after book two's Bower of Bliss episode, *Amoretti* LXXVI names the beloved's bosom "the bowre of blisse, the paradice of pleasure" (3). That bower appears cleansed, home only to "thoughts," free from even a single "sparke of filthy lustfull fyre" (LXXXIIII). As *The Faerie Queene's* House of Holiness redeems book one's House of Pride, so the proper colonial relation found in the *Amoretti* redeems Acrasia's bower, and the "lilly paps" of "the wanton Maidens" who tempt Guyon on his way there (2.7.66.6,1) become the beloved's "paps like early fruit in May" that safely harbor the lover's rash thoughts that boldly venture to rest between them (LXXVI.9). The beloved is not wanton; she lives in a "chast bowre of rest" (LXXXIIII.7). But the lover is still yearning; and unlike Guyon who passes through the allurements leading to the bower and looks "still forward right" (*FQ* 2.7.53), the lover's eyes are "hungry" and his thoughts have "wanton winges" (LXXXIII.1, LXXVI.11). And despite the desires of many critics to see the *Amoretti* end in mutuality, its lover is still complaining of his beloved's "cloud of pryde" and her "too constant stiffenesse" (LXXXI.7, LXXXIIII.12). As Anne Fogarty notes, at the beginning and the end of *A View of the Present State of Ireland*, "Ireland becomes emblematic of a pleasure which is both frustrating and endlessly enticing" (87). In *A View of the Present State*, then, Ireland becomes a Petrarchan beloved; and the semi-Petrarchan beloved of the *Amoretti*, Spenser's Irish love, his Ireland, submits, but never completely.

Although it accompanies the feminization of Virginia in English colonial discourse, the image of Ireland as a woman constructed in the *Amoretti* and English literary and historical texts was not an English colonial invention; it had a long history in Gaelic poetry before it was appropriated by the English, and in later years that same image was used by the Irish resistance. Ireland has been figured as a powerful bride, a virgin to be penetrated, a demanding mother, and a captive widow, among other feminized images. When Spenser came to Ireland with Lord Grey, Irish "bardic" poetry, "the refined product of a literary class that had practiced its art for centuries," had a tradition of treating Ireland as a woman and praising

Gaelic chieftains as her potential and fitting mates (Carney 6). Fearghal Óg Mac an Bhaird, writing for Cú Chonnacht Mág Uidhir [Maguire], Lord of Fermanagh (1566-1589), promises the imagined unified Éire: *"dob fhiú céile Cláir Da Thí/ Éiri ar a láimh dá léigtí."* ("if Ireland were allotted to him he would be a worthy mate for her") (*Duanaire* 22-23; Greene trans.). Another poem in Cú Chonnacht's poembook, possibly by An Giolla Riabhach Ó Cléirigh, personifies Ireland as a desiring woman impressed by Maguire's soldierly prowess:

> *Sí ag féin Éirne ionchosnaimh*
> *d'éis éirghe na ferchonsain,*
> *ben Da Thí ar tí an Mhanchoidhsin*
> *do-chí nach ní nemhchosmhail.*

Ireland has set her eye on that Maguire as she sees that it is not improper that she should be contended for by the soldiers of the Erne after the rising of that hero. (Duanaire 68-69, translated by Greene)

Although these poets feminize the land, they treat her as deserving and choosing a heroic ruler-husband. Their feminine land is an active partner, not a subjected victim; instead of submitting to a conqueror-husband, she takes a mate.

In 1589, Cú Chonnacht's son Hugh, the Maguire besieging Enskillen in 1594, succeeded him; Hugh died fighting the English. Tadhg Dall Ó Huiginn addressed a poem to him about his possibilities as a patron; its first stanza reads:

> *Léigfead Aodh d'fearaibh Éireann,*
> *lór don bhaisgheal bhairrséimhseang;*
> *leision Éire acht meise amháin,*
> *a seise, a céile compáin.*

I shall leave Hugh to the men of Ireland, they are enough for the white-handed one of the fine, soft hair; save myself alone all Ireland is his; he is her comrade, her companion. (Knott 1922, XXII.81, XXIII.54)

The words *"céile"* and *"compáin"* can both be translated as husband but they each have primary meanings of "equal match" and "companion" as Knott's translation indicates (*Dictionary of the Gaelic Language* "*céile*" "*companach - aich*"). The feminized Ireland is Hugh's spouse in this poem, and *he* is also her spouse, her partner, rather than her subordinate. James Carney has translated a piece of Eochaidh Ó hEoghusa's inauguration poem for Hugh:

Ireland is in the mood for love and has put on her wonderful raiment of green and purple and gold. The voices of her streams are low and she is wooing Hugh who is the second Naoise of the Ulstermen. It is for him she has put on her wonderful clothes and cast aside her mourning and enchantment. (19)

The poem pictures Ireland as a maiden bewitched and disfigured by the English; Hugh Maguire will bathe her in English blood and so release the spell and marry her.[25] But even this feminized Ireland who has been disempowered and left weeping by English conquest was before those terrible events "a happy maiden, merry and wise [who] had many suitors" (19). Unlike Morton's virginal New England, Hakluyt's expectant Virginia, or Spenser's submissive beloved, Ó hEoghusa's maiden Éire actively woos for herself. Hugh frees her, but she also casts her spell away to marry him.

In *A View*, Eudoxus asks Irenius whether the Irish bards' poems have any art, and Irenius replies that he has "caused dyverse of them to bee translated vnto [him], that [he] might vnderstand them" (97-98). In Irenius's opinion, the poems are graceful and comely, but they grace "wickednes and vice" (98). If we can infer from Irenius's claim that Spenser himself had been reading Irish poetry, then he would certainly have encountered this empowered feminized Ireland who populated the tradition everywhere as the patron's prospective mate. The submissive woman that Spenser's poetry creates answers this image with the "properly" feminine helpmate who recognizes her place and glorifies her conqueror. In the Irish poetic tradition, Spenser encountered both a cultural position to be envied and a feminized Ireland to be transformed. Much of his poetry can be read as instructing a patron to treat her poet in the manner of the Irish "bards,"

> whose profession is to sett forth the prayses and disprayses of men, in theire Poems or rymes the which are had in so high regard and estymacion amongst them, that none dare displease them, for feare to rvn into reproch, through theire offence and to bee made infamous in the mouthes of all men, for theire vearses are taken vpp with a generall applause, and vsuallie sounge at all feastes and metinges, by certaine other persons whose proper function that is, which also receyve for the same greate rewardes, and reputacion besides. (*View* 94)

But Irish poetry's subject matter, including its powerful, feminine Éire, was distinctly unpraiseworthy, and in his sonnet sequence to Ireland and his New English beloved, Spenser could transform the woman these poets constructed to praise into a willing, submissive sonnet mistress-mate.[26]

While Spenser's work constructed a feminized Ireland in order to justify a gendered domination, very soon that image became a tool for enemies of the Tudor conquest. In his examination of the "Gaelic Response to Conquest and Colonization," T. J. Dunne notes that in the 1650s Ireland appeared personified as a woman in captivity or in mourning, and in the early eighteenth century Ireland was imagined as a sad widow and an enslaved and helpless woman (26). Patrick Hanafin argues that "Ireland has in metaphorical terms been identified as female since early colonial time. It was an example of myth in order to compensate for colonial domination" (251). Although Hanafin ignores the indigenous positive roots of a female Ireland, both of these critics point to the use of an image of a feminized Ireland to mobilize resistance to English domination.

However, as Rick G. Canning notes in his analysis of Swift's *Injured Lady,* "the tension between flattery and inferiority implicit in this vision of the feminine had its price" (77); the image remained a powerful way to degrade the Irish. Declan Kiberd argues that in English discourse of the nineteenth century, "the Irish were to read their fate in that of two other out-groups, women and children. . . . The political implications were clear enough in that age of severely limited suffrage: either as woman or as child, the Irishman was incapable of self-government" (30).[27] Kiberd notes that in response to the country's long history of feminization, by the nineteenth century, "militant nationalists . . . called on the youth of Ireland to purge themselves of their degrading femininity by a disciplined programme of physical-contact sports" (25). This nationalist response calls for resistance to colonialist power but assents to the disempowered image constructed by that very dominance. By accepting that their [male] youth were feminized and that feminization equaled degradation, these nationalists confirmed a stereotype created by the settlers whose historical actions they reviled. The internalized oppression of this nationalist call to arms and the late-twentieth-century gender oppression that Hanafin argues derives from those stereotypes both demonstrate the power of the image of Ireland as subservient woman constructed in Spenser's *Faerie Queene* and the *Amoretti.*[28]

And ye high heavens, the temple of the gods,
In which a thousand torches flaming bright
Doe burne, that to us wretched earthly clods
In dreadful darknesse lend desired light;

And all ye powers which in the same remayne,
More then we men can fayne,
Poure out your blessing on us plentiously,
And happy influence upon us raine,
That we may raise a large posterity,
Which from the earth, which they may long possesse,
With lasting happinesse,
Up to your haughty pallaces may mount,
And for the guerdon of theyr glorious merit
May heavenly tabernacles there inherit,
Of blessed Saints for to increase the count.
            (*Epithalamion* 409-23).

Like Cymochles and his beastly companions in Acrasia's bower of bliss, *The Faerie Queene*'s knight of Justice, Artegall becomes a "womans slave" (5.5.23.5) in a woman's den. Acrasia's "yron mewe" haunts Artegall's text, which compares that knight to an injured falcon attacked by the kite-like Radigund after he pityingly spares her life:

Like as a Puttocke having spyde in sight
A gentle Faulcon sitting on a hill,
Whose other wing now made vnmeete for flight,
Was lately broken by some fortune ill;
The foolish Kyte, led with licentious will,
Doth beat vpon the gentle bird in vaine,
With many idle stoups her troubling still. (5.5.15.1-7)

Injured hawks or falcons were often put in mews to mend and to molt. The *Oxford English Dictionary* cites Sir T. Stafford, who, in 1623, writes presumably to a patron, "the faulcon your Lordship sent was so brused and ragged . . . [that I] have put her into a mieu" (*OED* "mew" 1). Just as in the poet's simile Artegall becomes a gentle female bird, in Radigund's mew Artegall will adopt "womans weedes" and serve "proud *Radigund* with true subiection" (5.5.20.7, 5.5.26.2).

As *A View* amply demonstrates, the knight of Justice's temporary fate—to resemble the Amazon other—was the planter's nightmare. To live with the colonized is always to risk becoming an image of the beast (Jones and Stallybrass 163).[29] Spenser writes that those old English in the pale, and therefore close to civilization, have retained their English identities, "but the rest which dwell aboue in Connagh and Mounster, which is the swetest souyle of Ireland, and some in Leinster and Vlster ar degenerate and growen to be as verie Patchokes as the wilde Irishe, yea and some of them haue quite shaken of theire Englishe names, and putt on Irishe, that they might bee altogeather Irishe" (*View* 84).[30] Fynes Moryson, who came to Ireland like Spenser in the service of a new lord deputy, writes in his 1617 *Itinerarie* that "the [Connaght] English family Bermingham [was] of old very warlike: but their posteritie have degenerated to the Irish barbarisme" (IV 190). Likewise in Ulster, "Monaghan was inhabited by the English family Fitzursi, and these are become degenerate and barbarous, and in the sense of that name are in the Irish tongue called Mac Mahon, that is, the sonnes of the Beares" (IV 190). If the Old English settlers' experiences caused them to "quite forgett theire Countrie and theire own names," and to "cast of" with those names their English "alleigeance" and term "themselves very Irish," what might become of the new men who would make their fortunes in Ireland (*View* 83-84, 85, 86)?[31] What might become of the planter-poet, an undertaker in the Munster plantation? Artegall's Amazon adventure illustrates this primal colonial fear: if one pities aliens, one may adopt their manners.[32] Part of Spenser's plan to exterminate the Gaelic Irish stemmed from the need to eliminate this temptation and this fear.

Degeneration threatened settlers, "servantes," "followers," and their families, as the old English in Munster took "on them Irishe habites and customes which could never since bee cleane wyped awaye but the Contagion thereof hath remayned still amongst theire posterities" (*View* 85-86). Both Moryson and Spenser warn that this colonial "contagion" infects not only those individuals who succumb, but most threateningly their progeny; in the English culture these men wished to perpetuate, gentry and noblemen's primary domestic concern was the purity and elevation of their families, their male heirs. This coda's epigraph, the blessing the poet begs from the "high heavens" in his *Epithalamion*'s second-to-last stanza, seeks to ward off such a fate from Spenser's desired "large posterity." The poet asks the powers above to bless his marriage with children and to grant those children the graces that Spenser thought his life so arduously demonstrated: possession of the land he had acquired from the infidels and elevation to the rank of the Protestant elect in heaven. He would have been cruelly disappointed. Not only did Spenser's descendants, as Henley notes, ironically lose all the land that he had appropriated for them by 1738, but a number of them would be excluded from the heavenly destiny of the religious tradition that Spenser so assiduously and violently promoted (211).

Like *The Faerie Queene*, the *Amoretti* is a distinctly Protestant text (Dunlop intro. in *Complete Poems* 589-90, Sinfield 66-68). Both poetic texts, like the *Epithalamion,* were clearly, for the poet, signs of his election and missionaries against the power of the whore of Babylon, who for Spenser lived with the Gaelic and the Old English who had adopted Catholicism along with Irish manners and clothing. But Spenser's own colonial situation proved to be as deeply dangerous, in his terms, as the poet feared. Spenser's eldest son, Sylvanus, married a Catholic woman in Ireland, and his early death ensured that the poet's grandsons Edmund and William were raised in the religious tradition their grandfather loathed. According to a letter by Cromwell on his behalf, William renounced his "Popish religion . . . since coming to years of discretion," and so he reclaimed his family's land, which had been seized by the English due to his mother's faith (Coleman 1894). But William's cousin Hugolin, Spenser's second son, Peregrine's eldest, fought with the Irish against the English in the war of 1689-91 and was convicted of high treason in 1694 (Henley 208-209). For Edmund Spenser, this would have been the cruelest colonial transformation.

Chapter 2

Bermuda's Ireland: Naming the Colonial World

Edmund Spenser worried that Old English settlers' experiences in Ireland had caused them to "forgett theire Countrie and theire own names" (*View* 64 (1970) 83-4). In *A View of the Present State of Ireland,* when Irenius finally introduces his program for Irish reformation he calls on the English to

> renew that old statute . . . by which it was commanded that whereas all men used to be called by the name of their septs according to their several nations, and had no surnames at all, that from thenceforth each one should take unto himself a several surname, either of his trade or faculty . . . whereby they shall not only not depend upon the head of their sept . . . but also shall in short time learn quite to forget his Irish nation. (*View* 155-56)

For Spenser a man's English or Irish identity rested in his English or sept (clan) name; name and "countrie" or "nation" were inextricably linked. While integrating with the native community could cause settlers to lose national identity and name, colonizers could use the integrity of name and nation to divorce the Irish from their traditional clan identification. Spenser's fear of English degeneration and his program for reidentifying the Irish point to the crucial place of names and naming in the new Atlantic world. As David Spurr argues, "the very process by which one culture subordinates another begins in the act of naming and leaving unnamed" (4). Colonization depends not only on military power, technology, politics, and economics, but also on acts of naming, and resistance to colonial power can manifest in naming and renaming. In this chapter, I contend that naming was a powerful transformative action in the mouths and texts of people in the emerging Atlantic world. This focus on naming follows Paul Carter's lead in his fascinating book on Australia's "spatial history," which, as he says, evokes "the spatial forms and fantasies through which a culture declares its presence" (xxii). Carter suggests that "spatial history begins . . . *in the act of naming.* For by the act of place-naming, space is transformed symbolically into a place, that is a space with a history" (xxiv). James Cook, in Carter's account, "proceeded within a cultural network of names, allusions, puns, coincidences, which far from constraining him, gave him . . . conceptual space in which to move" (7). Likewise, the first colonial namer considered in this chapter, Fynes Moryson, when he named Ireland in his early modern travel book, brought a particular English conceptual space to bear on the country he wished to dominate. Moryson's example illustrates how the imposition of English names on other people's land helped the English to dominate the new Atlantic world. As Spenser's worry shows, personal naming as well as place-naming indicates the vicissitudes of colonial transformation, as does the naming of groups and the assignment of epithets. Thus this chapter focuses broadly on colonial naming in the Atlantic world, looking at place-naming and also extending Carter's analysis beyond the naming of territory, settlements, and bodies of water, to personal naming and postcolonial renaming.

In understanding naming in the colonial world we must look not only at how places, people and things are named and what conceptual spaces and histories are thereby created, but also at how these names are used, by whom, and under what compulsions. For names exist only in the human world, where they can be reappropriated and also given from below. Naming can figure in resistance as well as in colonial domination. This is the same dynamic process of transformation that we have seen in the discussion of the feminization of Ireland in chapter 1. That feminization signified differently in the mouths of the Irish poets than it did in the mouth of an English colonist fearing his own transformation. Like Spenser's imagination of his Ireland as a resistant but finally submissive beloved, colonists' naming and renaming of land and people were powerfully transformative in the new Atlantic world; but like Spenser's feminization of Ireland, those names could be reappropriated and reimagined.

This chapter argues that transformative naming in the colonial world was not solely the colonizer's property or process, and also that naming revealed divisions between colonists and the faultlines in colonial settlement. Another of my examples, Shane O'Neill's (the Earl of Tyrone) act of naming himself in Ulster, shows that in the colonial world, naming was and is often contested. English explorers and colonists named their colonial world and so did poets like Spenser and colonists like Moryson; but Gaelic poets, Native Americans, and disgruntled returned colonists were also naming and transforming space. Names given by English colonists were taken up and used to other effect by Spanish colonists. English colonists competed to name territory in Virginia, and groups and individuals in the new Atlantic world had different and sometimes opposing stakes in the names of England's often tenuous colonial holdings. J. B. Harley observes about colonial naming generally that "the naming process was not entirely a history of naked imposition, nor of passive acceptance, nor yet of subsequent unified resistance to these offensive 'speech acts' on the map" (1990, 4).[1] In the names of—and the naming of—the colonial world on both sides of the Atlantic, we can see struggles over hegemonic definitions of Irish, English, and North American spaces and people. Colonial names and maps show colonial power and conflicts among colonists, as well as resistance by native and creolized settler communities, and they also show the "conceptual spaces" in which all these people and communities moved.

However, while naming is a contest, those colonial people and groups with the most economic, political, or military power are often the winners. Naming and mapping in colonial Virginia and in Bermuda, as this chapter shows, often reflected the domination of native people and land and the imposition of English conceptual space on places previously dominated by other people and ecosystems that had never known human settlement. And such naming in the early modern new Atlantic world still determines our understanding of the spaces created four hundred years ago. "Big Turtle Island" is a Native American name for North America (Dufrene 129), but the vast majority of Americans and inhabitants of the larger world have never seen or heard that name. Most of the world, in contrast, recognizes the name "the United States," even though, if that country honored its own treaties and court and legislative decisions, perhaps 35 percent of the continental United States would belong to "native nations" (Churchill 1996 49). Despite this chapter's analysis of colonial naming, in this book I depend on names—like "Virginia" and "Indian"—that denote power and acts that I actively criticize, because those names are the only names my readers will recognize. Place-naming can support other forms of oppression, and native inhabitants of postcolonial societies, as well as descendants of imported slaves, must often fight against hegemonic, oppressive toponyms that have achieved mystified and powerfully interpellative status over hundreds of years of domination. Another of this chapter's examples, Jamaica Kincaid's renaming of her postcolonial world, is an example of that fight, as is the struggle by Native American groups to change the names of sports teams like the Cleveland Indians and the Atlanta Braves.[2] Names can always be changed and native names recovered. But naming is not only a reflection of power and resistance, it is transformative in itself.

In his 1617 *An Itinerary Containing His Ten yeeres Travell through the Twelve Dominions of Germany, Bohmerland, Sweiterland, Netherland, Denmarke, Poland, Italy, Turky, France, England, Scotland & Ireland,* after a paragraph placing Ireland latitudinally and longitudinally, Fynes Moryson opens his description "of Ireland" with a discourse on that island's names:

> This famous Iland in the Virginian Sea, is by olde Writers called Ierna Inverna, and Iris, by the old inhabitants Eryn, by the old Britans Yuerdhen, by the English at this day Ireland, and by the Irish Bardes at this day Banno, in which sense of the Irish word, Avicen cals it the holy Iland, besides Plutarch of old called it Ogigia, and after him Isidore named it Scotia, This Ireland according to the Inhabitants, is devided into two parts, the wild Irish, and the English Irish, living in the English Pale. (4.185-86)

Moryson's opening has almost a postmodern relativistic feel, recording as it does not only the English name for the island but also its Gaelic name Erin (*Éire*) and supplying, as well, an appellation used by the "Irish Bardes," a source already discredited and vilified by Moryson's English contemporaries.[3] But Moryson's use of the bards actually indicates the domination implicit in his act of naming.

Moryson's contemporary New English poets wrote vituperatively about the bards. Although he envied their power and prestige, Spenser castigated these poets, who should celebrate virtue but instead praise rebels and encourage the wicked:

> But these Irishe Bardes are . . . so farr from Instructing yonge man in morall disypline, that they them selues doe more deserve to be sharpelie discipled . . . such lycentious partes, as these, tendinge for the most parte to the hurte of the English, or maintenance of theire owne lewd libertie, they them-selues beinge most desyrous thereof, doe most allowe, Besides these evill thinges beinge deckt and suborned with the gaye attyre of goodlie wordes,

maye easelie deceave and carrye awaye the affeccion of a younge mynde.
(*View* 95-96)

Spenser calls the bards deceivers of the young, infidels favoring "lewd libertie" over proper restraint and needing the discipline of their masters. In
his description published in 1610, another New English author, Barnaby
Rich, shares both Spenser's concerns and his disgust:

> There is nothing that hath more led the Irish into error, than lying histori-
> ographers, their chroniclers, their bards, their rhymers, and such other their
> lying poets; in whose writings they do more rely, than they do in the Holy
> Scriptures, and this rabblement do at this day endeavour themselves to noth-
> ing else, but to feed and delight them with matter most dishonest and shame-
> ful . . . the songs that they use to sing are usually in the commendation of
> theft, of murder, of rebellion, of treason, and the most of them lying fictions
> of their own collections. (340-41)

According to Spenser and Rich, Irish bards have abandoned the poet's duty
to "frame . . . polliticke vertues" (*FQ* Letter of the Authors 15-16). Rather
than fashioning gentlemen, these alien poets fashion criminals. Spenser's
mentor in poetics, Sir Philip Sidney, argued that "right" poets "imitate to
teach and delight, and to imitate borrow nothing of what is, hath been or
shall be; but range, only reined with learned discretion, into the divine consideration of what may be and should be" (*Apologie* 102). Seconding Sidney's belief in poetry's potential, Spenser decried the bards' use of the
poet's privileges of invention to achieve evil ends.

Moryson and Spenser shared a confirmed hatred for the Gaelic population. Like Spenser, who served as Lord Grey de Wilton's secretary from
1580 as Grey was suppressing the Desmond rebellion, Fynes Moryson went
to Ireland in 1598 as secretary to Charles Blount, Lord Mountjoy, the new
Lord Deputy and accompanied him as he suppressed the Tyrone rebellion
which destroyed Spenser's escheated castle. Moryson tells us that he will
write of his trip to Ireland "in another manner, then I have formerly done
of other Countries, namely, rather as a Soldier, then as a Traveler" (2.166).
Moryson, Spenser, and Rich all described the island as confirmed oppo-
nents of the Gaelic population and advocates of its bloody conquest. That
Moryson, then, would quote bardic authority for the island's name is surprising.

However, as Ashis Nandy cautions, "cultural relativism by itself is not incompatible with imperialism, as long as one culture's categories are backed by political, economic and technological power" (100). And Moryson's apparent relativism demonstrates the imperialism associated with naming and the power to name. Moryson's naming offers his audience an English colonialist understanding of the island. Even as Moryson translates his bardic name "Banno" for his audience into the English "holy Iland," he misnames the Irish literary tradition he cites; and he misnames a tradition deeply involved in naming the island on which it developed. While the word "bard" in Ireland actually denoted a "subordinate functionary" who might recite poetry, the *filid* or Irish poets traditionally named and renamed Ireland throughout their work (Knott 61). This practice is immediately visible even in the short passages of Irish poetry that I quoted in chapter 1. The court poet's repeated and varied namings were an integral part of Irish court poetry; the *filid* employed numerous epithets for the island, not one as Moryson's description asserts, each name reflecting "some item of ancient historic or mythological lore known and cherished by contemporary listeners."[4] In court poetry, the island was called "*Inis Fáil* [Island of Destiny], *Teach Tuathail* [House of *Tuathal*], *Fonn Feradhaigh* [Country of downpours]," "*Fiadh Fuinidh* 'land of the west'; *Clár* (*magh, tulach*, etc.) *na bhFionn* 'Plain (hill, etc.) of the Fair ones'" (Knott 69). *Duanaire Mhéig Uidhir* (The Poembook of Cú Chonnacht Mág Uidhir, Lord of Fermanagh 1566–1589) which contains the work of at least eleven *fili* has twenty-four poems which use fourteen different names for the island (Greene 288). Irish poets called the island by a series of kennings in which a significant part of the island denoted the whole. Knott suggests that "The most frequent of all these 'kennings' is *Breagha* 'Bregia', the Leinster territory so conspicuously associated with the rulers of Ireland both in ancient and more modern times, the territory wherein Tara stood and where the great prehistoric monuments of Dowth and Newgrange are still to be seen, as well as *Bóinn* 'the Boyne' [the Pale]" (Knott 69).

Moryson's use of "Banno" reveals his own interested act of naming. Knott doesn't find *Bán* [*o* _], which might have meant in 1617 "white, fair, bright; pure, holy, blessed," among the most common poetic epithets for the island. "Banno" also doesn't appear in Cú Chonnacht's *Duanaire*. Moryson's cited source, William Camden's 1607 *Britannia*, from which he gets the word "Banno," is truer to the tradition of multiple names than the *Itinerary*. Camden tells us that "the Irish bards celebrate in their songs *Tivolac, Totidanan*, and *Banno*, as much the most ancient names of the island" (217). Moryson misreads Camden when he says that "Banno" means "holy island" according to "Avicen"; Camden actually wrote, "for the rest *Biaun* signifies in Irish *holy*," and Camden then explains that the classical author Festus Avienus called the island "*Insula Sacra.*" The word "Banno" in itself does not contain the noun "island." By choosing the adjective epithet from among many used by the *filid* and, tellingly, by choosing it from among Camden's three possibilities, Moryson chose a word that might also mean "blank" when used to describe a page and might mean "untilled, waste, unoccupied" when used to describe land (*Dictionary of the Irish Language* [B]). It is also possible that Moryson may have seen a manuscript of Spenser's *View* in which Irenius tells Eudoxus, "this is the wretchednes of that fatall kingdome, which I thincke, therefore in ould tyme was not called amisse *Banno* or *Sacra insula,* taking sacra for accursed" (120). Moryson's bardic name serves his colonialist purposes, rendering the island a blank space awaiting occupation.

Misnaming the island and attributing that misnaming to the bards, Moryson erased Gaelic culture, its history and myth. What might appear as relativism to a twentieth-century audience is in fact part of the *Itinerary*'s project of denigrating the Irish. Moryson's location for Ireland has the same effect: placing the island "in the Virginian Sea," Moryson situated it in a body of water called after the English Queen, under whose name the late Tudor conquest of the island was coordinated. Moryson's "Virginian" is another misreading of Camden who calls the sea "*Vergivian*" and derives that name from Briton and Irish appellations (217). Here again Moryson's misreading is a symptom of his colonialist agenda. The English cult of Elizabeth attempted to replace the Catholic focus on Christ's virgin mother with a worship of Elizabeth as the virgin head of the English church. The Gaelic people whom Moryson refers to as "the wild Irish" were Catholic and would no more have wanted to participate in this replacement than to have glorified the English queen. Just as Moryson's text situates this island within the English Virgin's territory, it also encloses the island's bardic name within its iteration of ancient and classical appellations. Thus the "Irish Bard"'s name for the island has almost the same status in the discourse as the names used by the "olde Writers," "Plutarch," and "Isidore." And while the discourse acknowledges that the bardic name is as current as the English name (both are used "at this day"), like the island itself enclosed by "the Virginian sea," Moryson's act of naming is enclosed by its naming of the island as Ireland. Moryson calls the chapter "Of Ireland," not "of the island the English call Ireland." The chapter opens with the phrase "The Longitude of Ireland." The marginal gloss next to that line says simply "*Ireland*," and Moryson starts the sentence after his list of names with the phrase "This Ireland." The discourse, then, continually announces that the island's other names are a decorative display of Moryson's erudition; the island is really named Ireland.

Moryson's cultural network, displayed in his interpellating discourse, included the cult of Elizabeth, England's chronicle histories, and Moryson's classical education—Elizabeth made him a fellow at Peterhouse in Cambridge where he received his B. A., and he took his M. A. degree at Oxford—as well as New English descriptions of Ireland like Rich's and Spenser's. That network distinctly did not include, or rather credit, the pope's condemnation of Elizabeth, Gaelic and Old English claims to Ireland, Native American claims to the "New World," and Gaelic Catholicism, all of which would have led him to a very different act of naming. His travel account introduced "Ireland" to its readers as a land with a colorful naming history, but ultimately as England's Ireland.

Of course, the English were not alone in naming the new Atlantic world. Father Bernabe Cobo's *History of the Inca Empire* offers a seventeenth-century Spanish perspective on the meaning and power of naming in the colonial world.[5] Cobo argued:

> In my opinion, it would be impossible to extinguish and erase, from now until eternity, even the memory of just the names of the provinces and towns that we have founded in this New World. The conquistadores and settlers have been naming these places in honor of our nation and in memory of their provinces there or for other reasons and motives, always with the objective of making eternal the fame of our people in these new lands. For this reason, there is hardly a kingdom in Spain whose name has not been taken by now to this land and applied to the provinces that have been pacified and settled. (79)

For Cobo, naming is a natural part of pacification and settlement. The conquistadors' names commemorate their victories for eternity; those toponyms make the "New World" speak only of the Spanish and of their imperial European center. He proudly asserts that "now there are so many names of places in Spain in this land [Peru] that it seems as if that whole kingdom had been brought here" (80). In its place-names then, Peru is a transplanted Spanish nation.

At times English naming followed a pattern set by the Spanish in their earlier settlement of lands in South America. According to Cobo, the Spanish have named Peru for Spanish toponyms, for kings, for their "founders," and for the Peruvian places' characteristics. What Cobo means by characteristics is usually not aesthetics (in itself a relative category) or geographical features, but utility to the Spanish, as in Puerto Seguro "Safe Port," a safe port for Spanish ships; Buena Ventura "Good Luck," good luck for the Spaniards (perhaps signifying bad luck for the native population); La Plata "Silver," and Monte de Plata "Hill of Silver" (82). This is the type of naming practiced by James Cook, in Carter's account, when he named his "first [New Zealand] landfall" "Poverty Bay" because of its storms, its unfriendly natives, and its lack of desired resources (14-15). Early English naming in Virginia also followed that pattern with designations like "Cape Comfort." A Virginia colonist who became council President, George Percy, wrote that in 1607, a party of English "rowed ouer to a point of Land, where wee found a channell, and sounded six, eight, ten, or twelve fathom: which put vs in good comfort. Therefore wee named that point of land, Cape *Comfort*" (1687). The "safety," "luck," "poverty," and "comfort" in these names are clearly the colonists', since the bloody repulse of Cook and his men might instead be named triumph by the successful aborigines in the encounter. The comforting depth of water next to that piece of Virginia land enabled the English to take the entire area from its native population just a few years after Percy's landing.

Most of the naming Father Cobo describes, however, would not have been within Protestant English conceptual space, and, as Cobo forecast, it indelibly marked Peru with Spanish Catholicism. Thus Spanish Catholic naming directly declared a religious domination anathema to English Protestant expansion, and English naming challenged that domination. The Jesuit missionary happily noted that "We have given Christian names to almost all of the Indian towns within the Spanish dominions, giving the majority the titles of saints" (82). He was particularly satisfied that,

> Not less in this respect have been the demonstrations of devotion that our nation has made to the Holy Mother the Virgin Maria; in her honor sixteen towns are named with the titles of her holy mysteries; for her sweet name three are named; for the same, with other attributes . . . five; out of devotion for her Inmaculada Concepcion six are named; one after the title of her Purificacion Santisima and another with the title of her glorious Asuncion. (81)

During Hugh O'Neill, the second earl of Tyrone's, Ulster rebellion, as also in the earlier Desmond Munster rebellion, Irish chieftains looked to Spain for assistance against the English. The chieftains' devotion, at least practically, was to the Virgin after whom Spain named so much of its colonial dominions. The English naming of Virginia was a direct challenge to this system, although intriguingly it is the only English name always reproduced on all European mapping of the territory.

Perhaps *Virginia's* double valence made it legible within both Catholic and Protestant systems depending on who was reading and speaking. When Spanish people repeated the name, they named the land for Mary, regardless of the English namer's intentions. The Spanish use of the English "Virginia" reminds us that "words, expression, propositions etc. . . . change their meaning according to the positions held by those who use them" (Ashcroft, Griffiths, Tiffin 171). Already aware in 1653 that the use of colonial names is a clear indication of their power, Cobo commented, "there are a great many Indian towns that have no names other than the ones we have given them, and these names are used not only by the Spaniards but also by the Indians themselves" (82-83). In Cobo's analysis, Spain's power has erased all traces of Indian names. Since the Indians use Spanish names to understand their own settlements, they have completely acceded to Spain's power over them. It is not enough for colonists to give a Spanish name to an "Indian" town; to ensure their domination, the colonized must reenact that domination in their own acts of naming.

But names do not always serve dominant colonial interests, whether English or Spanish. When Tyrone took the name O'Neill to signify his control over Ulster, he was asserting power by directly disobeying English authorities who very well understood the force of that traditional name. In 1569 as part of its Act of Attainder against Shane O'Neill, the first earl of Tyrone, the English crown had declared:

> Forasmuch as the name of the O'Neill, in the judgements of the uncivil people of this Realm, doth carry in itself so great a sovereignty, as they suppose that all the lords and people of Ulster should rather live in servitude to that name, than in subjection to the crown of English: be it therefore, by your Majesty, with the assent of the lords spiritual and temporal, and the Commons in this present Parliament assembled, and by the authority of the same, that the same name of O'Neill, with the manner and ceremonies of his creation, and all the superiorities, titles, dignities, pre-eminences, jurisdictions, authorities, rules, tributes, and expenses, used, claimed, usurped, or taken by any O'Neill, as in right of that name, or otherwise, from the beginning, of any the lords, captains, or people of Ulster, and all manner of offices

given by the said O'Neill, shall from henceforth cease, end, determine, and be utterly abolished and extinct for ever. (*Irish History* 174)

In this Act, the crown was attempting to suppress the power contained in the name O'Neill, which they recognized could serve as an icon to attract rebels. By opposing "servitude to that name" to "subjection to the crown," they attempted to belittle Gaelic "judgement." The crown implied that servitude is a poor and ignorant form of subjection, just as a name is a weak substitute for a crown. But the Act's juxtaposition of "servitude" and "subjection" calls into question the distinction between the terms, while its acknowledgement of all of the name *O'Neill's* accoutrements and powers inadvertently equates name with crown. What is a crown after all but the ability to install "superiorities, titles, dignities, and pre-eminences," to take "jurisdiction," wield "authority," make "rules," and collect "tribute"? If the name O'Neill offered that power to an Irish rebel, then in essence it offered him a crown to counter Elizabeth's crown.

In 1569, the year England declared Shane O'Neill a traitor, Edmund Campion wrote that, in preparation for his rebellion against the English, O'Neill had "fortified a strong island in Tyrone, which he named spitefully 'Foogh-ni-Gall' [*Fuath na nGall* (editor's brackets)], that is, the 'hate of Englishmen,' whom he so detested, that he hanged a soldier for eating English biscuit" (Maxwell 173). Whether or not Campion's story is true of O'Neill or a story meant to condemn him in English eyes, that story and the crown's attempts to take away O'Neill's name show that names can be powerfully transformative in the mouths of native people. When North American Indians called plantain, one of the most persistent weeds accompanying England's biological invasion, "Englishman's Foot," they were naming the English as invasive, choking, and unwanted (Cronon 143, Crosby 1978, 19). When they called honey bees "English flies," they were insisting that what the English considered a domesticated cash crop was instead a nuisance animal, as pestilent as the English themselves (Crosby 1978, 15). As the Virginia Company's official sources were calling Virginia a paradise of commodities, the Indians were naming the effects of their wasteful consumption.

In 1988, at the other end of England's empire, Jamaica Kincaid furiously remembers the colonial interpellation of her childhood's physical world: "In the Antigua that I knew, we lived on a street named after an English maritime criminal, Horatio Nelson, and all the other streets around us were named after some other English maritime criminals. There was Rodney

Street, there was Hood Street, there was Hawkins Street, and there was Drake Street" (*A Small* 24). She notes that English colonial names powerfully formed the world she and her compatriots knew and understood. Those names were part of a system that led the oppressed descendants of slaves to celebrate the English empire and to consistently fail to recognize racism in their white rulers. But while Kincaid recalls her world being named in spite of her and her people's histories, she also performs a powerful renaming as she relabels Nelson, Rodney, Hood, Hawkins, and Drake as "criminals."

Each of these seafaring men was a British imperial hero; Queen Elizabeth knighted both Drake and Hawkins, one of England's first slave traders; and Hood, Rodney, and Nelson were all raised to the British peerage, Nelson and Hood as viscounts and Rodney as a baron. The 1878 edition of the *Encyclopædia Britannica,* published near the British empire's height, and itself an imperial document, records each of these men's imperial canonization.[6] The *Encyclopædia* says that "Stow speaks of [the slaver Hawkins] as a very wise, vigilant, and true-hearted man," and it lauds Drake for "his success and honourable demeanour." The *Britannica* praises to the sky the three later men, all responsible for England's late-eighteenth-century maritime dominance: Rodney for "masterly decision and confident boldness"; Hood for "thorough seamanship, and . . . a rare union of courage and decision with coolness and caution"; and Nelson for "the transcendent gifts which made him pre-eminent." The 1878 *Encyclopædia* is especially enamored of Nelson, who it says "was unrivalled, in an eventful age of war, for resource, daring, professional skill, and the art of winning the hearts of men." This entry's language encapsulates the document's imperial hubris: "men" are clearly defined here as not only the British seamen Nelson commanded, but also the text's implied audience, proud British citizens of the empire who see its holdings, like Kincaid's Antigua, as rightfully and happily their possessions. But Kincaid's act of naming spits in the face of this hubris, calling these men simply and eloquently "maritime criminals," and thereby revealing the interest behind her streets' names and the British creation of maritime heroes.

Father Cobo understood that a native population's assent to the names the conquistadors assigned to native towns signaled colonial capitulation. But Hugh O'Neill's assumption of the name O'Neill indicates that names can also figure in a different colonial dynamic, in this case when colonial capitulation turns to rebellion. While the late twentieth century lives with the results of early modern Atlantic world naming, names can be changed and can invoke resistance. Despite Father Cobo's belief in their permanence, Kincaid's courageous outrage indicates that the names whose imposition this chapter documents can always be challenged.

On November 18, 1618, the Company of Adventurers and Planters of the city of London for the first colony of Virginia issued the following instructions to Captain George Yardly, the colony's governor-elect:

> to put in Execution with all convenient Speed a former order . . . for the laying and seting out by bounds and metes[7] of three thousand Acres of land in the best and most convenient place of the territory of James town in Virginia and next adjoining to the said town to be the seat and land of the Governor of Virginia for the time being and his Successors and to be called by the name of the Governors Land. (*Records* III 99)

Yardly and his officers were to make that land, "the Lands formerly conquer'd or purchased of the Paspeheies," and adjacent lands into common land for the settlers. The company also instructed Yardly to "lay out by bounds and Metes" another three thousand acres to be "called the Companies Land" and occupied by the Company's tenants (99-100). In its conquest and purchase this land had become England's, but in its reduction to English acres, it entered the English spatial imagination which required measured property to understand the land. As it instructed the colony to measure and name Virginia land, the Company brought territories formerly unspecified in English terms under English control. And in so doing they attempted to erase Indian claims as they substituted an English spatial imagination of Virginia for Indian names for and understanding of the land.

The Company's instructions divided a territory called "James town" into areas called "the Governors Land" and "the Companies Land," bringing English social divisions, mercantile systems, and hierarchies into American territory. By dividing "James town" into common land, "the Governors Land," and "The Companies Land," the company declared that Virginia owed its existence in the final instance to the crown, that the whole belonged to James. It also announced that the people it imported would be under the rule of a governor whose preeminence was expressed in ownership of a large amount of land and whose dominance also rested in a title and not in a person—Virginia went through a number of governors quickly in its early years. The Company also deeded as much land to tenant farmers, calling that land the property of a group of merchants and noblemen investors based in London. It thereby re-created in Virginia a severely hierarchical English society and a political situation in which the crown was nodded to as the ultimate power but was also dependent on, as it authorized, monied interests.

In its acts of naming, the Virginia Company brought that land into an English cultural network, made it signify a specifically English rather than Indian space. The Company instructed Yardly to "reduce" its "plantations . . . into four Cities or Burroughs *Namely* the cheif City called James town [,] Charles City [,] Henrico [,] and the Burrough of Kiccowtan" (100). William Cronon suggests that in the case of New England, we can see the different "conception[s] of property" held by the English and the Indians in "the names they attached to their landscape." While the English named property after its owner,[8] "the great bulk" of New England Indian names for their landscape "related not to possession but to use" (65). Virginia Indian toponyms also follow that pattern, indicating, like New England Indian names, places where plants could be gathered, where animals could be caught, or people sweated; they also denote idiosyncratic geographical characteristics (Barbour 1971, Cronon 65, Heckewelder).[9] The Algonquian name "Kiccowtan" may mean "a person that heals, or where the sick are cured" or it may mean "great town" (Heckewelder 379 "Kiquotan," Barbour 1971, 288; Barbour 1981, 25).[10] If it signifies "great town" it must have commemorated a temporary greatness as the Powhatans had a "settlement pattern that was both dispersed and constantly fluctuating" (Rountree 1989, 58). In contrast, the English names, "James town," "Charles City," "Henrico," named an English system of ownership and permanent settlement. English relations under King James were organized around property and also around patrilineal inheritance, primogeniture, patronage and compliment. All of these pieces of the English "cultural network" were present in the Company's naming of its "plantations."

In contrast, the Powhatan Indians, that is those Indians the English called "Powhatan," were a matrilineal society, and Helen Rountree, whose sole sources are the English colonists who were looking for forms of their own social relations, can call Powhatan social divisions only "an incipient class system" (1989, 79). As Rountree notes, the English were continually looking for social distinctions but they "say nothing whatever about a real Powhatan 'aristocracy'" (1989, 143). The Powhatans had "little or no economic specialization" (Rountree 1989, 32); their lands were viewed as common unless they were being farmed, and they abandoned land to fallow after probably two years or so. Rountree suggests that "ownership of fields was apparently based on usufruct ["ownership" predicated on use]" (1989, 46). There is no evidence for their having a ruler with power even approximating the English monarch's; what evidence we have for their "emperor" Powhatan's power comes solely from English sources who were looking for a "king" with whom to deal. Powhatan's "empire" was very possibly formed in response to pressure from the English; that is, the English "found" what they were looking for, a transformed culture that could be read as approximating their own image (see my introduction note 12).[11] The social relations of English culture, recreated in the Company's names for Virginia territory, were surely foreign to the native inhabitants of that land, just as the social relations and relationships to the land indicated by Powhatan names were foreign, sometimes attractive, and sometimes threatening to the English settlers.

While the Indian name "Kiquotan" indicated either that there was at one time a large settlement on that piece of land or that one might go there to be cured, the English names showed that James was the supreme ruler of England and the territory that the Indians had occupied, and that his sons, Henry and Charles, would succeed to the crown and its claims. They also showed that the noble men organizing the Virginia settlement owed their compliments to their king and his sons, despite his sons' youth. Unlike the native groups who had lived, and were still attempting to live, on this land, English society was deeply stratified by rank, and its members ordinarily owed respect to rank above any other attribution And the English system mystified its rigidity in a culture of male friendship and compliment, also denoted in English naming. For example, in 1621 William Powell wrote from James City to Sir Edwin Sandys, "Not any waye moved with the power of your place, Right Noble Sir, although I hartelie wish such honnors might euer be so worthelie conferd: but as I must confess invited, naye incited by those inward beautyes, pietie, and pittye which do so loudlie speak you to the world more then man . . ." (*Records* III 436). Common men like Powell owed respect to their patrons and owed them also the mystification that they deserved that respect because of their natures not their rank; noble men like Sandys owed a similar respect to their king and paid complimentary tribute in their acts of naming.

In English eyes King James I's dominion over the land preceded the geographical face of Virginia—perhaps accurately, as it would change that face—and encapsulated the hundreds and plantations named for large-scale investors in the colony. The Virginia Company's charters from James granted them the land "To bee houlden of us our heirres and successors as of our Manor of East Greenwhich." As Robinson observes, the Manor at

East Greenwhich "refers to the residence of King James I at the royal palace of Greenwhich and was used as a descriptive term in many grants to indicate that the land in America was also considered a part of the demesne of the king" (11). James owned Virginia as he owned large pieces of England, and since property ownership, tribute, and compliment were the Virginia Company's and its settlers' ways of understanding colonial territory, the names of their king and his family soon dominated their maps of the Virginia landscape. In his letter reporting the adventures of the 1609 colonists who had been diverted to Bermuda and had finally built ships to carry them to the mainland, Strachey describes their arrival into Virginia as follows: "This is the famous *Chesipiacke* Bay, which wee haue called (in honour of our young Prince) Cape *Henrie*, ouer against which within the Bay, lyeth another Head-land, which wee called in honour of our Princely Duke of *Yorke* Cape *Charles*" (*A true reportory* 1748). The company, its cartographers, and those colonists not specifically employed as mapmakers were especially eager to name territory and features after the king himself, the center of the network of tribute and compliment. James issued the Company's three charters, and under its first charter he personally appointed the members of the governing Virginia Council from among the Adventurers (Craven 1957, 4). The English transformative fantasy was that these lands were his to give to his loving subjects. When Englishmen named Virginia geography after James, they both acknowledged his prior claim to the land, bid for his patronage, and marked the territory as English.

The English men and women in Virginia clearly understood the power of names in the colonial world. In 1619, their Generall Assembly petitioned the Treasurer, council and Company to say "they wilbe pleased to change the savage name of Kiccotan, and to give that Incorporation a new name" (*Records* III 161). The Company agreed to the change, issuing a broadside addressing "the foure ancient generall Burroughs, called *James* City, *Henrico*, *Charles* City, and Kicowtan, (which hereafter shall be called *Elizabeth* City, by the name of his Maiesties most vertuous and renowned *Daughter*)" (*Records* III 276).

Unlike James City, Henrico, or Charles City, Kicoughtan had a long history of Indian-English relations that had established the Algonquian name in English written records and English, as well as Indian, oral culture. Writing about the 1607 English foray into Virginia, George Percy described their arrival on the west bank of the river the English called "James":

> presently the Captaine caused the shallop to be manned, so rowing to the shoare, the Captaine called to [the Indians] in a signe of friendship, but they

were at first very timersome, vntil they saw the Captain lay his hand on his heart: vpon that they laid down their Bowes and Arrowes, and came very boldly to vs, making signe to come a shoare to their Towne, which is called by the Savages *Kecoughtan*. (1687)

This entrance initiated the town as the first port of trade between Indians and the English, and for about ten years they lived together in the area relatively peacefully (Hatch 95). Captain John Smith describes a Christmas among the "savages" where "we were never more merry, nor fed on more plentie of good Oysters, Fish, Flesh, Wild-foule, and good bread; nor never had better fires in England, then in the dry smoaky houses of Kecoughtan" (II, 194). But the English found this section of land very desirable, and they were troubled by the less than complete submission of Indians there and in adjoining areas. William Strachey described the 1610 English takeover of the town: "The nineth of Iuly [Sir Thomas Gates] prepared his forces, and early in the morning set vpon a Towne of theirs, some foure miles from *Algernoone Fort*, called *Kecoughtan*, and had soone taken it, without losse or hurt of any of his men" (1625, 1755). While the settlement was forcibly Englished, the name "Kicoughtan" in all of its variant spellings persisted in the documents for nine years and has remained in use for at least some of the area. In a letter to the Virginia Company from the Virginia Council after the 1622 Indian rebellion, "Kickoghtan" appears among the "few places" not abandoned by the English (*Records* III 612).

The toponym's persistence indicates its significance within the English conceptual world, but the Generall Assembly's efforts to discard the name show that its Indian associations were too strong to be completely assimilated even in the calm period before the rebellion. It is tempting to believe that the English tolerated the name "Kiccowtan" as long as they did because of the orthographical choices they made as they transcribed the Powhatan name.[12] As Margreta De Grazia and Peter Stallybrass suggest about variant spellings in Shakespearean texts, in texts from the early seventeenth century, "the field of the signifier is a field of difference—a field where 'true etymologies' are displaced by the orthographic play that allows for (indeed cannot prevent) the intermingling of signifiers and the coining of ever-new etymologies" (264). John Heckewelder, an early nineteenth-century German scholar of Algonquian languages rendered the Algonquian word as "Kiguatank" or "Kigŭétank." By transcribing the Algonquian name as "Kiccowtan," and by hearing in the foreign word the English sound that would allow that transcription, the English included in it the residents they preferred to the Indians: cows. In that transcription, which is only one of the choices the English made for that name, the Algonquian word entered an English "field of difference" and therefore looked as if it could signify within an English vocabulary as well as within their understandings of an Indian town, a town of Indians and English people, and a town taken from the Indians. Maybe, for a little while, they were just as happy to include a cow town among their burroughs.

As settlement progressed, however, the name was too much of a reminder of a blended English-Indian society; it seemed "savage" among the names of England's royal family that covered the land, and the settlers felt the need to change it into the king's daughter's name. And that daughter's name was given the land as part of the English system of mystified tribute. "Elizabeth City" is named, the document says, for James's daughter's "virtue" and "renown," not, as it surely was, as a complimentary bid for the royal family's patronage. The Company and colony's naming and renaming of territory in Virginia reveals an English conceptual world that conflicted with the conceptual world of the people whose native names the English either ignored or replaced. But English social divisions and internal conflicts were part of the conceptual world encoded in their names. The story of English mapping complicates a binary picture of colonial naming in Virginia.

30 Item [:] that ther be espetiall care taken both of generall and particular Survayes, whereby not onlie a true Mapp and face of the whole country costs Creeks riuers highe ground & Lowe ground &c. may bee exactlie discouerid: Butt also the Boundaries of the Severall Hundreds and Plantacions with the perticuler directions in them bee perfectlie sett forth from tyme to tyme maynetayned to peruent therby future differences that arise vpon questions of possestion, wherein also itt may be fitting and moste vsefull to posteritie, to Cast an Imaginarye eye and view, wher and which way the grand highewayes may bee like to strike and passe through the Dominions. (Instructions to the Gouernor for the Time Being and Council of State in Virginia July 24, 1621, *Records* III 477)

Naming of colonial territory is, of course, part of the process of mapping that accompanies every settlement by a colonial power. Maps are vital sites for understanding the place of naming in the new Atlantic world. European maps, like the one requested in item thirty of the Virginia Company's 1621 instructions to Governor Wyatt, were ways of understanding, claiming, and demarcating territory; they announced possession as they guided the settlers who would complete a conquest. They were an integral part of expansion. Item forty-four of those instructions to Governor Wyatt declares "as wee hold itt most necessarie that you provide for the generall safety and securing of your selues and estats together: So doe wee conceaue it a matter of exceeding great advantage & incouragement to discouer everie day farther by the sea Coast and wthin Land" (*Records* III 481). In the Company's discourse, both mapping and exploring are forms of "discovery." As Carter suggests, exploring and surveying can be "two dimensions of a single strategy for possessing the country. The map [can be] an instrument of interrogation, a form of spatial interview which make[s] nature answer the invader's need for information" (113). Whether colonial settlement follows exploratory "discovery" or outright violent conquest, the map is the result and the prophecy. J. B. Harley claims that across time and space,

> as much as guns and warships, maps have been the weapons of imperialism
> . . . Surveyors marched alongside soldiers, mapping for reconnaissance, then
> for general information, and eventually as a tool for pacification, civilization,
> and exploitation in the defined colonies . . . Maps were used to legitimize
> the reality of conquest and empire. They helped create myths which would
> assist in the maintenance of the territorial *status quo*. (1988a, 282)[13]

Certainly, in both Ireland and Virginia, English colonists used maps as tools of conquest, and native populations responded to them in those terms. According to John Andrews, "the reigns of Henry's successors witnessed a cartographical revolution [in Ireland] as mapmakers followed the English armies from one war-torn district to another" (5). Quinn records that the Irish population reacted to the mapmakers accompanying the Munster plantation by dropping stones on surveyors, one of whom described himself as "nearly starved because no one would provide him with food and shelter." A surveyor, Francis Jobson, recalled "being every hour in danger to lose [his] head" (Quinn 1966 28).[14] Jobson had been sent to Ireland to assist a man named Robins, "the chief measurer," but the Dublin council had great trouble persuading surveyors to serve the plantation. As Michael MacCarthy-Morrogh explains, "Reluctance to serve was unsurprising, for it was pointed out by Robins that a surveyor in England could expect . . . the aid of the locals, and the benefit of service in 'a quiet country.'" In Ireland, on the other hand, Jobson was poorly paid, and "'the people for the most part discontented with the course to be observed'" (61).

These reactions to mapmakers demonstrate that native populations fully understood the consequences of their territory being mapped by prospective colonial rulers: their land would be controlled and transformed by the mapmaking colonists. Edward Waterhouse, in the English colony, both asserts that the Powhatan Indians' 1622 rebellion was demonic and entirely unjustified and acknowledges, without remorse, that "the dayly feare . . . possest [the Indians] that in time we by our growing continually vpon them, would dispossesse them of this Country, as they had beene formerly of the West Indies by the Spaniard" (22). Waterhouse views the Americas as populated by an undifferentiated population of Indians, and his comment is also directed against the Spanish. But his comment reveals a native Virginia population that, like the Irish people in Munster, understood what the English desired as a result of their mapping and discovery.

Item thirty of the Virginia Company's 1621 instructions eagerly anticipates those consequences, and it also confirms Carter's understanding of mapping as a process of invention. The Company's discourse seems to postulate an existing buried or masked "face" in Virginia, what it calls "a true Mapp," that may be "exactlie discouerid." When this surface is discovered, the map will have reached truth; but the map's purpose is not to uncover a truth but to marry topography to boundaries—which were matters of land grants and therefore invented—and to imagine or invent future truths: the highways that will admit travel and commerce, that will dominate the land. The process of mapping is all of this at once: "exact discovery," imagination of the land as possession, and imagination of pathways for trade. Since the "true" geographical map exists within "the Boundaries of the severall Hundreds and Plantacions," that true map is not prior to possession but a part of possession. The "questions of possession" that the Company imagines may arise in the "future" reveal the Company's understanding that the English will own both the land and the future. "Future differences" may arise between Englishmen who will be the "posteritie" to whom the grand roads will be of use. The map erases Indian land claims as it bequeaths the land to English owners and supplies what the makers hope will be permanent boundaries.[15]

The Company's instructions ask that "a true Mapp and face of the whole country costs Creeks riuers highe ground & Lowe ground &c. may bee exactlie discouerid." As it includes the word "costs" in its list of the geographical features of the country, it reveals another desired result of the map. "Costs" certainly signifies "coasts" in the list, but it also signals the money interests involved in this map. The English drew maps of Virginia not only to divide the land for settlement but also to convert the land into currency. From the beginnings of its operation, the Company sold the equivalent of stock in the country; when subscribers bought shares in London, they were buying pieces of the mapped countryside, and the map made the division of the land into pounds and shillings possible. When the Virginia Council paid Sir Thomas Dale seven hundred pounds for his services in Virginia but ordered that money to be "received in land distribution," they were declaring that land in Virginia was money (Robinson 17). To shareholders and adventurers, the land represented an investment. Virginia had coasts and it also had costs; to recoup those costs, the land itself had to be convertible to capital. Mapping and naming the country realized the capital in the land, not only in the country's potential commodities.

As G. N. G. Clarke suggests, "nomenclature, as one aspect of the look of maps, becomes a primary ingredient of the visual dimension of possession: a verbal pattern through which culture speaks itself *onto* the land; renaming as it wipes clean one history and rewrites, as it renames, its own history onto the surface of the map (and land)" (456). As early as 1608, Robert Tindall's "Draughte of Virginia" named the large river entering the territory, on which the English located Jamestown, "King James his river" (see figure 2.1). The present day name of that river, "the James," first appeared on a printed map in 1635; Coolie Verner notes that the four "major Virginia rivers appear together correctly named on the 1671 Ogilby copy of the 1635 map 'Nova Terrae Marie Tabula'" (Verner 15). Although Tindall's chart may not "correctly" name the river according to its present-day appellation, it reveals what the "James" meant to his 1608 subjects. The river belonged to James because he was king, and it belonged to his English subjects because, in their subject status, they were members of his body. Thus Tindall drew this river in 1608 as a part of an embodied England.

Tindall also attempted to embody some of the country in his own image, naming two points on his map "Tindall's Shoals" and "Tindall's Point," but neither of these names appear on present-day maps of Virginia (Verner 7). While James's name was a fantasy backed with economic power, Tindall's was a personal fantasy that, while it shared an English understanding of possession, lacked the worldly support that James's name had in abundance. Tindall's attempt to name for himself indicates that an individual colonist might aspire to assert himself outside of the network of tribute and compliment. But it was his naming for King James that Tindall ultimately contributed to the map of the new Atlantic world.

Another early Virginia map, probably drawn in 1610 by a surveyor employed by James, called the river simply "the King's" (Brown 1, 456). And "the Kings Riuer" is what Strachey calls it in his 1610 *Reportory*

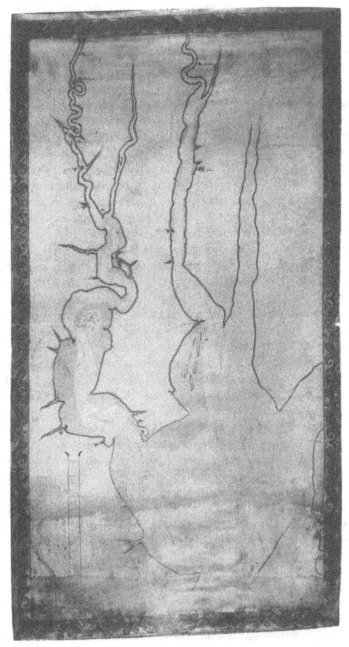

Figure 2.1—Robert Tindall's "Draught of Virginia" Cotton MS Augustus I, Vol. II. No. 46. Reproduced with permission of the British Library.

(1752). Percy describes an act of naming the river in 1607: "the foure and tweṅtieth day wee set vp a Crosse at the head of this Riuer, naming it *Kings Riuer,* where we proclaimed *Iames* King of *England* to haue the most right vnto it" (1689). The anonymous surveyor's 1610 map's name for the river is even more revealing of an English world view than Tindall's, since the surveyor deemed it completely unnecessary to specify which king "owned" the river, although English documents at the time acknowledged, and probably even created, Native American kings. "The King" could only be King James.

But another name given to "the James" river in the period reveals that authority and power was still contested in Virginia. Captain John Smith's "Map of Virginia" named the same river "Powhatan" in 1608. Smith's name for the river, after the Indian ruler Smith called an emperor, "Powhatan," survived on maps for some time, as atlas compilers used maps engraved from or using Smith's 1608 "Map." "Nova Virginia Tabvla," a map published in Mercator's Atlas of 1636, for example, situates Jamestown on a river it calls "Powhatan flu" (II, 437; see figure 2.2), and "Nova Anglia Novvm Belgivm et Virginia," another map in the same volume, puts the settlement on what it calls "R. Pawhatatan" (II, 441).[16] The temporary impasse in naming of the river—between "James," "the King," and "Powhatan"—indicates more than one struggle in the colonial world. An English worldview, predicated on possession, centered on ranked individuals and believing in God-given favor, was replacing a Powhatan worldview based in dwelling and use. The impasse in naming is also a sign of a contest between English colonists: between John Smith's elevation of a Powhatan "emperor" whom he could control and understand and the Virginia Company's and the King's subject-employee's debts to James as the authorizer. That Continental mapmakers preferred the name "Powhatan" over the name "James" shows that English possession of Virginia was still contested in the larger European world.

The historical record tells us that Powhatan Indians had mapped the river we now call "the James." Strachey says that the river was named "*Paspiheigh,* which wee haue called the Kings Riuer" (1752). But if Barbour is right, "Paspahegh" means "At the mouth (outlet)" and was therefore not the river's name but the name for the place on the river where the English located themselves (1971, 296). That we have permanently lost the aboriginal name for that river, and especially that the only native name the English record gives for it is Smith's, an English colonist, affirms Harley's broader observation that "conquering states impose a silence on minority or subject populations through their manipulation of place-names. . . .

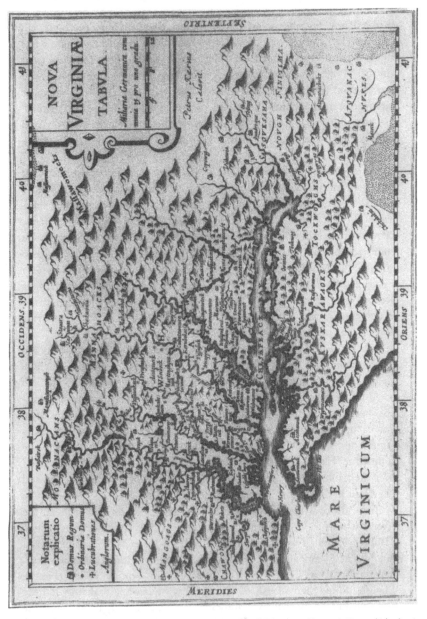

Figure 2.2—*Nova Virginiae Tabvla* in Gerard Mercator, *Gerardi Mercatoris Altas or A Geographicke description of the Regions, Countries and Kingdomes of the World.* Trans. Henry Hexham. Amsterdam: Henry Hondi, 1636. Reproduced with permission of the John Carter Brown Library at Brown University.

Whole strata of ethnic identity are swept from the map in what amounts to acts of cultural genocide" (1988b, 66). The English sources tell us that Powhatan himself took his name from a settlement (Rountree 1989, 7). The name probably meant "Medicine-man's (shaman's) village or hill" (Barbour 1971 297; see also Rountree 1989, 11), although it is worth noting that this is virtually the same meaning the white sources seem to have assigned to "Kequotan." If we can trust the etymology of "Powhatan," then a native leader took the name of a place made famous by a curer or priest, and subsequently an Englishman—Smith—took that name and gave it to a river. When Smith named the river, he used his English understanding of naming: that settlements and geographical features must belong to someone prominent and be designated by their personal name, and that his naming of the river for Powhatan gave him a special link to the emperor that he had supported and created by making claims on his behalf. Then, while Smith was attempting to gain prestige through his relationship with a native ruler, the Company's representatives were challenging his designation in the name of their desired direct relationship with their monarch.[17]

Although the toponym "The James" may not have triumphed on maps until the late 1630s, it carried more significance with the Company then did the name Powhatan. On August 24, 1621, the Virginia Company granted a commission to Arthur Guy and Nicholas Norburne for a voyage to Virginia "there to land and sett on shore all the said persons goodes and provisions soe shipped and deliuer them at James Citties in James Riuer in the Kingdome of Virginia" (*Records* III 499). Likewise, Edward Waterhouse, in his declaration written to assuage English fears after the 1622 "massacre," tells his audience that "at the time of this Massacre there were three or foure of our ships in *James-River*" (17-18).

Military might assured that James's name would triumph, but its triumph may also be attributed to the name's ideological significance for the English. In the tradition of English and Scottish kings, James was named for his ancestors in the kingship and for a culture hero. James was the name of five previous Scottish kings who were all, like James I of England, named for the biblical hero Jacob, the meaning of whose name is given in the King James Bible as "he that supplants." The author and collector of the famous expansion-oriented compilation of travel and exotica stories, Samuel Purchas, who collected his stories with an eye to converting nations to Protestant Christianity, exclaims in his prefatory letter to the Reader, "*God hath shewed his Word unto* our IACOB (THE DEFENDER OF HIS FAITH) *his Statutes and his iudgements unto this* ISRAEL *of Great Brittaine*" (par. 5).

The Genesis stories of Jacob are origin stories and therefore naming stories: Rebekah's womb is said to hold "two nations" in the twins Esau and Jacob (25:23). God names Jacob Israel, and Jacob names the places he encounters after his relationship with God. Intimately familiar to the Protestant adventurers in London and Virginia, Jacob's stories, invoked by their king's name, must have authorized colonial naming and also English expansion in Virginia. Read in the Protestant tradition as types and therefore as prophecy, the collection of stories accruing to the name Jacob would have justified every English possession and theft.[18] In naming the land after James, the English, using Genesis as a master narrative of colonialism, could identify the Indians with Jacob's dispossessed brother Esau, who in the words of the King James Bible "was [only] a cunning hunter, a man of the field; [while] Jacob was a plain man, dwelling in tents" (25:27).[19] In that story, Jacob steals Esau's inheritance and blessing, and Esau is left to live off the land and to be watered by the sky. Ultimately undisturbed by Jacob's treachery, Isaac blesses him: "that thou mayest inherit the land wherein thou art a stranger" (28:4). And God tells Jacob "thy seed shall be as the dust of the earth, and thou shalt spread abroad to the west, and to the east, and to the north, and to the south; and in thee and in thy seed shall all the families of the earth be blessed" (28:14).

James's name on the land was prophetic to the English, who were most unlikely to see the irony in God's blessing. The bitter irony for the Indians was that the English colonists' meagerly successful attempts at converting Indians to this tradition fulfilled its prophecies—at least in the English stories. The 1622 Powhatan rebellion that killed about 350 settlers could potentially have wiped out English settlement in Virginia; had all of the settlers been killed, it is unlikely the Virginia Company or the crown would have strenuously attempted to settle that as-yet-unprofitable part of North America. But, Edward Waterhouse explains, the English were saved by an Indian convert: "Such was (God bee thanked for it) the good fruit of an Infidell conuerted to Christianity; for though three hundred and more of ours died by many of these Pagan Infidels, yet thousands of ours were saued by the means of one of them alone which was made a Christian" (20-21). Although the English had not succeeded by 1622 in converting much of the native population to Christianity, their efforts had transformed at least one Indian into a loyal Protestant, believing in the superiority of a religious system that authorized the dispossession of his people in its king's name.

As English settlers in Virginia mapped and named the land and features after their king, they discovered Virginia in the image of their culture, a culture that believed that God authorized possession. They reaffirmed their culture's hierarchies and its parceling out of privilege. But names on English maps of Virginia also reveal a far from unified understanding of the place of individual settlers in the new Atlantic world. The modern world's loss of the native name for the river now called "The James" exposes England's domination of the Virginia Indians; but the conflicts over the central river's name among Englishmen show that an abstract understanding of colonial naming as unified comes under pressure from individuals with a personal project who might challenge the established system in their own names.

In 1609, in a diversion from the Virginia mainland colonization made famous by *The Tempest,* Sir George Sommers, William Strachey, Sir Thomas Gates, Captain Newport, and a group of other colonists bound for Jamestown in the *Seaventure* were shipwrecked on the islands we now call the Bermudas. Their experiences led the English to begin settling those islands in 1612. As Jeanette Black notes, they "were the earliest English colony to be completely surveyed and mapped. The small land area—only about twenty square miles—made a complete survey feasible, and the comparatively rapid growth of the population in the early years of settlement rendered it necessary" (151). Contributing to the feasibility of completely mapping and surveying the territory was the lack of any native human population: no prior land claims, no one dropping stones on the surveyors' heads, and no previous names, aside from the stubbornly persistent name Bermuda, apparently for an earlier Spanish explorer, Juan Bermudez (Craven 1990 15).[20] The islands were a blank slate for English naming, and the marks the English made on that slate reveal not only their understanding of North American colonial territory, but—as in naming on the Virginia mainland—English social relations. Intriguingly, English naming in Bermuda reveals as well English colonial desires for Ireland, a part of the Atlantic world that might seem entirely unrelated to these previously unpopulated islands.

Initial acts of naming in Bermuda took two related forms. The settlers named land for property owners, some of whom were the early inadvertent explorers. They also named territory to honor those original Englishmen who set foot on the land and who made the land hospitable for their companions and the coming settlement waves. Strachey relates that "Sir

George Summers in the beginning of August, squared out a Garden by the quarter (the quarter being set downe before a goodly Bay), vpon which our Gouernour did first leape ashoare, and therefore called it (as aforesaid) *Gates his Bay*" (*True reportory* 1739). As it is extremely unlikely that Sir Thomas Gates, who was traveling to Jamestown as the new governor of Virginia, was in reality the first of the men on his ship to step on the bay's beach, what was commemorated in that naming was not an act by a man, but that man's prominence within his company of men and women. Again English colonial naming denoted rank within English metropolitan culture above other considerations such as priority or effort. In Shakespeare's *Henry V,* the victorious king reads "the number of our English dead" after the battle of Agincourt:

> *Edward* the Duke of Yorke, the Earle of Suffolke,
> Sir *Richard Ketly, Dauy Gam,* Esquire;
>   None else of name: and of all other men,
> But fiue and twentie. (TLN 2821-2825; 4.8.102-106)

As in Shakespeare's play-world, Gates's common companions had names in their private world, but in the public sphere of commemorated actions they remained nameless. Sommers's power to name came from his rank, and he named the bay after a man with a title, not after those people whom the fictional king Henry terms "other men": men who might have acted in the same manner to possess or settle the colony, but whose status did not rate either the prose or the cartographic historical records.[21]

In those records, Sir George Sommers became the settlement's most important hero; all the early accounts credit him with saving the ship-wrecked group from destruction on the islands. While Gates survived to profit from his early, forced, foundational moment (the shipwreck that caused him to "leap ashore"), Sommers died trying to feed the starving Virginia colony with provisions from Bermuda. Captain John Smith tells the dramatic story of Sommers's heroism as he attempted to return to the islands to gather food for the mainland colony:

> Much foule and crosse weather he had, and was forced to the North parts of Virginia [perhaps what we now call the coast of Maine], where refreshing himselfe upon this unknowne coast, he could not bee diverted from the search of the Bermudas, where at last with his company he safely arrived: but such was his diligence with his extraordinary care, paines and industry to dispatch his businesse, and the strength of his body not answering the ever

memorable courage of his minde, having lived so long in such honourable services, the most part of his well beloved and vertuous life, God and nature here determined, should ever remaine a perpetuall memory of his much bewailed sorrow for his death: finding his time but short, after he had taken the best course he could to settle his estate, like a valiant Captaine he exhorted them with all diligence to be constant to those Plantations, and with all expedition to returne to Virginia. In that very place which we now call Saint Georges towne, this noble Knight died, whereof the place taketh the name. (II 351)

Complimenting himself as well as his subject, Smith's hagiography ends with an act of naming that at once commemorated a great colonial service to His Majesty's plantations and claimed the islands for England in opposition to Spanish claims and naming practices. Smith, who always insisted on both his title "Captain" and his own "diligence" in maintaining the Virginia plantation, lauds Sommers as "like a valiant Captaine." Thus, Smith made Sommers' actions metaphorically increase Smith's own status; "like a captain" becomes the ultimate accolade, complimenting a man whose actual rank, "knight," made him better than a captain in the English status system. As he tried to transform its terms, Smith also paid tribute to that system by calling Sommers "this noble knight." So without actually challenging English ranking of men, Smith managed to elevate his own status by attributing the nobility of a knight to his acting "like a valiant captain."

For Sommers's extraordinary act of service, defying his health for his compatriots, the settlers called their principal town "Saint Georges towne." Sommers's personal name's consonance with that of England's patron saint allowed the English settlers to canonize George Sommers and simultaneously to name the town for England. The name "Saint Georges towne" seems to resemble all of the Spanish place naming for saints that Father Cobo lauded: "in order to appreciate the high regard, esteem, and filial love that we the Spanish nations have for the Patron of Spain, it should be known that we give nineteen towns of these Indies the title of Santiago; in honor of San Juan we give sixteen towns his name; two have that of San Pedro and one that of San Pablo . . ." (81). But in reality, "Saint Georges towne" marks a distinctly Protestant appropriation of Catholicism's canonizations. Whereas Cobo argued that the Spanish named towns "Santiago" to show their reverence for their patron saint, the English gave the town their patron saint's name to honor an uncanonized individual. They revered not primarily the Church's saint but their own heroic colonial creation. In addition, however, the patron saint's name stood for England in the English and European systems, so in that saint's name they could oppose Spanish claims to the same territory.

Smith followed his story of the naming of "Saint Georges towne" with a sentence in which he recorded the rechristening of the entire territory: "But [Sommers's] men, as men amazed, seeing the death of him who was even as the life of them all, embalmed his body and set saile for England, being the first that ever went to seeke those Ilands, which have beene ever since called Summers Iles, in honor of his worthy memory" (351). However, despite the happy pun on summer/sommer, which the propaganda-minded English very soon tried to make the most of, they were finally unable to change the islands' name. The impetus for this unsuccessful attempt at renaming was presumably England's colonial contest with Spain. "Bermuda" as a name was clearly Spanish, a reminder to the English, and to Spain and the larger world that the first men to step on the islands' soil were Spanish, not English. The English tried valiantly in a number of textual venues to get their new preferred name to stick. But even their own efforts betray the task's impossibility.

A February 12, 1612 (new dating) letter from John Chamberlaine in London to Sir Dudley Carleton, who was then serving as James's ambassador to Venice, mentions the attempted name change as part of the news from England about the Company's doings in America:

> There is a lotterie in hand for the furthering of the Virginia viage, and an under-companie erecting for the trade of the Bermudes, which have changed theyre name twise within this moneth, beeing first christned Virginiola as a member of that plantation, but now latley resolved to be called Sommer Illands, as well in respect of the continuall temperat ayre, as in remembrance of Sir George Sommers that died there. (*Letters* I 334).

In his characteristic slyly humorous epistolary style, Chamberlaine pokes mild fun at the islands that cannot decide what to call themselves and at the Company, the absent subject of this naming process. His letter illustrates the Company's desire to name in order to assert possession against Spanish claims and, perhaps, to familiarize the islands for their colonists; the letter also indicates the difficulties that would eventually block that renaming. The 1612 colonists worried constantly about a Spanish invasion. Both Smith and Nathaniel Butler[22] relate two stories about scares at supposed sightings of Spanish fleets, although Smith derisively lampoons the colonists' temerity. According to Butler, the islands' first governor, Richard Moore, spent most of his time fortifying them against incursions by foreign ships. And Butler attributes the Company's resupplying of the islands so quickly "by reason of some distrusts they tooke of the Spaniards soudaine supplanting of it in its birth (as not likely to endure patiently such a thorne in his West Indies sides)" (27); As governor, Butler himself charged the colonists in 1619 to fortify the island, "man" the fortifications, and to "be soldiers," for, he said, "if it should so fall out that any soudaine breach happen betweene England and Spaine (and who knowes hoe sone this may be,) ther is not any place that it will breake out vpon soner than vpon this" (196). The Company was determined to replace the islands' Spanish name even before Sommers provided a martyr's patronym, and their original choice "Virginiola" was a clear attempt to link the islands with their tenuous but relatively more established holdings in Virginia and perhaps even to shore up those holdings by adding territory. Butler says that the original unintentional colonists, when they returned to London, inspired the Virginia Company, "who rightly apprehended the aydefull vicinitie of the place to that colony of thers" (17). But Chamberlaine's syntax reveals how well-entrenched the name Bermuda was; rather than telling Carleton that the Company was forming a new wing to "trade in the Sommer islands, formerly called the Bermudas," he writes that the trade is "of the Bermudas," demonstrating that he could rely on Carleton's knowledge of the islands under the name "Bermudas."

Chamberlaine could rely on Carleton's previous knowledge of the islands under their Spanish name because the "Bermudas" had an established legendary presence in the European travel literature. Because of their position in the Atlantic, they were extremely difficult to navigate around safely, and because of their frequent storms, mariners had long feared the islands and considered them haunted (Quinn 1989). Silvester Jourdain in *A Discovery of the Barmvdas, otherwise called the Ile of Divels,* published in London in 1610, explains that the *Seaventure's* landing on the islands was "most admirable. For the Ilands of the Barmudas, as euery man knoweth that haath heard or read of them" were "euer esteemed, and reputed, a most prodigious and inchanted place, affoording nothing but gusts, stormes, and foule weather; which made euery Nauigator and Mariner to auoide them, as Scylla and Charibdis; or as they would shunne the Deuill himselfe" (8-9). Strachey set out to dispel this prejudice in his 1610 narrative: "I hope to deliuer the world from a foule and generall errour: it being counted of most, that they can be no habitation for Men, but rather given ouer to Deuils and wicked Spirits" (*True repertory* 1737). Smith made those stories into a sign of England's exceptionality: "for you have heard, it hath beene to the Spainiard more fearefull then an Utopian Purgatory, and to all Seamen no lesse terrible then an inchanted den of Furies and Devils, the most dangerous, unfortunate, and forlorne place in the world, and [Sommers' company] found it the richest, healthfullest and pleasantest they ever saw" (II 348). This legendary history was solidly in the way of the islands' renaming, for it had given them general notoriety; and that renaming would even have unnamed two English children, who being born to members of Gates's group after the 1609 shipwreck were called "Barmuda" and Barmudas" (Strachey, *True reportory* 1746, Smith II 349).

James I's June 29, 1615 charter to the newly formed Bermuda Company, while it opens by granting them a patent for "All those Islands . . . formerly called by the name of Bermudas or the Bermuda Islands now called the Somer Islands," repeats that phrase or a variation of it so often that the document mentions the name "Bermudas" seven times before finally starting to call the islands "the Somer Islandes" (Lefroy 1877 83-98 *passim.*). Thus, while the charter participates in the English attempt to change the toponym permanently, it also testifies to the name's previous significance. And although in their texts and business dealings the Company valiantly alluded to the islands only as the Sommer islands, other significant texts about the islands use both names prominently. Even Smith's 1624 *The Generall Historie of Virginia, New-England, and the Summer Isles,* despite its hagiographic narrative of the renaming, titles its fifth book's section on the islands *The Generall Historie of the Bermudas, now called the Summer Iles.*

The islands' climate, to which its legendary history alludes, was also a significant barrier to popular acceptance of the name Summer Islands. Purchas writes that Bermuda "is also called the *Iland of Deuils";* at least one sea-monster was spotted there, "but this name was giuen it not of such Monsters, but of the monstrous tempest which here they haue often sustayned" (Lefroy 1877 102). George Sommers's poetic epitaph, recorded by Smith encapsulates the name change's irony:

Hei mihi Virginia quod tam cito præterit Æstas,
Autumnus sequiur, sæviet inde et hiems,
At ver perpetuum nascetur, et Anglia læta,
Decerpit flores florida terra tuas.

In English thus:

Alas Virginia's Summer so soone past,
Autumne succeeds and stormy Winters blast,

Yet Englands *joyfull Spring with joyfull showers,*
O Florida, *shall bring thy sweetest flowers.* (II.351)

But as Strachey informed his readers, "these Ilands are often afflicted and rent with tempests, great strokes of thunder, lightning and raine in the extreamity of violence" (*True reportory* 1738). Since the "Summer Islands" were liable to make settlers and mariners feel "stormy Winters blast," the propagandistic memorial toponym was naturally doomed.

In 1615, Richard Norwood completed his first survey of the islands and he surveyed them again between 1616 and 1618. Although a printed map was registered at Stationers Hall in January of 1621/22, it has not survived, but John Speed published an engraved version of Norwood's map in his 1627 atlas, *A Prospect of the Most Famovs Parts of the World . . . Together With all the Prouinces, Counties and Shires, contained in that large Theater of Great Brittaines Empire* (Black 151, Speed XI; see figure 2.3). In that atlas the map is called *A Mapp of the Sommer Ilands once called the Bermudas,* and it appears within a decorative frame below which is a band of text titled "The names of the now Adventurers, *viz.* this yeare 1622 so neare as wee could Gather" with a decorative emblem on its left side and the Virginia Company's seal to its right. Everything about this map, from its layout to its naming of specific features, argues for England's possession and shows the islands subsumed in an English cultural world.

On the engraved map, Norwood's surveyed islands appear in the visual context of other European and especially English possessions in America; thus the map renders the islands as impossibly large and located next to and spanning Virginia and New England. The cartouches located at the top left hand and bottom right hand of the map give separate scales for the map of the islands and the distances outside of the islands, but the effect of the engraved map, if a viewer does not carefully consult the scales, is to magnify enormously the islands' area and therefore their accessibility and attractiveness, while simultaneously minimizing their distance from the mainland; even according to the upper-left-hand scale, the Sommer Islands' "Ireland" island is located less than two hundred and fifty miles from Roanoak Virginia and "St. Georges Iland" is less than two hundred miles from the Northern New England coast.[23] Modern maps locate the islands over seven hundred miles off the coast of Virginia and, of course, nowhere

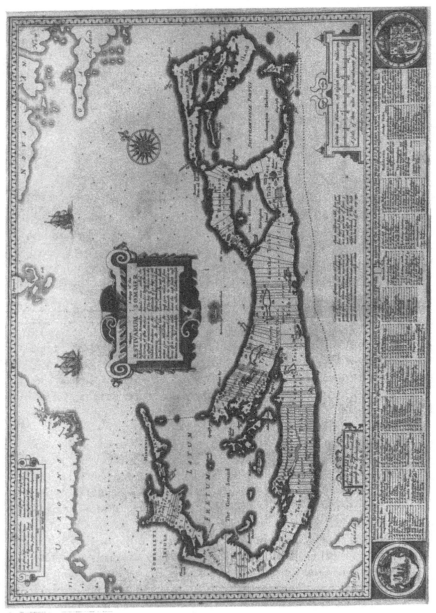

Figure 2.3—Engraved version of Richard Norwood's map of Bermuda. John Speed, *A Prospect of the Most Famovs Parts of the World . . . Together With all the Prouinces, Counties and Shires, contained in that large Theater of Great Brittaines Empire.* London, 1627. Reproduced with permission of the Folger Shakespeare Library.

near Maine. Although two small pieces of the Spanish colony, Hispaniola, appear also impossibly close to the bottom-left-hand corner of the mapped islands, the prominence of the English possessions at the map's top two corners and middle announce the islands as England's and as related to England's other colonies. William Boelhower suggests that maps evoke a "narrative world" that encloses their readers and defines their understandings of territory. He comments that "part of this narrative world even involves the spatialization of letters, their differing dimensions, the variety of their characters, and, of course their hierarchical grades" (492).[24] On the top right hand and left hand of Norwood's map respectively, Virginia and Nova Anglia are set in all capitalized italics; thus the English possessions draw the eye and significantly frame the islands. On the bottom-left-hand corner, "Hispaniola" is engraved in smaller script letters with only the "H" capitalized. The map thereby acknowledges but minimizes the Spanish presence in this part of the "New World."

Clarke suggests that borders, images, cartouches, and fonts are "a series of interrelated indexes which bind [maps] within a series of ideological assumptions as to the way the land is viewed" (455). In the case of this engraving of Norwood's map, these features all insist on England's possession, yet even this map cannot fully instantiate the Sommer Island Company's desired renaming of the islands. On the large cartouche at the map's center, *A mapp of the Sommer Ilands once called the Bermudas,* "Sommer" appears in large bold capitals while the rest of the title is set in mixed italics less than half the size of the patronym. Thus the map participates in the Company's attempt to rename the islands. However, it also alludes to the islands' legendary history by framing the left end of the band of adventurers' names with an emblem of a wrecked ship on a shield held by a ferocious lion flanked by the legend "Quo fata ferunt" ('where the fates take us'). That tag presumably refers to the fateful 1609 shipwreck that enabled the colonization, the shipwreck that *The Tempest* refers to, and that is also referenced in the list of colonizers' names that appears directly to the emblem's right. The map also addresses another instance of "providential deliverance" in a small text set in the bottom center of the framed map, which reads (in English), "About midsomer 1616, five persons departed from these Ilands in a Smal open boat of some 3 tunn and after 7. weeks arived al safe in Ireland [the island next to England], the like hath scarce bene heard of in any age." Both text and emblem, then, can be seen to promote the English efforts as especially blessed, although ironically, the text takes a story of five men who rebelled against the tyrannical but sanctioned governor, Daniel Tucker, and transforms it into an officially approved miracle. Running from Tucker's brutal administration and injunctions to work, the five men had launched themselves into the sea pretending that they were going fishing.[25] The text connects this story with the islands' patronym by setting its tale of miraculous deliverance in "about midsomer," rather than giving a month (the Latin reads "*Circa solstitium æstivum*" with the Latin title of the map as *Mappa ÆSTIVARUM . . .*). But neither emblem nor text can avoid their narratives' connections to the proverbial destructiveness of "the Bermudas." Although both speak of English deliverance, they speak also of the terrifying storms and wrecks that made that deliverance so unlikely.

Smith's history of the islands credits Norwood's 1618 survey with civilizing the English settlers who had previously spent their time squabbling over government and laboring to avoid work. According to Smith, when Governor Tucker arrived in the colony in 1616 and Norwood began his survey, the governor found "the Inhabitants both abhorring all exacted labour, as also in a manner disdaining and grudging much to be commanded by him" (II 362).[26] But when Norwood completed the survey in 1618 and the land was carefully divided, "this which was before as you have heard, but as an unsetled and confused Chaos, [received] a disposition, forme, and order, and [became] indeed a Plantation" (II 370). The engraved map from that survey shows the islands carefully divided into shares within "tribes" named for their noble sponsors back in England, called "adventurers" in the documents: "Hammilton Tribe," "[Sir Thomas] Smiths Tribe," "Devonshire Tribe," "Pembroke Tribe," "Pagets Tribe," "Warwick Tribe," "Southampton Tribe," "Sandys Tribe." Those patrons' names cover the land, dominating recently established toponyms, although a number of those toponyms were also for those patrons: "Pagets port," "Sandys narrow," "[Sir Thomas] Smith's Ila.," etc . . .

Most of the tribes' names were hereditary English titles; Smith and Sandys were only knights, but each of the others was a lord or an earl. Only Smith and Sandys actively governed the Virginia and Sommer Islands Companies (in London), but their activity was the smallest consideration when the Company divided the land and named the tribes. Butler's text explains the procedure: "First therfore, eight of the chiefest persons and deepest adventuerers wer assigned to haue ten shares apeece in euery tribe, answerable to the quantitie of their adventures, and euery one of them (for honnours sake) to haue the tribe called by his owne name" (105). None of the adventurers actually came to the islands, as Tucker implicitly laments in a July 14, 1616 letter to Nathaniel Rich, a principal adventurer in Southampton tribe: Tucker urges, "Sir, I should much rejoice to see you,

Sir Robert Riche or some of the Cheefest Adventurers, in these ilands. It is a very pleasant voyage and I will assure you, you will thincke your tyme well spent and your labours well bestowed" (*Rich Papers* 8). But Tucker assured Rich in vain; Sir Robert Riche never visited the Bermudas. Thus, like the naming of Jamestown and Cape Henry, the islands' map's names microcosmically denote a society of prominent Englishmen still located in England, and those names reflect a reified English system of rank.[27]

Naming of natural features and built "improvements" on the islands followed the same pattern as the naming of the tribes. "Paget's port," although originally established by Governor Moore (an untitled gentleman) and subsequently improved by Tucker, was named "in honnor of the Lord Pagett, whoe [was] one of the company" in England (Butler 24-25). Likewise, Butler relates that Moore laid "the foundation of a large redoubt upon a hill to the east of St. George's towne, the which he afterwards brought to good perfection, beinge at the present called Warwick Fort" (29). And Butler wrote to the Company that he had worked hard to restore the "burnt redoubt in the King's Castle . . . and [he has] bin bold to call it Devonshyres Redoubt" after England's Lord Devonshire (215). Again rank determined English colonial naming above considerations of effort. Rather than recording the labor that established the settlement or even commemorating the directors of that labor, the Sommer Island map records its home culture's rank order.

This system was so accepted by the colonial government's representatives (and indeed can still seem so natural to us) that they continually reinvented it, and in that reinvention they illustrate the condition of a hegemony's maintenance. So it is not that these governors and the nameless men they commanded built all of the islands' named forts only to have these forts named in spite of their builders. On the contrary, those governors and their anonymous workers were as likely themselves to name the forts for nobility in order to participate in and perhaps benefit from the system of tribute and compliment that I have already described, just as they were unlikely to have named a fort for an untitled man working on it, however important his efforts. So Butler relates that, in the course of fortifying the islands, Moore "rayseth a fashionable redoubt in Coopers Iland, and calls it Pembroke Fort" (39).

Like Tindall in his draught of Virginia, Moore did try to commemorate his own efforts; Butler described him

> busied about the rayseinge of a foure-square frame of timber vpon a high
> hill ouer toppeinge the towne of St. Georges to the westwards to serue for

the discouery of shippinge vpon the coast; the which worcke the gouernour accounted for his masterpeece, and was earnestly affected to haue it carry his owne name, and to be called Moores mount. (28)

But the islands' inclement weather countered Moore's dreams and knocked down the only architecture to bear his name, as if somehow nature supported the nobility's dream of naming the colony "for honnour's sake." Butler also related that Governor Tucker named a piece of what was to end up as general or company land after himself; Norwood, Butler wrote, began his survey "at the east end of the islands, which boundeth vpon that generall land nowe called Tuckers Towne (being so tearmed by the Gouernour Tucker himselfe vpon his erectinge ther of two or three cottages of leaues and plantinge the liue-tenant of the castle and his wife vpon it)" (104). But Butler's parenthetical comment on the naming of Tuckers Towne lampoons the former governor's quest for honor. Its few and unsubstantial buildings and its meager imitation of the adventurers' planting prerogative make his eponymous town a joke. Where the naming of the tribes and the islands' fortifications embeds them in rank, privilege, and unsullied honor, Tucker's naming signifies his dishonorable character.

Because the islands had no native population and therefore no prior names, it may seem unremarkable that the English settlers covered them with noble patronyms and the names of the first inadvertent explorers. After all, in their naming of the "Sommer Ilands" the English were not appropriating another culture's land or language. But English naming in these islands reveals its culture's fundamental preoccupations, what Antonio Gramsci would call its "common sense," or what David Turnball calls its "forms of life." That common sense determined that whoever actually worked at settlement, whether in building or government, was less important than whoever held a higher rank in London and might be able to provide capital for further settlement. That common sense also determined the basic system of naming on a deeper level. The vast majority of names on the engraved map are English men's names with a few exceptions: the settlers named a shore point "Spanish Point" because they found a cross there left by Spaniards from a stranded galleon. They also named a bay "Tobacco Bay" for a small tobacco growth there which had presumably been introduced by another shipwrecked crew (Quinn 1989, 12, 18). In addition, Speed's engraved map shows perhaps six relatively insignificant places named for animals: Hog bay, Long bird Iland, White hearne bay, Cony Iland, Whalebone bay, and Gurnett's head. This latter point may well have been named not for the fish, "Gurnett or Gurnard," but for that fish's place in English popular discourse;[28] as the *OED* notes, the term "*Gurnet's head* [was] used allusively with reference to the disproportionate size of the fish's head." The dictionary offers a number of references for the term from plays written for the public theater, revealing the fish's general and common comprehension as metaphor. In contrast to this meager naming for animals, not one Sommer Islands' toponym refers to a plant or tree. In general the islands' toponyms demonstrate a pointed lack of interest in the native flora and fauna, despite the crucial part played by nature in the colonists' initial and continuing survival.

As Harley suggests, "that which is absent from maps is as much a proper field for enquiry as that which is present . . . silences should be regarded as positive statements and not as merely passive gaps in the flow of language" (1988b 58). Norwood's map as engraved for Speed is a rich text, covered with noble names; however, as well as concealing the labor of the nameless people building the settlement, its toponyms ignore the natural abundance that made the settlement originally feasible and desirable. In 1610 Jourdain wrote that despite the islands' legendary history, the original shipwrecked crew found a veritable natural cornucopia there: Sommers, he said, went "and found out sufficient, of many kind of fishes, and so plentifull thereof, that in halfe an houre, he tooke so many great fishes with hookes, as did suffice the whole company one day. And fish is there so aboundant, that if a man steppe into the water, they will come round about him" (11-12). Sommers's men found hogs, left by earlier boats, that had multiplied and waited to be eaten, and, Jourdain wrote, "there is fowle in great num. vpon the Ilands, where they breed, that there hath beene taken in two or three houres, a thousand at the least" (1610 13). Gates and Sommers's men found birds that laid their eggs among men, herons they could beat "downe from the trees with stones and staves" and "other small birds so tame and gentle, that a man walking in the woods with a sticke, and whistling to them, they wil come and gaze on you, so neare that you may strike and kill many of them with your sticke" (14). Jourdain wrote that "the Country yeeldeth diuers fruits, as prickled peares, great aboundance, which continue greene vpon the trees all the yeare. . . . And there is a tree called a Palmito tree, which hath a very sweet berry, vpon which the hogs doe most feede . . . which occasioned us to carry in a maner all that store of flower and meale, we did or could saue for Virginia" (15-16).

Like Jourdain, Strachey praised the islands' fishing: "Wee haue taken also from vnder the broken Rockes, Creuises oftentimes greater then any of our best *English* Lobsters; and likewise abundance of Crabbes, Oysters, and Wilkes. True it is, for Fish in euerie Coue and Creeke wee found Snaules,

and Skulles in that abundance, as (I thinke) no Iland in the world may haue
greater store or better Fish" (*True reportory* 1740). He described "a kinde of
webbe-footed Fowle . . . of the bignesse of an *English* greene Plouer, or
Sea-Meawe" that appeared in great quantity on the islands in Fall and Winter. In Strachey's account, these birds come to be killed when the English
call them:

> I haue beene at the taking of three hundred in an houre, and wee might
> haue laden our Boates. Our men found a prettie way to take them, which
> was by standing on the Rockes or Sands by the Sea side, and hollowing,
> laughing, and making the strangest out-cry that possibly they could: with the
> noyse whereof the Birds would come flocking to that place, and settle vpon
> the very armes and head of him that so cryed, and still creepe neerer and
> neerer, answering the noyse themselues: . . . and so our men would take
> twentie dozen in two houres of the chiefest of them; and they were a good
> and well relished Fowle, fat and full as a Partridge. . . . There are thousands
> of these Birds, and two or three Ilands full of their Burrowes, whether at any
> time (in two hours warning) wee could send our Cock boat[29], and bring
> home as many as would serue the whole Company. (1741)

In addition to touting these amenable birds, like other observers, Strachey
remarked on the turtles that surrounded the islands and offered an alternative food source when the colonists' semi-domesticated hogs failed them:
"true it is the Hogs grew poore, and being taken so, wee could not raise
them to be better, for besides those Berries, we had nothing wherewith to
franke them: but euen then the Tortoyses came in againe. . . . One Turtle
(for so we called them) feasted well a dozen Messes, appointing sixe to
euery Messe" (1741). Even when rudimentary domestication failed, the
natural world provided for the shipwrecked colonists.

All of the primary texts about the early years of settlement—Strachey,
Butler, Jourdain, Smith—treat the islands' plants and animals at length; those
descriptions are at once a real record of the crucial role that the islands'
"abundance" played in saving Sommers's people and then sustaining human
life on the islands, and at the same time propagandistic lures to future (and
perhaps hungry) settlers. But those descriptions do not carry over into the
actual process of naming the islands. "Hog bay" and "White hearne bay"
are small inlets whose presence on the map is barely legible in comparison
to the toponyms "Harington Sound," "Southhampton Harbour," and
"Pagets port." While the colonial texts frequently mention acts of naming
for noble and prominent men, none of the texts talk about naming land,

harbors, or waters for nature, emphasizing how negligible the practice was. Butler's text comments that the Governor began to repair "the fort commanding the towne, called Warwick Castle" and with all his improvements "it might appreare, so much the more worthy of the noble appellation it was distinguished by" (231). But rather than talking about naming the islands after their natural "abundance," the narratives speak of naming the flora and fauna after the English landscape. Butler says, "another smale birde ther is, the which, by some ale-hanters[30] of London sent ouer hether, hath bin termed the pimplicoe, for so they imagine (and a little resemblance putts them in mind of a place so dearly beloued), her note articulates" (4). Butler's comment initiates his running critique of common drinkers who impeded the settlement of the islands, but it also records a typical practice of naming the natural world for the familiar London scene. While the English happily named the islands' fauna after their own "*English*" haunts and animals, they distinctly did not memorialize the natural world on their maps. Their naming practices distinguish an English conceptual world from the world of the Indians they encountered in Virginia who named their spaces and themselves for the foods the land offered to them.

Although naming for people rather than animals or nature was so commonsensical for the English as to be absolutely unquestioned, the narratives of the colony's settlements demonstrate, as Peter Hulme has noted, that the dream of living in the colony and naming the islands "for honnour's sake" was not universally shared by the English men in Bermuda (1986, 103–104). All of those narratives teem with stories of discontented and rebellious lower orders, and many of them record direct verbal challenges to any English authority based in the system of rank and honor that determined so much of English colonial naming. Strachey tells the story of a "Gentleman" named Henry Paine who when called to stand guard struck his commander and when threatened with Governor Sommers's wrath spoke so blasphemously against the system that he had to be shot. Strachey writes that

The said *Paine* replyed with a setled and bitter violence, and in such vnreuerent tearmes, as I should offend the modest eare too much to expresse it in his owne phrase; but the contents were, how *that the Gouernour had no autoritie of that qualitie, to iustifie vpon any one (how meane soeuer in the Colonie) an action of that nature, and therefore let the Gouernour* (said hee) *kisse, &c.* (*True reportory* 1745)

And Paine was not alone in forgetting his place in the transplanted society. In 1615 Governor Moore, disgusted that the Company's promised rewards had not arrived, left the islands under the command of six governors, whose rules were described by both Smith and Butler as drunken, chaotic revels. Butler described a crew of men, during that time, finding what they believed was a wrecked treasure ship; celebrating their (illusory) newfound wealth and prominence, they shouted "the treasour is found at the Flemish wrack, the treasour is found: we are all made men, made men." Illustrating De Grazia and Stallybrass's point about the orthographic play in early modern English, Butler wryly comments "Madd men indeed." Butler's text narrates one man's dreams as particularly risible:

> among the which merry gang I must nott forgett a charitable smith, who being told by one of the company (betweene iest and earnest) that in respect he was one of the chiefe of the gouernours fauorites and a good bowler, it could not be but that his share in this purchase would amount to noe lesse than the makeinge of him a gentleman, if not with an improuement of a ladyship for a wife, "And then (quoth he) you will not knowe vs poore men." (68)

Although Strachey's story describes a gentleman's deadly punishment for disobeying a social superior, while Butler, less ominously (for his subject), invites the reader to laugh at the aspiring smith, both stories functioned to warn readers against ambitions above their station and visions of the world turned upside down; at the same time both accounts expose the system's faultlines. Paine clearly threatened the quotidian social order more seriously, so much that he could not be directly quoted and his "vnreuerent" mouth had to be shut forever; the (unnamed) smith seems less dangerous as he was being fooled even by own his companions. Like the drunken butler Stephano in *The Tempest* who dreams of being a king, this poor smith was alone in his belief that he would rise to the rank of gentleman and even marry a noblewoman. Butler's text invited its readers to laugh along with the smith's companions at the extent of his ambitions, which are amusing in themselves; that is, the text presumes that those ambitions are inherently impossible, only believable by a clown or a fool. The smith is "charitable" in the *OED*'s sense of "inclined to think no evil of others, to put the most favourable construction on their actions"; he has no critical faculties, and Butler's readers are to laugh at the smith's trust in his "friends."

However, the very depth of the hilarity the text demands, its overdetermined character, indicates that the smith's fantasy was ultimately as threatening as that of Paine, the gentleman. The colony had to execute Paine because as a gentleman he was expected to uphold the system that afforded him his rank; instead of acting as the system's agent, he undermined it completely, not because he denied that Sommers, a knight and the anointed governor, had authority over him—after all, Paine the gentleman might have become a knight—but because he denied the knight's authority over any member of the colony, in Strachey's words, "how meane soeuer." Paine's invitation to the governor "to kiss" his ass reduced Sommers's authority to the lowest level and required that Paine be silenced. But the smith had to be even more effectively silenced, branded as uncomprehending and an object of hilarity to all, because his rise would have embodied a real and threatening social disruption—that money would so transform the poor that they would become identical to the gentleman, so undistinguishable that they could marry into the nobility.

These social threats to the established order of rank certainly existed in England but the colonial world became an intensified microcosm of them. Robert Rich (the younger cousin of the Robert who would become the second Earl of Warwick in 1619), his cousin and his brother Nathaniel's representative on the islands, wrote in May of 1617 to Nathaniel that a cooper named Dysone "hath given out very baise speches unto divers that I spoke noething a tall for the people" (*Rich Papers* 27). When Butler arrived as governor he convened a grand jury, and in his charge to them, he felt obliged to address the problem of challenges to the established order:

> Ther is another ill-fauored vice and custome rauiginge amongst you, which by all meanes is to be suppressed: and it is an itche of tounge and a kind of base delight to depraue and slaunder the actions and good names one of another. . . . And if the deprauations extend to men of place and command, very seuere and sharpe punishments are inflicted: and with good reason for howsoeuer it is true, that wordes are but wind (as they saye), yet experience tells that oft times they proue the fire-brands of all other ills; for *a verbis ad verbara* [from words to lashes]: it may be that some busie heades and factious discontents haue bin the breeders and sowers of this ill seed among you: but let them take heed least, at length it produce hempe, to make a rope to hang themselues. (186)

In this charge, Butler attempted to enlist his appointed judges to defend those in power from the words of the discontented. This must have been a delicate and difficult task, since Butler's own text reveals the depredations and tyranny of those, like Tucker, who had been "men of command" on the islands. By linking slander against the high with neighborly slander (a mode possibly experienced and maybe even perpetrated by every man on the jury), by branding the slanderers as "factious discontents," and therefore dangerous to the colony's harmony, and by threatening capital punishment, which at least one slanderer had already suffered, Butler could hope to induce the colonists to police themselves and thereby uphold the English system of rank and authority.

Butler's charge shows that Bermuda, the small, easily surveyed plantation, was also a potential hotbed of colonial transformation—a fantasy land for the most disruptive English who might there become transformed unrecognizably. Butler's charge also shows that colonial naming on the islands was as much an effort to suspend colonial transformation as it was an effort to transform territory. Covering the islands with the names of men of place was a way for colonial governors like Moore and Butler to pay tribute to men of place and thereby hopefully garner patronage, but it was also a way to attempt to instantiate permanently a system that the colonial experience on the islands showed was constantly threatened by the transformative cauldron that was the isolated colonial experience—where gentlemen could aspire to have their posteriors saluted by noblemen, and poor men could fantasize a world in which they could attain the power and status of wealthy noblemen.

One of the islands' toponyms, however, signified a more collective English fantasy of domination over its perpetually troublesome enemies to the west, the Irish, whom Fynes Moryson attempted to name as submissive. Sometime before Norwood drew the first of his maps, the small thin island that reaches north, completing the islands' northeasterly curve on their western side, was named "Ireland." Perhaps the naming was thought of as a joke, but that name holds all the hopes for an English relationship with its western neighbor that hundreds of years of military attempts had yet to achieve. Ireland island's position in relationship to the colony's main island may have suggested the toponym, but if so the name would have been as or more appropriate for the island called "Somerset iland," a much larger piece of land than "Ireland" which sits more directly adjacent to the main island, as *Éire* sits in relationship to England. Actually no piece of land in the group of islands is related to the main island as *Éire* is to England, but that seems to be precisely the name's point. The smaller "Ireland" is a more appropriate container than "Somerset iland" for English dreams since that "Ireland" sits dominated and yet isolated on the islands' map. In the colonial fantasy world denoted by its name on Norwood's map, the English could imagine an Ireland many times smaller proportionally than *Éire* was or is to England. *Éire,* the land near England, is actually close to England's size and therefore physically as well as politically and religiously threatening (if those categories can be separated in the early modern period). In contrast, the Sommer Iland's "Ireland" is proportionally tiny, and the map's typefaces emphasize and exploit this size differential; although the map clearly labels the small island "Ireland" and supplies its Latin name "*Hibernia,*" the island's neighbor Somerset iland's Latin name *Somerseti Insula* is set in much larger type, making that island—named like "St. Georges" for George Sommers, the islands' and England's patron saint—physically dominate "Ireland."

Ireland's name on the map, given simply as "ireland" without the qualifier "island," is itself a sign of its conceptual significance for its cartographers and audience. The map names many other islands, from large islands, including "Somerset Iland," "St. Georges Iland," and "St. Davids Iland" to very small islands, such as "Pearle Iland" and "Daniels Iland"; Norwood's map labels every one of those islands as an island. Only "Ireland" and the island labeled "Nonsuch" sit on the map without the label "Iland." Perhaps that is because "Nonsuch" may also allude to a place: "*Nonsuch* near Epsom in Surrey" (OED);[31] thus actual places resemble their namesakes completely. But the effect of leaving off the designator "Iland" is such that what one sees on the Sommer Ilands map is not "Ireland Iland" but "Ireland," insignificant in relation to land designated as belonging to important English noble families.

Norwood's map completely surrounds the island "Ireland" with English possessions. Flanked to the west by the much larger "Somerset Island" and the even more prominent Virginia, and to the northwest by New England, Ireland sits enclosed and unthreatening. Because of its location the island was completely unfortified, unlike the large and small islands on the northeast side of the group, and that lack of visible fortifications on the map renders Ireland still more dominated and defenseless. Since the Sommer Islands' settlers built those fortifications largely in response to real and imagined Spanish threats, and *Éire* was a real jumping-off place for a threatened Spanish invasion of England, naming the small, surrounded, unfortified, and unthreatening island "Ireland" symbolically reduced Spanish danger to the home country.

The map uses three symbols to signify habitation: a small house, a larger house with visible windows and a connected tower, and turrets, which signify the fortifications so assiduously built by Moore and neglected by Tucker in Butler's account. These houses and crenellated towers—all of which look European and so were purely iconic in a settlement that used the available leaves to roof its dwellings—speak of the islands' inhabitation and defenses (the map also shows scattered ordinance symbols at strategic coastal points). But Ireland island is represented as "socially empty space," free, therefore, of human barriers to possession, and unprotected, distinctly not the condition of England's metropolitan neighbor (Harley 1988a 284). *A Mapp of the Sommer Ilands, once called the Bermudas* performs a colonial transformation on *Éire* and that island's relationship to England, reducing the recalcitrant kingdom and colony to the surrounded, minimized, and empty space of New English desires.

In its names and symbols, Norwood's map of the Bermudas imagines the islands as a perfected British Isles. The Sommer Islands are a sun-drenched, ordered paradise: well-fortified, stable, controlled by "natural" masters, and in proper hierarchical relation to one another. English lords dominate the main islands, which themselves naturally dominate the smaller, unprotected Ireland island. But the map's names, while they themselves denote massive colonial transformations, especially of the island's natural world, cannot entirely contain the colonial transformations envisioned by the disempowered in the new Atlantic world. The next chapter will show how the London stage became a crucial space to display colonial transformations and to investigate their effects on the metropolis.

Chapter 3

The New Atlantic World Transformed on the London Stage

At the same time as English noblemen were transforming Bermuda into their home by proxy, English poets were using colonial expansion into the Atlantic world as an opportunity to comment on and attempt to transform their homeland and its imperial ambitions. It was the London stage that reached a critical mass of English people and displayed the new Atlantic world and its colonial subjects, and on that stage England's colonial exploits and native subjects spoke to the condition and transformation of English culture. This chapter focuses primarily on transformative visions on the English stage, particularly in Ben Jonson's dramas. However, Ben Jonson is by no means the only English playwright from the early seventeenth century who used the emerging Atlantic world as a touchstone on the London stage. Indeed, as my introduction indicates, most early modern drama is involved with producing English subjects who understood themselves in relationship to that new world. Recently scholars have focused on how Shakespeare's plays treat Ireland and the related questions of Irish and English identities. *Henry V* in particular, because of its Irish character MacMorris and its overt reference to Essex's Irish military campaign, has attracted critical commentary that recognizes the play's implication in English colonial policy.[1] And Shakespeare's interest in Ireland is not limited to the history plays. Where *The Tempest* has long been seen as the location for Shakespeare's "New World" interests, David J. Baker argues that Ireland is the "necessary (but unnamed) place for [that play's] dislocations" (1997b, 81); and Andrew Hadfield suggests that "the ghostly presence of Ireland haunts many of Shakespeare's works" (1997, 52).[2] In addition, Gordon McMullan has argued that John Fletcher's plays *The*

Island Princess and *The Sea Voyage* uneasily explore "the desires and demands" of Virginia adventurers.[3]

Both canonical and less widely read early modern plays speak to England's developing role in the new Atlantic world. In his effort to worm the truth out of his prince, *The Duchess of Malfi's* Bosola uses the proverbial danger of the Bermudas, which the Bermuda Company worked so hard to rewrite, as the ultimate metaphor for the perfidy of a ruler who values lineage over personal integrity: "I would sooner swim to the Bermoothas on / Two politicians' rotten bladders, tied / Together with an intelligencer's heart string / Than depend on so changeable a prince's favor" (III.ii.266-69). Written probably in 1613, Bosola's metaphor, which convinces the Duchess of his sincerity, is evidence that the Company's propaganda had not triumphed, and his invocation of the "Bermoothas" also speaks to the place of those islands as an imaginable outland for the English theater audience. The same English audiences were accustomed as well to the appearance of derogatory references to Ireland and Irish people on the London stage.[4] Thomas Dekker's 1630 play *The Second Part of the Honest Whore* features an Irish footman whose dialect provokes this comment from an Italian gentleman: "Is't not strange that a fellow of his starre, should bee seene here so long in *Italy*, yet speake so from a Christian?" (I.i.23-24). Because Bryan is Irish he is identified as a pagan, and his inability to speak intelligible English means that his mouth is full of English foul language: "My Lady will haue me Lord sheet wid her" (I.i.20). When *The Second Part of the Honest Whore's* Infelice, the hero Hippolito's wife, tells him that she has cuckolded him with Bryan, Hippolito exclaims, "Worse than damnation! A wild Kerne, a Frogge, / A Dog: whom Ile scarce spurne" (III.i.148-49). Although the falsely accused Bryan vows to leave for Ireland, he returns later in the play as a walking dialect-joke in a scene in a linen-draper's shop. Dialect jokes like Dekker's were a regular feature of Irish characters on the English stage, and they also appeared in court masques, as we will see in chapter 4. And court masques that lampooned Irishmen were imitated in plays. John Ford's play *The Chronicle History of Perkin Warbeck* (printed in 1634) puts a masque of "four wild Irish in trouses" on the public stage (III.ii.111). In characters like Bryan, who populated the London stage, the stereotypes and insults from colonial texts such as Spenser's *View* walk in front of London audiences and are made real for those audiences, some of the members of which had economic or personal interest in the larger Atlantic world.[5] Beyond the canon of Jonson's plays, the London stage as a whole was a site of commentary, display, and intervention in that emerging world.

However, Jonson's corpus offers an especially accessible and complex site for an investigation of the new Atlantic world on the early modern English stage, since in his plays Jonson sees the colonial world talking back to and describing England's emerging imperial identity.[6] Because Jonson saw himself as the guardian of English manners and values, his texts speak powerfully to and of England's colonial concerns, for in the emerging Atlantic world the realms of English identity and England's colonies were becoming irrevocably entangled. This chapter looks to Jonson's plays for signs of—and as agents of—the emerging new Atlantic world and the transformations that were producing that world. The plays Jonson set in England are home-making fantasies that envision England in terms of its colonial spaces and as a colonial space. In this chapter, I examine the evolution of the characters Ben Jonson crafted as explorer-poets;[7] and I show how Jonson used these figures and the worlds he created for them to display and contend with these transformations. I suggest that in order for Jonson to aspire to what Richard Helgerson has called poet-laureate status in England, he had to become the interpreter of empire. Jonson was the poet-playwright of England's colonial efforts at home. His texts concern themselves with the impact of empire at home, with how the burgeoning efforts to become an empire encompassing Ireland, Virginia, and Bermuda would shape England, with how the English would comport themselves as colonists and how they would enjoy the fruits of their colonial ventures. His English plays evince and display a lexicon of otherness; they are also deeply involved in the inevitably coupled project of defining Englishness. Those plays display the colonial implications for English identity, implications that became reality in the turmoil of the Bermuda settlement's early years. Because Jonson's texts see English identity as troubled, because they define many technically English people as savages, and because they are metadramatically aware of their divided audiences, they reveal England's own colonial transformations: the contentious transformations of London and Englishness itself that came along with English efforts to expand into and control territory in the new Atlantic world.

As the previous two chapters have indicated, Jonson's plays and the London stage as a whole were only one site through which English people were negotiating their places in the new Atlantic world; and his plays make most sense when read in the context of that larger negotiation process. Therefore this chapter reads them as part of a textual world that includes Virginia Company propaganda, especially its lottery announcements, but also some of its Declarations of the colony's condition. That textual world includes, as well, John Mandeville's often-printed travel fantasies, which were read as fact for hundreds of years, Jonson's own poetry, a Michael Drayton poem, and John Donne, William Crashaw, and Patrick Copland's sermons written for the Company. Again, these are only a selection of the myriad early modern literary texts that spoke to the emerging new Atlantic world. I read these texts together through the postcolonial lenses of Gerald Vizenor and Mary Louis Pratt's theories, which suggest the constructed nature of colonial fantasy and the possibilities for resistance. Jonson's theatrical negotiations within the new Atlantic world should both remind us of how it was made and alert us to other possible worlds.

In 1605, Ben Jonson collaborated with George Chapman and John Marston[8] on a play for the Children of Her Majesties Revels entitled *Eastward Hoe*. Dedicated to the city of London, that play satirizes a thirty-pound knight, Sir Petronel Flash, who wastes his money and dreams of recouping his losses and avoiding his creditors by fleeing to Virginia. Sir Petronel may actually be even cheaper than a thirty-pound knight; one gentleman accuses him of having bought his knighthood from a page for four pounds. This knight and his dissolute and lowly friends, Seagull, a ship's captain, and the adventurers Scapethrift and Spendall, whose names define them, dream of a Virginia that marries Sir Walter Raleigh's fabled South American Manoa to Sir Thomas More's 1516 Utopia:[9]

> *Seagull.* Come boyes, *Virginia* longs till we share the rest of her Maiden-head.
> *Spendall.* Why is she inhabited already with any English?
> *Sea.* A whole Country of English is there man, bred of those that were left there in 79. They haue married with the Indians, and make 'hem bring forth as beautiful faces as any we haue in England: and therefore the Indians are so in love with 'hem, that all the treasure they haue, they lay at their feete.
> *Scapthrift.* But is there such treasure there Captaine, as I haue heard?
> *Sea.* I tell thee, Golde is more plentifull there then Copper is with vs... Why man all their dripping Pans, and their Chamber pottes are pure Gold; and all the Chaines, with which they chaine vp their streetes, are massie Golde; all the Prisoners they take, are fetterd in Gould.
> (IV:III.iii.14–31)

Performed for the first time in 1605, after two unsuccessful English attempts at establishing a colony in Virginia, *Eastward Hoe* delineates and ridicules contemporary and even archaic English fantasies about Virginia. Seagull's reference to Virginia's maidenhead clearly echoes Raleigh's famous comment in his 1596 *Discoverie of the Large Rich and Bewtiful Empyre of Guiana* that "Guiana is a Countrey that hath yet her Maydenhead" and is waiting for England to ravish her (115); and the play's dialogue's lavish Indians who craft golden toilets and chain their prisoners with precious metal are direct descendants of Utopia's people who do the same to show their contempt for avarice and worldly delights. Certainly Chapman, Jonson, and Marston would have expected at least some of their audience to understand the adventurers' Virginia fantasies as antiquated fictions.

Unlike the aspiring smith in Nathaniel Butler's Bermuda narrative, Sir Petronel Flash, his lapsed apprentice buddy, Quicksilver, and their lowly cohorts Seagull, Scapethrift, and Spendall never make it to their promised land. The play shipwrecks them on the Thames at Cuckold's Haven, and instead of landing in their prospective colonial refuge they land in the counter (debtor's prison), a destination slightly less humiliating and dangerous than that threatened by the constable in the play who takes them up on shore as masterless men. The constable had pressed them for military service in the Low Countries before their true identities were revealed. Their fantasy of escape from responsibility disappears as the play makes them pay for their metropolitan crimes with humiliating servitude. In the counter, Flash and Quicksilver, along with the usurer Securitie who helped to finance their trip, become converted to Christian piety and renounce their economic crimes.

Seagull, Spendthrift, and Spendall dream of a ludicrous fabled Virginia that Chapman, Marston, and Jonson could safely lampoon on the London stage, but their Virginia fantasies are not limited to recycled material from sixteenth-century colonial dreams. *Eastward Hoe* also speaks to colonial fantasies of social advancement that, as we have seen in chapter 2, had enough substance to threaten the colonies in Virginia and Bermuda established just a few years after the play first appeared on the stage. In *Eastward Hoe*, Seagull's golden chamber pots and fetters have the same status as Sir Petronell Flash's gulled bride Gertrude's dreams of rescue from the penury in which Flash has left her. Gertrude believes that "A *Fayrie* may come, and Bring a Pearle, or a Diamonde. . . . Or, there may be a pot of Gold hid o'the backside, if we had tooles to digge for't" (IV:V.i.83-86). Like Gertrude, Seagull lives in a world of fairy stories and legends. But Seagull also dreams of a Virginia that surpasses quotidian London in a more potentially attainable fashion. He declares,

Then for your meanes to aduancement, there, it is simple, and not prepos-
terously mixt: You may be an Alderman there, and neuer be a Scauinger; you
may be a Noble man, and neuer be a Slaue; you may come to preferment
enough, and neuer be a Pandar, to riches and fortune enough, and haue
neuer the more villanie, nor the lesse wit. (IV:III.iii.48-54)

It is in response to this inducement that Spendall expresses his willingness
to embark: "Gods me! And how farre is it thether?" (IV:III.iii.55). The
implication that in England noblemen might be construed as slaves was
evidently perceived as dangerous, for when *Eastward Hoe* was revised in
response to a complaint that a passage in the same scene satirized the Scots,
the reviser changed the words "Noble man" to read "any other officer."[10]
Despite the fact that the foolish Seagull speaks these words, whoever cen-
sored and revised *Eastward Hoe* realized that they could be taken seriously.
The unnamed charitable smith in Bermuda, who dreamed that his share in
a treasure ship "would amount to noe lesse than the makeinge of him a
gentleman, if not with an improuement of a ladyship for a wife," clearly
took claims like Seagull's seriously (Butler 68).

As early as 1605, Jonson, Chapman, and Marston understood the lure of
social advancement in the colonies. They saw that dangerous English social
fantasies might be active in the imaginations of men who envisioned colonial space as a place to transform society. Like Seagull, the smith in
Bermuda believed that away from English social control, a man might live
in a transformed England; a country with all of the same hierarchies, but
an England in which a lowly born man might have access to the benefits
of those hierarchies, privileges previously shared only among the well-
born. In addition, Jonson, Chapman, and Marston, in the safe space of
Seagull's colonial fantasy, embed a salient and powerful critique of the
human cost of those hierarchies. It seems in Seagull's eyes that in England
a man's social advancement requires that he sell his human dignity and
freedom. This is certainly not a novel critique; in fact, *Utopia*, the text that
undergirds Seagull's other arguments about Virginia, makes the same argument through its traveler Raphael. But in Seagull's fantasy, Jonson, Chapman, and Marston expose that vision of colonial transformation to the
London theater audience. The colonial adventurer becomes a mouthpiece
for a vision of a perfected England. While, like the unnamed smith in
Bermuda, Seagull is a laughing-stock, his complaint might be heard, and
clearly was heard, above the laughter.

As well as potentially advocating English reform through the vehicle of
colonial desires, *Eastward Hoe* warns that colonial adventuring might depre-

date England's economic base: the land that supported its nobles and gentlemen and the resources of its citizens. The play warns against England's transformation into fodder for the as-yet-totally unprofitable Virginia. Sir Petronel Flash marries Gertrude for the sake of the land she inherited from her grandmother: "a pretie fine Seate, good Land, all intire within it selfe" (IV:II.ii.143-44). Sir Petronel intends to sell that land through his buddy Quicksilver to the usurer Securitie in exchange for the means to sail for Virginia. For Securitie's delectation, Quicksilver characterizes that land as "Two hundered pounds woorth of wood readye to fell. And a fine sweete house that stands iust in the midst an't" (IV:II.ii.146-48). Securitie is a loathsome character who sees every person and thing in terms of profit and will not even lend his victim, Sir Petronel, the use of his stable, but as Securitie himself indicates, the moving force behind the decimation of England for profit that he envisions is Sir Petronel's pipe dream of adventure to Virginia. As if he were some sort of nightmare version of Virginia propaganda, Securitie laments that "we haue too few such knight aduenturers: who would not sell away competent certainites, to purchase (with any danger) excellent vncertainties? your true knight venturer euer does it" (IV:II.ii.171-74). Again, this sentiment comes from the mouth of a degraded character; however, the play approves of Securitie's anti-adventuring stance, since its adventurers' Virginia really is a ridiculous investment that sucks Sir Petronel and Gertrude, his victim, dry. Likewise, Quicksilver has robbed his former master Touchstone, a goldsmith, of "some small parcels . . . to the value of 500. Pound . . . to furnish this his *Virginian* venture" (IV:IV.ii.236-38). Since that "venture" is an empty hole into which England's vital landed and commercial resources disappear, *Eastward Hoe* tells its London audience that colonial desire endangers England as the play also points to social problems at home.

Although *Eastward Hoe* means its audience to laugh at the (sea)gulls who believe in a land of gold, in 1605 that fantasy about Virginia, promulgated by early explorers like Raleigh, had as yet little English competition. When the Lady Elizabeth's men revived *Eastward Hoe* for the court in 1615, these lines must have sounded even funnier, as publicity by the Virginia Company had placed the settlement firmly in the public imagination and its image had undergone substantial revision. English people now not only knew that Virginia had no gold, they knew also that the project of settlement would require hard work and would be accomplished, to some extent at least, against the resistance of its native population. *Bartholomew Fair* played at about the same time as that revival for the same court audience; by that 1614 performance, the English colony in Virginia was alive but small and struggling. By the time Ben Jonson wrote *The New Inne* in 1629, about three thousand English people lived in the rapidly expanding Virginia colony. During the period of time between these two plays, England was beginning perforce to reexamine itself as a trans-Atlantic colonial power and to constitute itself in relation to its American settlements. The 1620 *A declaration of the state of the colonie and affaires in Virginia . . .* , one of the many Declarations issued by the Virginia Company over that crucial fifteen year period, states that the colony "hath as it were on a sodaine growne to double that height, strength, plenty, and property which it had in former times attained" (A3,1). In those fifteen years England became a colonial power outside of its tenuous contiguous empire in Ireland. *Eastward Hoe, Bartholomew Fair,* and *The New Inne* each show an England implicated in its colonial expansion, and the differences between the plays reveal a developing metropolitan English vision of colonial control.

Bartholomew Fair's fair is the most public of spaces. To its London audience, the play offers the fair with its puppets as all the world. The foreign colonized in the cast and the play's Induction's references to England's newest possession—Virginia—both signal the fair as England extended, an imperial England at its birth. According to the play's denigrating Stagekeeper, the play's fair resembles the Virginia colony as much as it resembles London's Smithfield: "When it comes to the fair, once: you were e'en as good go to Virginia, for any thing there is of Smithfield" (9-10). As much as they despise the Stagekeeper, the Bookholder and Scrivener who displace him also remind us of that resemblance; in the Scrivener's articles for a contract between the audience and author, he offers the audience "censures" in the play, lots of criticism that he compares to the lots that can be bought to support the Virginia project. When he presents Smithfield's puppets in place of *The Tempest's* "servant-monster," he offers the opportunity to see this Smithfield as critics have so long seen Shakespeare's *Tempest's* island: as a place to observe a representative England colonial other. It is as if *Eastward Hoe's* crew of cozeners and gulls had landed in Smithfield rather than at cuckold's haven and produced their Virginia at the fair.

Eastward Hoe and *Bartholomew Fair* show Jonson displaying English "savagery." Witness, they say to their audiences, drunkenness, lewdness, dialect, tobacco smoking, cozening, and the need for citizen and aristocratic control;[11] look and you will see not only the Bermuda straits of London, where Overdo says one will find quarreling, "bottle-ale" and "tabacco," but the larger Atlantic world in need of an explorer to settle it in the proper English mode. When Wasp tells Bartholomew Cokes in *Bartholomew Fair* that his tenants are "a kind o' ciuill Sauages that wil part with their chil-

dren for rattles, pipes and kniues" (VI:III.iv.37-38), Jonson's diction draws
on and shapes his London audience's notion of Indians. Jonson's vision of
"civil savages" depended on a variety of sources: travelers' histories, Virginia company propaganda, his audience's understanding of savagery. Wasp
could be quoting from any number of colonial narratives; by 1614 the
notion of the "New World" savage with no comprehensible notions of
value and with an insatiable hunger for trinkets had attained a canonical
place in the colonial and exploration literature.[12] Jonson's offhand use of it
implies that his public theater audience also knew this topos well. When
Jonson calls Englishmen savages, he indicates that he sees the emerging
Atlantic world as one world, degraded and begging to be perfected.

Ben Jonson's sometime friend[13] Michael Drayton depended on a similar understanding of English savagery when he wrote an epistolary poem
addressed to George Sandys, who was serving in Virginia as treasurer for
the colony. Sandys produced Virginia's first published literary effort, a translation of Ovid's *Metamorphoses,* the quintessential account of transformation. Drayton's poem encourages Sandys to pursue his literary efforts on
Virginian soil:

And (worthy *GEORGE)* by industry and use,
Let's see what lines Virginia will produce;
Goe on with *OVID,* as you have begunne,
With the first five Bookes; let your numbers run
Glib as the former, so shall it live long,
And doe much honour to the *English* tongue:
Intice the Muses thither to repaire,
Intreat them gently, trayne them to that ayre,
For they from hence may thither hap to fly,
T'wards the sad time which but to fast doth hie
For Poesie is followed with such spight,
By groveling drones that never raught her height,
That she must hence, she may no longer staye.
(Drayton III 207-208 lines 37-49)

In Drayton's epistle, action in the American colonies becomes an example
for the metropolis, the colonial atmosphere a safer and happier place for a
distinctly *English* project, here the production of noble poetry. Under the
guise of a description of colonial life, the English poet offers a critique of
English life and an object lesson for its improvement. Drayton's bitter
poem, which compares his treatment at the hands of English censors to a

"shipwrack" (20), explicitly praises Sandy's translation project, while it dispraises the state of English poetry—written by "groveling drones"—and laments Drayton's own reception as an English poet. Offering Virginia as a refuge for the muses when English "Base Balatry" (79) drives them out, Drayton implies a vision of a perfected England, where "noble" poetry is valued over the "base," where poets may write freely without fear of the censor, where Drayton himself would get his well-deserved patronage from James I.

However, although Virginia provides safety for the poet and the project of advancing the *"English* tongue," the poem finds its native human inhabitants beneath description. Continuing the poem's fictional epistolary frame, Drayton asks Sandys to reply, and in that reply to educate him about Virginia's resources, its physical nature, and to keep him informed about the colony and its governor's health and progress—in short, to write another of the Company's public declarations privately for him:

> If you vouchsafe rescription, stuffe your quill
> With naturall bountyes, and impart your skill,
> In the description of the place, that I,
> May become learned in the soyle thereby;
> Of noble *Wyats* health, and let me heare,
> The Governour; and how our people there,
> Increase and labour, what supplyes are sent,
> Which I confesse shall give me much content;
> But you may save your labour if you please,
> To write to me ought of your Savages.
> As savage slaves be in great *Britaine* here,
> As any one that you can shew me there.
> (III 207-208, lines 93-104)

In response to Drayton's own public "letter" to the Company's treasurer, Drayton asks for a private declaration of the colony's health, framed exactly like those numerous "Declaration[s] of the State of the Colony in Virginia," but with a twist. Like those official declarations, Sandys's "rescription" should laud Virginia's commodities, praise its beauties and promote the colony's progress; but unlike those official declarations, Sandys's reply should make no mention of Virginia's native inhabitants, who appeared often in those Company accounts—alternately offering the settlers food and comfort, welcoming conversion to satisfy English people's pious ends, or unsuccessfully resisting (and thereby justifying) English encroachment.

Drayton's poet-persona asks Sandys not to describe Virginia's "Savages" because the poet finds enough human curiosities at home: "As savage slaves be in great *Britaine* here, / As any one that you can shew me there" (103-104). As in Fynes Moryson's account of Éire's names (chapter 2), in Drayton's poem we find an apparent early modern relativism that, under scrutiny, reveals itself as a more insidious repression. At first sight the poem seems to be resisting any essential othering of Virginia's native population; if England's "savage slaves" can substitute for those natives in Sandys's account of the exotic, distant land, then Virginia's Indians must be, at the worst, familiar Englishmen, another version of something found at home. Equating Virginia's "Savages" with England's "savage slaves" might seem to elevate the Indians to a higher status than they were offered in many English accounts, which envisioned the natives as more brutish than animals (see Bach 1997 and introduction). Such a reading, however, underestimates the brutality of Drayton's satirical attack on his own homeland. Actually, Drayton's bitter satire degrades Virginia's native inhabitants by comparing them with the English people Drayton's poem savages. Drayton employs a litany of degenerate and demonized otherness to describe his English savages. He calls them "groveling drones," "that bestiall heard," and equates them with "the stiff-neckt *Jewes,*" "Th'unlettter'd *Turke,* and [the] rude *Barbarian*"; they are worse for English poetry than "all the old *English* ignorance before," unleashing a ravenous "blind Gothish Barbarisme" on Drayton's times.

Not just a collection of the poet's worst possible epithets and accusations, Drayton's list also mirrors the series of accusations and comparisons settlers used to describe Virginia Indians in accounts like the one Drayton's poem requests from Sandys. Rather than elevating Virginia's Indians to civilized status, Drayton's epithets degrade them, associating them with the worst people and attributes conceivable under Drayton's progressivist Christian schema. Like the Jews, England's and Virginia's "savages" refuse to condone and live by the truth; like the Turks, and both the Goths and the unlocatable but beyond-the-borders barbarians, they are illiterate and without manners—two charges clearly related, as Norbert Elias's history of manners argues. Instead of championing progress, as Drayton and other poetry-loving Englishmen do, these "savages" represent a step backward for Englishmen, from their earlier innocent ignorance toward a chosen beastly, sluggish existence. Drayton's satirical linking of slurs, like the William Strachey sonnet to the lords of the Virginia council that I examined in the introduction, illustrates how in this period, oppressive stereotypes were linked and mobilized in each others' service.

This dynamic of oppresive stereotypes used to reinforce each other runs through Jonson's poetry as well as the plays this chapter treats. So Jonson's "Ode Allegoric," written to preface Hugh Holland's 1603 *Pancharis*, celebrates Holland's poetry's power to charm "the rout" (unspecified ignorant rabble), as well as to charm Irish rebels in the pale and "the kern and wildest Irishry"; in this poem, Holland's poetry works in the service of conquest, and Irish colonial conquest equals control of ignorant Englishmen and women. The English "rout" and Irish kerns react identically to poetry's civilizing force. Jonson's poem envisions a poetic transformation of uncivilized people both at home and abroad; and like Drayton's epistolary verse, it links the degraded English population to barbaric Irish wildness. In Jonson's English plays, this same dynamic recurs, as within the explicitly Virginian space of London's Bartholomew Fair, representative Irishmen run wild among London's other "rout." Like his "Ode Allegoric," Jonson's plays also develop theories of colonial control and resistance to that control.

Bartholomew Fair's "civil savages" include and conflate all of the populations eluding English citizen and monarchial mastery: the Irish (in these representations an undifferentiated mass), gypsies, English economic criminals, lustful masterless men aspiring to mastery, even cuckolding wives. Although the plays generally work to distinguish the gulls from the cozeners,[14] both categories of people elude proper domination and beg (in the imaginations of their dominators) for civilization. Gerald Vizenor describes American Western movies as "vicious encounters with the antiselves of civilization, the invented savage" (7). This formulation should remind us that none of the play's "savages" represent even the actual London underworld,[15] much less the real world of Virginia Indians, Irish people in London, or gypsies anywhere in the early seventeenth century. But these "antiselves" and their transformations on stage represented that real world to the London audiences that would fund more material transformations across the Atlantic world. They were the representations that helped to convince the crown to send populations from London prisons to Virginia, where most of them died. The crown in that case was hoping for a partial mirror image of *Eastward Hoe*'s final transformation. In that play, Virginia becomes the counter and inside the counter the criminals (cozeners and cozened) become pious model citizens. When King James ordered one hundred prisoners shipped to Virginia in 1619, he hoped that Virginia would serve as a real-life version of the play's transformative counter.

Bartholomew Fair's induction refers to the lotteries, licensed in 1612, that ran until 1621 and provided the chief support for the Virginia Company's enterprises.[16] The lotteries were in every sense a public project, depending on the broad-based support of audiences like Jonson's, literate and illiterate. The Virginia Company issued at least eleven publications to advertise the lotteries, including the five extant broadsides, most with simple texts that could easily have been read aloud at taverns, inns, and other public places. The company depended on lotteries to raise up to 8,000 pounds a year.[17] A lot in the "running lottery" that took place in Smithfield, Southwark and other locations cost twelve pence, twice as much as the cheapest seat at Blackfriars, where *Eastward Hoe* and *The New Inne* played, and probably twelve times as much as the least expensive seat in the Hope, where one could have seen *Bartholomew Fair;* but the smallest prize was two crowns, and prizes were as large as 4,500 crowns, a fortune at the time. The price of a lot would have been beyond the means of an apprentice, but not outrageous for many other Londoners, and the lure of such prizes must have attracted many who could not really afford the price. One argument for the suspension of the lotteries was their effect on the budgets of the "common sort." Although the "Standing" lotteries in which a lot cost two shillings six pence did not succeed as well as the less expensive lotteries, they were heavily promoted, and that promotion contributed to a widespread public awareness of the Virginia Company's projects.

Since the Virginia Company depended both on a large subscription by Londoners to its lottery and on a continual influx of settlers from the city to Virginia, it needed to generate awareness of its goals and enthusiasm for its project. Unlike his masques, which I will examine in chapter 4, Jonson and his collaborators' representations for the public stage, despite their civilizing displays, existed in a matrix of public images of the Virginia settlement rather than in line with any official version of the project. During the period in which Jonson wrote for the public stage, the Virginia Company had placed its project more and more in the public eye. The London Company, composed of London merchants, represented the project to Londoners as a civic duty in their lottery advertisements: "hoping that the inhabitants of this honourable Citie adventuring even but small summes of money, Would have soone supplied so little a summe appointed to so good a Worke" (*By His Maiesties Councell for Virginia*, A). Their discourse connects the honor of the city and, by association, of its people with the success of the lottery and thereby the Virginia settlements. In its 1620 *Declaration of the State of the Colony*, the Company published lists of adventurers, both individuals and companies like the clothworkers, the barbersurgeons, and the fishmongers (*Records* III:307-340). The Company's publications postulate the support of all of London's communities.

Jonson's plays and popular London opinions troubled the identity that the Virginia Company attempted to promote between a virtuous London that would support the colonial endeavor and a pure Virginia colony that would civilize the Indians around it. In opposition to the purified homology promulgated by the Company, rumors posited an equivalence between London's underworld inhabitants and dangerous presences in the settlements—both Indians and "corrupted" Englishmen. Like the Company's propaganda, rumors identified London and the colony; but unlike that propaganda, rumors identified the two only by incorporating the colony's wildness into representations of the metropolis and metropolitan wildness into representations of the colony. Both the Company's defensive stance in its propaganda and the slang that Jonson picks up and uses in his plays respond to the rumors current in London that the settlement abounded with thieves and scoundrels—the population of *Eastward Hoe* and *Bartholomew Fair*—and that the "wild" Indians and the settlers were becoming indistinguishable. A 1610 broadside, *A Publication by the Counsell of Virginea, touching the plantation there*, attests that "some few of those unruly youths sent thither, (beeing of most leaud and bad condition) . . . are come for *England* againe, giuing out in all places where they come . . . most vile and scandalous reports."[18] The text calls these youths "lasciuious sonnes, bad servants," "ill husbands," and "an idle crue." We can hear in their rhetoric the cast of characters in Jonson's plays: the crew headed for Virginia in *Eastward Hoe*, and the fair's salespeople and attendants. As Joseph Sigalas notes about *Eastward Hoe*, "every character linked with the play's Virginia voyage is either dishonest, drunk, incredibly gullible, or all three" (89). Both London and Virginia looked like a new world, addicted to tobacco, drinking, gambling, and other forms of "savagery."

Central to these representations of English antiselves is their lack of control over bodily sensation, their inability to stop eating, drinking, smoking, gambling and swearing, inabilities that the plays continually link to lack of sexual chastity and lack of respect for chaste male-female relations. These antiselves's lack of bodily integrity indicates their need for mastery.[19] The gulls' conversation in *Eastward Hoe* feminizes the Indians of Virginia, who are imagined as breeding machines for English men, but it also indicates the gulls' own interest in sex with Indian women. Likewise, *Bartholomew Fair*'s stage Irishman Whit is a bawd, continually offering up women to Ursula's customers and attempting to suborn all women into breaking their marital oaths. Whit reinterprets Win-wife and Quarlous's "respectable" searches for wives with dowries as searches for sexually available women: "Gi' me tweluepence from tee, I vill help tee to a vife vorth forty marks for't, and't be. . . . And shee shall shew tee as fine cut 'orke for't in her shmock too, as tou cansht vishe i'faith; vilt tou haue her, vorshipful Vin-vife?" (III.ii.7-12). Along with Whit's inability to pronounce an English "w" comes an understanding that all women are whores (cutwork lace was a badge of prostitutes) and that all men desire whores and do not care if they marry women available to other men. Thus he informs Littlewit's wife, Win-the-Fight, that lying with twenty men is an entirely justifiable matter: "Tish common, shweet heart, tou may'st doe it, by my hand: it shall be iustified to ty husbands faish, now: tou shalt be as honesht as the skinne betweene his hornsh, la!" (IV.v.45-47). Again, Whit's Irish inability to pronounce the word "honest" properly is the indication of his Irish inability to believe that there are sexually honest women.

Bartholomew Fair's induction implies that its lewd and uncontrolled fair is also Virginia; the play will show Jonson's audience what their money supports; even more importantly, it presupposes that they are familiar with the colony that the lottery finances. Captain John Smith, contemporaneously writing his experiences as a Virginia settler and colonial leader, represents both the English men in foreign space and the native people he encounters as an English underworld: as criminals, cozeners, gallants, and prostitutes—[20]the denizens of Jonson's *Bartholomew Fair*. The tirelessly self-promoting Smith thereby demonstrates that he alone comprehends the settlement's difficulties and can control both groups as a knowing English master. Likewise, for Jonson, the metropolis figures the colonized and behavior toward the colonized figures the metropolis. In Jonson's 1616 play *The Devil is an Asse,* Manly remarks of the foolish gallant Fitz-dottrel "Old *Africk,* and the new *America,* / With all their fruite of Mon sters, cannot shew / So iust a prodigie" (VI:I.v.8-10). Fitz-dottrel demonstrates his monstrosity in his misunderstanding of value. Cokes's tenants, according to Wasp, will sell their children for trinkets; Fitz-dottrel will sell his wife's time and her honor for a cloak worth thirty pounds. Neither the tenants nor Fitz-dottrel comprehends worth outside what Jonson sees as the debased market. Both can be understood in the same terms as the monsters outside the natural order that Smith found in Virginia.

In both Smith's text and Jonson's plays such monstrosity begs for reformation, and all of these texts portray that transformative process, although with varied success. In the 1629 play *The New Inne,* Jonson rewrote the chaotic fair as a controlled colony, and *Eastward Hoe*, *Bartholomew Fair*, and *The New Inne* each have a character who sets out to amend the play's miniature society. The earliest reformer, Touchstone, who plays multiple roles as *Eastward Hoe's* goldsmith, Quicksilver's master, and Sir Petronel's father-in-law, is no discoverer, simply a good citizen, scorning his subordinate's fantasies; Justice Overdo, *Bartholomew Fair's* inept reformer, aspires to be a world-explorer (he calls himself a "Columbus"); and *The New Inne's* Host, Lord Frampul, is a fully developed explorer-colonist. Lord Frampul is Jonson's vision of a successful colonial leader, able to oversee his world and subdue it in the interest of a perfected social order.

The Host of Jonson's *New Inne*, "The Light Heart," has seemingly lived two lives. As the play unfolds, he gives two accounts of his past; a past in which he left his family and his role as their patriarch, the Lord Frampul, and assumed the identity of Goodstock, the inn's Host. Early in the first act the Host declares that, as the tale goes, the Lord Frampul left "to turne Puppet-master." His guest Lovel continues the story: "And trauell with *Yong Goose*, the Motion-man." The Host replies: "And lie, and liue with the *Gipsies* halfe a yeare / Together, from his wife" (VI:v.61-64). According to the Host, popular opinion maintains that Frampul ran away with a traveling theater. But the end of the play tells quite a different story. When the Host reveals himself to his newly found family he tells his audiences, without disclaiming his old tale, that

> I am Lord *Frampull*,
> The cause of all this trouble; I am he
> Haue measur'd all the Shires of *England* ouer:
> *Wales*, and her mountaines, seene those wilder nations,
> Of people in the *Peake*, and *Lancashire*;
> Their Pipers, Fidlers, Rushers, Puppet-masters,
> Iuglers, and Gipseys, all the sorts of Canters
> And Colonies of beggars, Tumblers, Ape-carriers,
> For to these sauages I was addicted,
> To search their natures, and make odde discoueries!
> (VI:V.v.91-100)

At the beginning of the play we learn that Frampul left to be a puppet-master, at the end that he left to be an explorer—one who observes and discovers the nature of "savages."

As Frampul concludes this revelatory speech, he turns to his newly rediscovered wife and reinforcing the tenor of that speech calls her a "she

Mandeuile." He implies that he is constitutionally a Mandeville: not a man of the theater at all but an explorer. The mid-fourteenth-century text that bears the name Sir John Mandeville was printed in England six times between 1582 and 1639, the span of time in which the English turned from sporadic adventurers and pirates in the Americas to determined American colonists. The two major collections of travel and exploration materials issued as propaganda for the early British trans-Atlantic settlement and exploration efforts, Richard Hakluyt's first edition of the *Principal Navigations,* published in 1589, and Samuel Purchas's collection in 1625, both reprinted the Mandeville text. Mandeville was the ur-traveler; Mary B. Campbell claims that in his text "Mandeville-the-narrator emerges . . . as a pure wanderer and travel as an activity in and of itself" (149). The text is a collection of earlier travelers' tales and purely imaginative fictions as ludicrous as Seagull's Virginia, but Mandeville's travels were clearly accepted in early modern England as "true travels."[21] *Mandeville's Travels* became a model for exploration literature and the name Mandeville an iconic name—an explorer, as Jonson's construction suggests. In Jonson's diction, "a Mandeville" is a species that can have a female member as well as a male.

Lord Frampul as a Mandeville would seem to be a different story than Lord Frampul as a motion-man. A puppet-master is not an explorer; to explore and discover is not to run a traveling theater. One could resolve this discrepancy by pointing out that "Pipers, fiddlers, rushers, puppet-masters" were among the people of the "wilder nations" that the Mandeville Frampul has explored. Still, watching or exploring puppet-masters among others is not becoming a puppet-master. Of course the audience first learns of Frampul's adventures merely from Lovel's hearsay. And as Shakespeare reminds his audiences via Rumour's prologue to *2 Henry IV,* stories change in transmission; so perhaps the story of Frampul as puppet-master evolved from his association with men of the theater. Such an explanation seems unlikely, however, since Jonson does not give the Host an aside designed to put the audience right in the first account, although he has used that device just a few lines previously to inform his audience of Ferret's good guess at Lovel's "mouldy passion" for Frances, Lord Frampul's daughter.

I would argue that we do not need to resolve the two stories and that indeed Jonson does not need to reconcile the two accounts, because in his late theater Jonson's reformer-playwright character *is* the explorer. The space of the theater that Goose the motion-man would inhabit and create is the space of the discoverer and explorer—Mandeville's world. The Host's inn, the Light Heart, is the Host's theatrical/colonial space; his—and the possessive is crucial here—England, England's world, his world. Anne Barton notes that "like the Smithfield of *Bartholomew Fair,* although in very different ways, the inn at Barnet lays claim to inclusiveness. It provides, as the Host himself points out, a metaphor for the world" (399):

> Where, I imagine all the world's a Play;
> The state, and mens affaires, all passages
> Of life, to spring new *scenes,* come in, go out,
> And shift, and vanish; and if I haue got
> A seat. to sit at ease here, I' mine Inne,
> To see the *Comedy;* and laugh, and chuck
> At the variety, and throng of humors,
> And dispositions, that come iustling in,
> And out still, as they one droue hence another.
>   (VI:I.iii.128-36)

Here the Host not only gives voice to the topos of theater as world, he also lays claim to that world as a theater that he has constructed and produced: "I'mine inn." What seems merely a banal rehearsal of "all the world's a stage" is also a potent claim to ownership and control of that stage and therefore of the world. The Host, who later compares himself to the playwright Jonson,[22] claims that he has imagined the inn's world—that world is a play he has written, even a humors play, Jonson's first form. The Host's "I" controls the world as a stage: "where *I* imagine" and "*I* have got" (128, 131; my emphases). That "I" then is essentially the same as the "I" that will claim an identity as Mandeville, the I(eye) addicted to watching savages. The Host, Goodstock, is what Mary Louise Pratt calls "the seeing-man . . . he whose imperial eyes passively look out and possess" (7). Such a man affects an innocent pose but simultaneously asserts possession. So the Host, accused by Lovel in this scene of keeping a lowly inn for "Iovial Tinker[s]," "Rouge[s], Baud[s], and Cheater[s]," responds that his inn is his theater, cleansed by his astute observation.

Unlike *Bartholomew Fair's* fair, the Host's inn offers an uncontaminating spectacle. If we unpack the similitude that Barton notes, *The New Inne* resembles *Bartholomew Fair* in more than just its assertion of inclusiveness. In fact, one can read *The New Inne's* world/theater as a revision of that other play's world/fair. The Host is the puppet-master of his inn/theater/world; in his character Jonson rewrites aspects of at least three characters in *Bartholomew Fair:* Littlewit, the author of that play's puppet-show who leads the way to the fair, Lanthorn Leatherhead, the motion-man who interprets the puppet play and keeps its tent, and Adam Overdo. We can see the "he-Mandeville," Lord Frampul, as an ultimately triumphant explorer who has "measured" the world and brought it home, and also as a more successful overseer of his Smithfield than *Bartholomew Fair's* foolish Overdo, who would like to be, as he claims, a "*Columbus; Magellan; or our countrey man Drake* of latter times"[23] (VI:V.vi.38-39).

And as it rewrites the explorer and playwrights of *Bartholomew Fair, The New Inne* rewrites the objects of their exploration and the attributes of the earlier play's world. The earlier play contains two characters, Northern and Puppy, perhaps from Lancashire, the Northern England that the Lord Frampul has explored and whose dialects were objects of fun on the London stage. At both the inn and the fair one can indulge, as these "foreign" characters do, in the wildness of tobacco and drink and quarreling, all vices associated with the colonies in the colonial literature.[24] However, where the fair's wildness is central to its identity, the inn's "civil savages" are relegated to the basement. Smithfield's denizens take on disguises, but in the inn the disguised characters are revealed as always already domesticated, already part of the noble English family. Like the fair, the inn contains a theater as well as constituting one: but the inn's central show is a court of love and valor. Both the inn and the fair house a stage-Irish character, the prototypical other to England; *Bartholomew Fair* has the Irish bawd Captain Whit, and *The New Inne* has the Irish Nurse Shelee-nien.[25] But where Whit panders married women, the *Inne's* Irish Nurse (Frampul/Goodstock's wife in disguise) "sells" Frampul his real daughter as an adopted son. *Bartholomew Fair's* degenerate (potentially) cuckold and bastard-producing pandering is rewritten in *The New Inne* as the confirmation of proper patriarchal linkage.

In *The New Inne*, Shelee-nien Thomas, Frampul's wife in her antiself disguise as a drunk Irish nurse, evidences the same "lewd" interpretation of male-female relations that audiences had seen in *Bartholomew Fair's* Whit. The nurse's chief characterization as an Irishwoman involves her addiction to drink; the Host attributes all of her behavior to drunkenness: "Goe aske / th'oracle / O'the bottle, at your girdle" (VI:IV.iv.343-44). When Beaufort offers to kiss her charge, Laetitia, the nurse's essential Irishness, encapsulated in her drunkenness and her dialect, leads to a Whit-like interpretation: "I, 'tish a naughty practish, a lewd practish, / Be quiet man, dou shalt not leip her, here" (VI:III.ii.115-16). Beaufort indicates the joke in her dialect: "Leape her? I lip her, foolish Queene at Armes" (117). Love, elevated in Lovel's discourse into neoplatonic piety, in "Irish" becomes lewdness; a kiss becomes sexual intercourse. Just as the Nurse's "Irishness" interprets the court of love as a spectacle of lust, at the fair, Whit's presence makes Ursula's booth and the puppet show into whorehouses. But in the Inn, the drunk, "wild-Irish" woman becomes the tame English wife and mother (VI:II.vi.26).

As *The New Inne* tames Irish lust and inebriation, it also rewrites *Bartholomew Fair*'s other wildness. At the end of its travels through the fair, *Bartholomew Fair*'s audience experiences a puppet-show that parodies classic stories of love and friendship: *Hero and Leander* and *Damon and Pythias*. Lovel's performances in the inn's theatrical "court"—platonic disquisitions on the perfect form of true love and the pure reason of true valor—are the exact antitheses of the puppets' rantings on bawdry and sordid betrayal. Love, the puppets say, is a matter of "a pint of sherry" and "treading" a goose; valor is breaking "heads with a pot." In the Inn, however, love is separated from lust and valor transcends messy violence. When challenged by Zeal-of-the-Land Busy, Smithfield's puppets reveal themselves as genderless, while Laetitia-Frank, hired as a player in the Host's theater, switches genders so often that she becomes strangely puppetlike. Even before Laetitia-Frank plays her real sister's servant, she looks like her father's puppet, mouthing the Host's sentiments in Latin and silently attending her own disposition at the other character's hands. But the Smithfield puppets are genderless bawds, while Laetitia becomes a properly obedient, noble daughter. *The New Inne*'s theater of love and valor rewrites the puppet-show and tames its "savage" values.[26]

While the Host tells his newly found wife at the play's end that he has "coffin'd [himself] aliue in a poore hostelry" (VI:V.v.105-106), in actuality he has made that hostelry into a receptacle for the specimens from his explorations. The men in his cellar are the "canters," "gypsies" and "beggars" of his Mandevillean exploits. The Host's drawer, Pierce, calls the cellar's denizen's "halfe-beasts," and the Host calls them "brute ghests." Sir Glorious Tipto, brought to the inn by the Lady Frampul, refers to the cellar men as both "pioneers" and "mine-men." "Pioneer" certainly signifies here foot-soldiers who "march in advance of an army" to dig trenches (*OED* "pioneer"). The *OED* mentions also a figurative usage of the word at the time of Jonson's play to designate a worker in exploration. Mine-men recall the colonial treasure found in South America and wished for in all subsequent European American exploration. The cellar's mine-men are the detritus of colonial expansion.

Tipto, who is very attracted to the men in the cellar, is also a devotee of anything foreign, especially Spanish. As we have seen in the discussion of

Bermuda's names in chapter 2, the Spanish signify continually in England's colonial literatures as a touchstone; English explorers invoke the Spanish example either as a bad precedent or as the pattern for successful colonization. Just as England's Irish exploits were ever-present to Virginia colonists, the Spanish were never far from England's self-image as American colonizers, and surprisingly often, given their current rivalry on the political scene (and England's fears about Spanish intervention in Ireland), the Spanish were seen as direct ancestors and models for the English trans-Atlantic colonial effort.[27] In the 1622 text *A declaration of the state of the colony in Virginia,* written after the Indian rebellion in Virginia, Edward Waterhouse urges the English to take advantage of the situation: "So the Spaniard made great use for his owne turne of the quarrels and enmities that were amongst the Indians" (D4v, 24). If the English demonized the Spanish, they also regarded them as successful overseas entrepreneurs who could provide examples for English colonial effort. Tipto uses the Spanish to control his mine-men. He commands those cellar denizens and demands their respect for the Spaniards who he says, carry "such a dose of [*grauidád*] in [their] lookes, / Actions and gestures, as it breeds respect, / To [them], from *Sauages,* and reputation / With all the sonnes of men" (VI:IV.ii.38-41). The Host calls Tipto's "whoop Barnaby" and his "hoop Trundle," two of the cellar's crew, "his two Tropicks." Hereford, Simpson and Simpson gloss tropics here as "turning points, with a quibble on the two circles of the celestial sphere." But "tropics" might also signify the two circles of the earthly sphere. Directed by the gypsy Fly, the Host's cellar houses Smithfield's underworld, its "civil savages," even to its chamberlain (chamberpot) Jordan, a reincarnation of *Bartholomew Fair*'s Jordan Knockem and its own Bartholomew, Bartholomew Burst, "citizen, courtier, gamester," and now fittingly a "merchant adventurer."

But for all its wild inhabitants, for all that his cellar contains the tropical underworld and its adventurers, the inn itself is also almost the Host's home. When all the fictional identities in the play's world are exposed in the fifth act, the Host has been living there with most of his family. The Lady Frances Frampul, the visitor who exploits the inn as theater, is really Lord Frampul's eldest daughter, and the "boy" pretending to be Frances's sister is really her sister. The drunken Irishwoman Shelee-nien Thomas is not Irish at all but instead the Lady Frampul, an English noblewoman whose breeding Jonson assures his readers of in his argument to the play: "a virtuous gentlewoman, Sylly's daughter of the South." This noblewoman begins the play as one of the Inn's exotic specimens, and ends as the *heimlich* wife and mother. *Bartholomew Fair*'s stage-Irish bawd Whit speaks more lines than Shelee-nien Thomas, but Thomas plays a more structurally significant part. And although Overdo invites Whit home along with the other Bartholomew birds at the end of that play, *Bartholomew Fair* does not indicate that he will reform and be bawd no more. *The New Inne*, however, explicitly recuperates She-lee Nien Thomas as Lady Frampul and restores her husband and family.

As Julie Sanders points out, compared to a tavern, an inn was a relatively prestigious public venue. But like the fair, inns were public spaces by definition, as Lovel's list of their possible occupants indicates. Ferret, Lovel's servant, displays his understanding of inn discourse for his master, Goodstock, and the audience:

> *Ferret.* A wench, i'the Inn-phrase, is al these;
> A looking-glasse in her eye,
> A beard-brush with her lips,
> A rubber with her hand,
> And a warming pan with her hips.
> (VI:I.iii.9-13)

The Host's response indicates his role as savage-tamer: "This, in your scurrile *dialect*. But my Inne Knowes no such language" (VI:I.iii.14-15). Unlike the fair, full of northern dialect and bawdy language, Goodstock's inn speaks a standard English that does not admit bawdy interpretation. The Host's adopted son (actually Laetitia) speaks Latin, not inn or fair speech. When Lovel offers to make Goodstock's son his page, the Host replies that he will keep his son in the inn since as a nobleman's page he would learn to pander, gamble, "mount the Chambermaid," and steal. With this formulation, the Host reverses Ferret and Lovel's assumptions: that an inn houses perfidy, a nobleman's house refinement. Goodstock's inn is a thoroughly domesticated public space, more decent than a nobleman's private household.

In *Bartholomew Fair* and *The New Inne*'s logic, it follows that one can fashion a clean, controlled metropolitan area in the same manner as a clean, controlled colony. Overdo finds himself overwhelmed (as his name implies) by the enormity of the otherness he encounters; he cannot properly interpret it and he therefore has no authorial control over it. He hides his identity to make his "discoveries" but is himself mistaken for a thief, and he "discovers" goodness in the worst of the fair's rogues. Frampul on the other hand has authored his play; he has explored and played the "gypsy" and brought the exotic home. This is why Overdo's claim to be the Columbus or the Drake of his world is so laughable and Frampul's stance as a Mandeville so apt, especially in that Frampul's underworld is as fictional as Mandeville's copious otherness. As a later development of Smithfield, *The New Inne* is a domesticated colonial space.

Bartholomew Fair displays a wild space out of doors; the play moves out of a home into the market and fair's liminality, but *The New Inne* domesticates the fair's wildness, transforming it into the idealized English home.[28] In the earlier play, even Littlewit's home, where the play begins, partly belongs to the fair as he has presumably authored the puppet-play inside its doors. And the play's final movement into Overdo's home for a feast is unconvincing. The audience does not see that household scene, and if it did that home populated as it now would be with all the cast of the fair would no longer look like home. As Jonathan Haynes suggests, "the feast would be the genuine festive experience, but it is hard to imagine it in the light of the still unresolved (and perhaps unresolvable) divisions the play has opened up" (1992 138). Performed at the Hope theater, *Bartholomew Fair* actually took place in a wild space—in an amphitheater on the bankside used primarily for bear-baiting, stinking of animality as the play's induction notes.[29] The induction draws attention to the smelly theater; then the play brings its audience into the pungent world of the fair and never convincingly escorts it out. In contrast, *The New Inne* locates its gypsies below stairs: not only indoors, but also underneath the Host's private domesticated theater. They become literally the underworld. The later play transforms the puppet-show of lust and betrayal at the fair's limits into a courtly philosophical disquisition on love and honor in the inner gallery upstairs, and the play's fifth act convincingly restores the nobility, transforming wildness into comforting and familiar household space. In addition, Jonson set *The New Inne* in the removed clean space of the country[30] rather than in dirty London; and it was performed in the relatively luxurious indoor Blackfriars theater.

Don Wayne proposes that *Bartholomew Fair* "acknowledges that a regrounding of the social order [based on the market] is indeed under way in England, and the play offers an account of the principles upon which the new system of social organization will be constituted" (115). Wayne sees that emergent social order as predicated on market principles and contractual obligations rather than the "shared assumptions" of religion. I would add that this new order had everything to do with England's expansion into the world market. The increased capitalization of England and Europe[31] and England's encounters with inassimilable difference were transforming the ways in which English people saw England and London itself.

In the early seventeenth century a disreputable section of London came to be called the Bermudas. In "An Epistle to Sir Edward Sacvile, now Earl of Dorset," in the course of thanking Dorset for his graceful patronage, Jonson satirizes ungrateful supplicants:

> Their very trade
> Is borrowing; that but stopt, they doe invade
> All as their prize, turne Pyrats here at Land,
> Ha' their *Bermudas* and their streights i'th' *Strand*.
> (VIII: *The Vnder-wood* XIII. 1.79-82)

Jonson uses the name "Bermudas" to signify doubly. That section of London, like the islands off America, is morally and even physically dangerous, a haven for pirates, where unsuspecting visitors might well metaphorically be shipwrecked, robbed, or cheated out of their goods; London's Bermuda is a den of iniquity. In *Bartholomew Fair*, Adam Overdo's perception of the district agrees with that of the "Epistle" when, disguised as a raving madman, the Justice warns Smithfield visitors against the evils of Bermuda's primary cash-crop, tobacco:

> Looke into any Angle o'the towne, (the Streights, or the *Bermuda's*) where the quarelling lesson is read, and how doe they entertaine the time, but with bottle-ale, and tabacco? . . . Thirty pound a weeke in bottle-ale! forty in tabacco! and ten more in Ale againe. Then for a sute to drinke in, so much, and (that being slauer'd) so much for another sute, and then a third sute, and a fourth sute! and still the bottle-ale slauereth, and the tabacco stinketh!
> (VI: II.vi.76-86)

Like other debased districts harboring degenerate gallants, London's Bermuda offers futile, profligate, and physically disgusting pursuits. The Bermudas and the Streights are sinkholes for money, invitations to waste one's resources on self-consuming goods: the obsessive vagaries of fashion, stinking tobacco disappearing into smoke, ale drunk endlessly and spat out again in quarreling saliva. On stage and in English imaginations, the underworld of London became its Bermudas redeemable by exploration, while Virginia became a new England whose Indian and base English inhabitants could be imagined as treacherous cozeners.

When Jonson rewrote *Bartholomew Fair* as *The New Inne*, he reorganized all the resulting exotica into the English family. What did the Englishman find when he set out as an explorer in imitation of Mandeville but the wonders of London, the exotica of home—these, if not quite assimilable, were still comprehensible and controllable. In *Bartholomew Fair*, Jonson displays public space as the colony, and in *The New Inne* he shows his audiences the successful colonization of the same. In the earlier play gentlefolk go out into the revels, while in the later the nobility bring their revels indoors and relegate the objects of their interest downstairs.

In at least one crucial way, this movement toward control mimics the resolution of the much earlier *Eastward Hoe*. Although that play's adventurers project their outdated and ridiculous colonial fantasies onto actual colonial space, they themselves resemble both the fair's denizens and the inn's cellar residents. Drunken, prodigal, lustful, and negligent, Quicksilver, Flash, the gull, the spendthrift, and the scapethrift, even though they do not get to Virginia, and although they can only conceive of a mythical golden paradise, themselves represent the "antiselves" that popular rumor imagined populated the settlement. And their various projected or actual destinations—the Thames' banks at Cuckold's Haven, the counter, the Low Countries—are versions of colonial space for *Eastward Hoe*. All are destinations for the city's masterless men. *Eastward Hoe* also hints at the place of Ireland in this gallery of potentially colonized spaces. It was not uncommon for ships bound for Virginia to shipwreck or be windbound close to home, not on the Thames but on Ireland's nearby shores.[32]

Written for the city (meaning its dominant guild members), *Eastward Hoe* thoroughly redeems its wasteful gallants, leaving Quicksilver and Petronel composing pious ditties in an effort to reform all of the counter's prisoners. The play's counter turns into a theater within the play, like the fair's puppet show and the inn's play-court. Along with Touchstone, the play's moral voice, the play invites its audience to watch Quicksilver's repentant performance, and to reintegrate the wastrels into the city. Likewise, in *Bartholomew Fair*, Jonson marks the space of the fair as the space of the colonies as the space of the theater. Those other to the English are visualized and controlled through the frame of theater in the theater. If we read *The New Inne* as a development of Jonson's own theatrical practice, the play becomes the counter and *Fair* extended. *Bartholomew Fair*'s author remained offstage, and the world of the fair remained imperfectly colonized; it had only an Adam Overdo as its Drake, who despite his constant talk of discovery resembles more the ineffectual disguised ruler than an incarnation of a bold adventurer. In the later play, the Host has settled England's others in his inn to watch over them. But those others remain strange and displayed until the metamorphoses of the fifth act make the inn into the Host's actual home, the space of the English family. The Irish drinking woman becomes the Lady Frampul. The unruly, addicted, wild woman transforms into the English noblewoman.

As I have shown, *The New Inne* rewrites *Bartholomew Fair*'s rhetorics of othering and similitude as a rhetoric of homemaking. Other colonial discourses at the time invoked these rhetorics concurrently. The large broadside that declares the drawing of the lottery in 1615 (our 1616) provides a case in point (*Three Proclamations*; see figure 3.1). Two columns of text frame[33] a list of prizes; over the text, at the far left and right of a panel of images arranged symmetrically, two Indians appear holding bows in conventional, unthreatening poses.[34] In the center, the two sides of the Virginia Company's seal frame pictures of prize money bags, trophies, and a man drawing lots out of two large baskets. The Indians look down toward a poem printed on either side of the figure drawing lots:

> *Once, in one* State, *as of one* Stem
> Meere *Strangers from IERVSALEM,*
> As Wee, *were* Yee; *till* Others *Pittie*
> Sought, *and brought* You *to* That Cittie.
> *Deere* Britaines, *now, be* You *as kinde;*
> Bring Light, *and* Sight, *to* Vs *yet blinde:*
> Leade Vs, *by* Doctrine *and* Behauiour,
> *Into one* Sion, *to one SAVIOVR.*

Addressing the lottery's players, the Indians ask them to buy lots and so to bring them into the light of Jesus as at one time other Christians brought Londoners' putative ancestors. These mock Indians, these antiselves, invoke an idea current in England's colonial discourses at least since Thomas Harriots's 1590 *A Briefe and True Report of the New Found Land of Virginia*, a text that ends with pictures of mythological ancient Britains "to showe how that the Inhabitants of the great Bretannie have bin in times past as sauuage as those of Virginia" (75). The same idea animates William Crashaw's 1610 sermon for the Virginia Company. Crashaw exhorts the Company to convert the Indians as the English were once converted: "another necessity in nature and reason lieth vpon vs: for the time was when wee were sauage and vnciuill . . . Did we receiue this blessing by others, and shall we not be sensible of those that are still as we were then?" (C4v).[35] The apposition of one state and one stem in the first line of the 1615 lottery broadside's poem marks a direct equivalence between these Indians and ancient Englishmen; these Indians need only a process of conversion, which the poem represents as a journey into Jerusalem, to become early modern Englishmen.

A Declaration for the certaine time of dravving the great ſtanding Lottery.

VVelcomes.

To him that firſt ſhall bee drawne out with a Blanke	100. Crownes.
To the ſecond	50. Crownes.
To the third	25. Crownes.
To him that euery day during the drawing of this Lottery ſhall bee firſt drawne out with a Blanke	20. Crownes.

Prizes.

1. Great Prize of	4500 Crownes.
2. Great Prizes, each of	2000 Crownes.
4. Great Prizes, each of	1000 Crownes.
6. Great Prizes, each of	500. Crownes.
10. Prizes, each of	300. Crownes.
12. Prizes, each of	200. Crownes.
100. Prizes, each of	50. Crownes.
400. Prizes, each of	20. Crownes.
1000. Prizes, each of	8. Crownes.
1200. Prizes, each of	6. Crownes.
4000. Prizes, each of	4. Crownes.
1000. Prizes, each of	3. Crownes.
1600. Prizes, each of	2. Crownes.

Rewards.

To him that ſhall bee laſt drawne out with a Blanke	25. Crownes.
To him that putteth in the greateſt number of Lots vnder one name or Poſie	400. Crownes.
To him that putteth in the ſecond greateſt number	300. Crownes.
To him that putteth in the third greateſt number	200. Crownes.
To him that putteth in the fourth greateſt number	100. Crownes.

If diuers be of equall number, then theſe Rewards are to be diuided proportionally.

Addition of new Rewards.

The Blanke that ſhall bee drawne out next before the Greateſt Prize, ſhall haue	25. Crownes.
The Blanke that ſhall bee drawne out next after the ſaid Great Prize, ſhall haue	25. Crownes.
The Blankes that ſhall be drawne out immediately before the 2. next Greateſt Prizes, ſhall haue each of them	20. Crownes.
The ſeuerall Blankes next after them ſhall haue alſo each of them	20. Crownes.
The ſeuerall blankes next before the foure Great Prizes, ſhall haue each of them	15. Crownes.
The ſeuerall Blankes next after them ſhall haue alſo each of them	15. Crownes.
The ſeuerall Blankes next before the ſix Great Prizes, ſhall haue each of them	10. Crownes.
The ſeuerall Blankes next after them ſhall haue alſo each of them	10. Crownes.

Imprinted at London by Felix Kyngston, for William Welby, the 22. of Februarie. 1615.

Figure 3.1—*A Declaration for the certaine time of drawing the great standing Lottery.* Reproduced with permission of the Society of Antiquaries, London.

In addition, this broadside displays Indian "antiselves" and clearly marks their difference and exoticism by placing them on a page with a Jacobean gentleman and by associating them with the turtles at their feet. Turtles are the endpoint of Harriot's list of the commodities to be found in Virginia: "There are many *Tortyses* both of lande and sea kinde . . . they are very good meate, as also their egges. Some have bene founde of a yard in bredth and better" (21); Francis Perquin writes from Jamestown in 1608 "I have sent to Madame *your wife* a pair of tortoises" (Alexander Brown I 176).[36] John White's famous drawings of American natives, flora, and fauna include a loggerhead turtle, a common box tortoise, and a diamond-back terrapin (plates 54, 55, and 56). Over the picture of the box tortoise, someone has handwritten, "A land Tort [which] the Sauages esteem aboue all other Torts" (plate 55). And as we have seen in chapter 2, the reports from the Bermuda colony describe the wonderful turtles who (in English eyes) arrived to feed hungry colonists. Turtles were a "New World" commodity and a sign of its alien abundance. While the lottery broadside defines the Indians as exotic commodities, it also characterizes them as like the English, introducing the Indians as essentially Englishmen before their acceptance of Christ. And the broadside represents that acceptance as a domestication, a movement from being strangers to being at home.

The lottery broadside alludes to a transformative conversion that should be familiar from my analysis of Jonson's plays. *The New Inne* also shows strangers collected from the wild who become grafted to the stem of "Goodstock," the English family. Shelee-nien, a speaker of broken English and a representation of drunken Irish savagery, begins to speak as the English noblewoman. The Indians in the broadside ask for such a conversion, allying themselves with a fantasy of a previously pagan England. Absent in the broadside but fully present in the play is the figure of the author or the explorer. The colonist, present by implication in the council's seal, even in the figure drawing lots, is not directly represented; his words issue from Matahan and Eiakintomino's mouths. The "blind" antiselves themselves ask to be made all-seeing Englishmen, their transformative destiny.

In 1622 the preacher Patrick Copland published *Virginia's God be Thanked* . . . , a sermon on the previous year's successes in Virginia. To this text are "adjoyned some Epistles, written first in Latine (and now Englished) in the East Indies by *Peter Pope* an Indian youth, borne in the bay of Bengala, who was first taught and converted by the said P.C." (see figure 3.2). Decontextualized, a modern reader would find the letters unremarkable; they resemble the thankful letter written by William Powell, the Virginia colonist, that we looked at in chapter 2; any reader of Renaissance letters has seen them many times before. Peter Pope has composed the customary letters to a patron: flattering epistles thanking the patron copiously for favors bestowed. "For the present," Pope writes, "I have but little, which I may render for your great liberality towards mee (and to returne nothing at all, were altogether a signe of an vngrateful mind) vnlesse it be this small Paper-gift" (A2). Like *The Tempest*'s Caliban, Pope has been given the English language, but unlike Shakespeare's "servant monster," Pope uses the gift to praise rather than to curse. The three letters embrace the humanist project of education as well as the linked projects of colonization and trade. Pope calls his patron Martin Pring both by his military title, "Commander of the Sea Nauy of the East-India Company," and "most Illustrious Mecoenas" (Av). The letters display Pring's colonial authority as well as his personal patronage; in Pope's epistles the colonial subject mouths the colonizer's words; and in them, the colonial subject appears undifferentiated—the text attaches a message of thanks by an East Indian for his instruction in the colonizer's language to a sermon of thanks for success in Virginia. Like the Richard Hakluyt translation of de Soto's travels I discussed in the introduction, Copland's text relies on the assumption of an undifferentiated Indian, here essentially the same across the almost twelve thousand miles between Virginia and Bengal. Copland means his readers to take one Indian subject for another and, from hearing English words emanate from one Indian's mouth, to infer the future success of conversion efforts in both places. In chapter 4, we will see many more of these undifferentiated Indians in court masques and Lord Mayor's pageants.

In Peter Pope's letters to his patron, in the Virginia Company's lottery broadside, and in Jonson's plays, we can see the production of antiselves and their transformations in the service of colonial efforts. Figures like Pope, the broadside's Matahan and Eiakintomino, *Eastward Hoe*'s Quicksilver and his gulls, *Bartholomew Fair*'s Cokes, Whit, and the fair's crew, and *The New Inne*'s Fly display savageness, and in their destinies audiences could see savageness transformed into manageable and normative Englishness. These representations comment on home; and, more tellingly, like Drayton's poem, they produce a potential perfected home. Jonson's plays, particularly although not exclusively, produce that perfected home through theatrical transformation. But the discourse of colonialism as action on a stage was not confined to actual London stages; and it was also not solely a trope of poetry. On November 13, 1622, John Donne preached a sermon to the Virginia Company upon the eighth verse of the first chapter of the acts of Apostles: "Beloued," he says, "you are Actors vpon the same Stage too: the

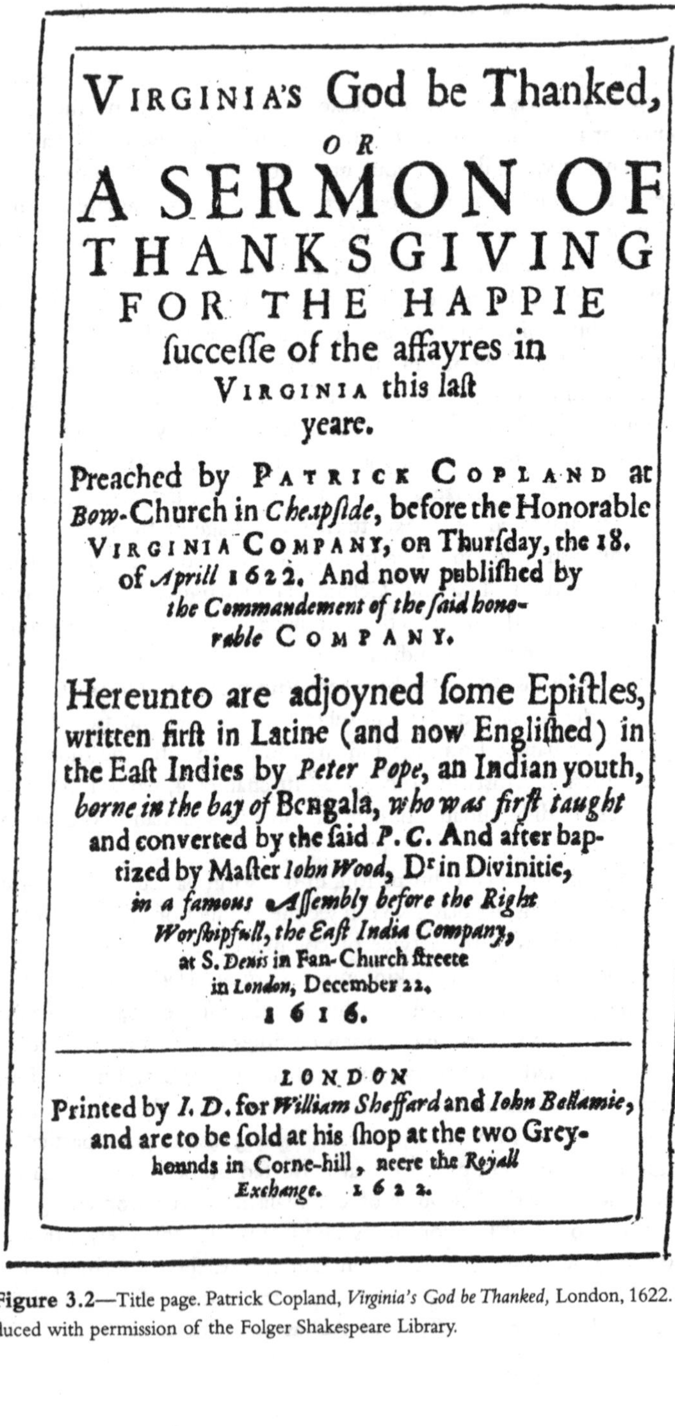

VIRGINIA'S God be Thanked,
OR
A SERMON OF
THANKSGIVING
FOR THE HAPPIE
succeſſe of the affayres in
VIRGINIA this laſt
yeare.

Preached by PATRICK COPLAND at
Bow-Church in Cheapſide, before the Honorable
VIRGINIA COMPANY, on Thurſday, the 18.
of Aprill 1622. And now publiſhed by
the Commandement of the ſaid honorable COMPANY.

Hereunto are adjoyned ſome Epiſtles,
written firſt in Latine (and now Engliſhed) in
the Eaſt Indies by Peter Pope, an Indian youth,
borne in the bay of Bengala, who was firſt taught
and converted by the ſaid P. C. And after baptized by Maſter Iohn Wood, Dr in Divinitie,
in a famous Aſſembly before the Right
Worſhipfull, the Eaſt India Company,
at S. Denis in Fan-Church ſtreete
in London, December 22.
1616.

LONDON
Printed by I. D. for William Sheffard and Iohn Bellamie,
and are to be ſold at his ſhop at the two Grey-
hounds in Corne-hill, neere the Royall
Exchange. 1622.

Figure 3.2—Title page. Patrick Copland, *Virginia's God be Thanked*, London, 1622. Reproduced with permission of the Folger Shakespeare Library.

vttermonst parts of the Earth are your Scene: act ouer the acts of the Apostles" (A4v). Paradoxically, however, when we recognize it, this conception of colonialism may leave some space to comprehend something other than the colonist's discourse. For theater is always a volatile and open transformative space, hence Ben Jonson's own distrust of the medium. As much as it attempts to shape its audience, it depends on audience response.

Likewise, English, as Caliban suggests, always leaves a space to criticize the conqueror. As Vizenor notes, "English, a language of paradoxes, learned under duress by tribal people at mission and federal schools, was one of the languages that carried the vision and shadows of the Ghost Dance, the religion of renewal, from tribe to tribe on the vast plains at the end of the nineteenth century. . . . English, that coercive language of federal boarding school, has carried some of the best stories of endurance" (105-106). Even as Pope's letters bear witness to his transformation into a proper humanist and colonized subject in his masters' eyes, they also indicate his facility with a language that could potentially liberate him and his fellow subalterns, transforming them into subjects who could speak with each other and, therefore, possibly against their masters and mistresses. Interestingly also, while Copland's use of Pope to promote the Virginia Company's efforts exposes the oppressive foundational assumption of the undifferentiated Indian, it also points the way to the liberating future Vizenor lauds. While the East India Company and the Virginia Company both attempted to indoctrinate the "Indians" they encountered by teaching them English, they were also enabling coordinated efforts at liberation within colonial territories through intergroup communication. In a different manner, the theater itself offered potentially liberating identifications to its audience, as well as offering visions of oppressive transformations.[37]

For all of *The New Inne*'s gestures toward making a home out of a once-savage space, and for all that the Host has mastery over discovery, *The New Inne*'s world remains imperfectly colonized; it remains a testimony to the impossibility of representing the absolute control of alterity to a theater audience itself far from homogeneous.[38] Even in Blackfriars there were cultural differences, and not all audience members would agree on the attractions of disquisitions on godliness. In *Eastward Hoe*, Quicksilver's pious, repentant song may have cheered guild masters, but a wink by the performer may also have signaled to those mastered men in the audience that his reformation was not complete, or was even a ruse. At the end of *The New Inne*, the Host leaves his theater to the gypsy Fly, and presumably when the master leaves, the downstairs revels will move upstairs. *The New Inne* ends in the bourgeois accommodation (here attributed to the aristocracy), a series of loving marriages, and the establishment of individual homes: the new social settlement in early modern England that would also constitute the possibility of settlement in Virginia. By 1629 the Virginia Company had started to import women into the colony, and the colony's government had started to divide the land into counties. The company realized that Englishmen would immigrate and settle only when the social structure and political structure of the "New World" resembled or could be made to resemble the old.[39] *The New Inne* ends with the redemption of wildness into civility, but on the English stage, the Host leaves the colonial theater to the "civil savage."

Despite its *Alchemist*-like ending, *The New Inne* appears to have been a failure. As we will see in chapter 4, Jonson could show the complete transformation of wildness to an ideal Englishness in his masques, but the public stage audience had no patience for such closure. Jon Lemly notes of the play that "in some key respects it resembles a masque" (138). Perhaps *The New Inne* was too close to a masque to succeed on the London stage. The play's last act represents and anticipates an ideal transformation, but the public stage, with its demands for improvisation and its love of playing, was perhaps not the perfect space for that particular fantasy. As London reimagined itself in terms of its colonies, its public stages could not successfully represent colonial space controlled. The fair, the market, and the colonies were perhaps still too inviting as spectacles, not always comprehensible by the imperial "I".

In Ben Jonson's 1626 play, *The Staple of News*, Peni-boy Canter, another disguised father, discussing the fitness of lodging the Lady Pecunia in a tavern, asserts that "the blessed / *Pokahontas* (as the *Historian* calls her) / And great Kings daughter of *Virginia* / Hath bin in womb of a tauerne" (VI:II.v. 121-24). In the Intermean following the play's second act, Tattle, one of the group of gossips serving as the audience within the play, complains about Jonson's use of the Indian woman's name: "Pokahontas, surnam'd the blessed, whom hee has abus'd indeed (and I doe censure him, and will censure him) to say she came foorth of a Tauerne, was said like a paltry Poet" (VI:II.Inter. 42-45). Of course, given Jonson's contempt for audience commentary and given the gossip's previous malapropisms, Tattle is wrong. Not only did the real Pocahontas indeed lodge at a London tavern—the Belle Sauvage Inn—but Pecunia is also an apt type for the Indian

"Princess" as represented by Jonson's "historian," Captain John Smith. Jonson's character Pecunia, called in the play "the Infanta of the Mines," is an allegorical composite picture of colonial treasure.[40] And Jonson uses this character to instruct his audience in the proper way to desire and possess the "New World's" riches.

The Staple of News is a play about the transformation of an English subject, a prodigal heir, Peni-boy Junior, into a proper English son; and that transformation involves his ability to manage an imperial fortune discreetly. Junior's inheritance comes from his father, presumed dead but actually in disguise as a canter, a beggar who knows thieves' jargon. However, the Peni-boy fortune, while possessed by Englishmen, derives from the larger world, especially the mines of the East and West Indies. I opened the book with a discussion of one of the *Two Gentlemen of Verona*'s anticipated education. Just as Protheus would be made a proper gentleman by discovering islands, Peni-boy Junior's interactions with Pecunia, the material result of discovery, will define his English gentility. By comparing Pecunia, "the Infanta of the Mines," with "Pokahontas," the Indian woman brought to London in 1616 as the transformed English princess Rebecca, *The Staple of News* domesticates colonial treasure and identifies it as the undifferentiated colonial booty of a representative English family.

Fifteen years earlier, in Jonson's 1609 play *Epicoene*, Sir Amorous La Foole tells Clerimont, a gallant, that another foolish character, Sir John Daw, keeps a "boxe of instruments," drawing implements with which he has taken the portrait, or as La Foole calls it, strikingly, drawn a "map" of "NOMENTACK, when he was here" (V.V.i.23). Namontack, one of Jonson's "great King" Powhatan's advisors, went with the Virginia Company's Captain Newport to England in 1608. Morose, *Epicoene's* curmudgeon, dreams of reducing his nephew, Clerimont's friend Dauphine, to a begging knight, like *Eastward Hoe's* Sir Petronel Flash, without even a hope of "repair[ing]" himself by going to "*Constantinople, Ireland, or Virginia*" (V:II.v.127-28). In this early Jonson play we have a record of the display of Virginia Indians in London, a practice employed by the Virginia Company to generate interest in the then ailing settlement effort; we see also one of that colony's prime sustaining forces—disappointed youth and younger brothers looking for a fortune. By the time of *The Staple*, the walking colonial allegory could become a focus of dramatic action in England. The colonial subject whose display provides local detail in the earlier play is the center of action in the later. While Pocahontas was also displayed by the Company, in *The Staple of News* her possession and its meaning for Peniboy Junior's character takes center stage; he becomes the figure whose destiny can be determined only in relation to his dealings with the imperial market's goods.

No one knows why the Virginia Company chose the Belle Sauvage Inn to house Pocahontas during her brief, fatal visit to England. Surely the name must have seemed appropriate, as Pocahontas was England's belle savage of the day. As Jonson's reference to Namontack attests, Pocahontas was only one of a number of Indians brought over from Virginia, but she was the first the Virginia Company could display both as a princess and as a convert. The Company's choice of the Belle Sauvage Inn is provocative for this chapter's concern with theatrical transformation, as that inn had served in the 1570s and 1580s as a theater. In 1616 Pocahontas lodged in an English inn that had been a theater; nine years later, in 1624, Ben Jonson staged an allegory of her at the nearby Blackfriars theater. And during her short English sojourn, Pocahontas attended one of Jonson's masques for King James and his court.[41] Pocahontas herself, as chapter 5 and the epilogue will discuss at length, is an apt figure for colonial transformation. And it is telling that her transformation into a princess and into a figure for display happened in the context of London theaters and then became fodder for *The Staple of News* on the London stage.

These conjunctions—Pocahontas on display at an inn that served as a theater, Pocahontas as a displayed guest at the royal theater, and Pocahontas as object of theatrical discussion and play on the London stage—point to the central role of theater and theatrical metaphor in public consumption of England's imperial expansion. The display of English "antiselves"— Irish, Northern, Virginian, East Indian—on London's stages enabled audiences to envision a transformed new Atlantic world and to negotiate their places within that world. In a sense such displays turned public and private theater audiences into "seeing-men" and "seeing-women," innocent observers who could also claim a sort of possession, or at least an understanding, as the stage turned Virginia into comprehensible London spaces and reassured those "respectable" audience members of those spaces' controllability. London's theaters became a place to witness faux "contact zones," what Pratt calls "the space of colonial encounters, the space in which peoples geographically and historically separated come into contact with each other and establish ongoing relations, usually involving conditions of coercion, radical inequality, and intractable conflict" (6). As Pratt suggests, "A 'contact' perspective emphasizes how subjects are constituted in and by their relations to each other. It treats relations among colonizers and colonized . . . not in terms of separateness or apartheid, but in terms of copresence, interaction, interlocking understandings and practices, often within radically asymmetrical relations of power" (7). *The Staple of News* is a prime example of a faux contact zone on the London stage, for while Peni-boy Canter and the audience determine his heir's suitability by his relations with "his" imperial treasure, that personified treasure acts only in relation to male desires on the stage; she is not a character, even in the slight sense that Junior and his father are characters. Rather, the Lady Pecunia, like the Pochahontas she glancingly represents, acts her role as treasure to be possessed completely, without the possibility of escape.

Despite *The Staple's* simplistic colonial allegory, what makes early-seventeenth-century London stages such a compelling and vital site for understanding contact in the new Atlantic world is both their audience dynamics and the complexity of their spectacles. Jonson's plays display contact between geographically and historically separated people, and since they played in London's public and private theater settings they played for audiences that were themselves differentiated and even potentially aggressively inclined toward one another. *The Staple of News's* Lady Pecunia, for example, shares the stage with Peni-boy Canter, a wealthy gentry father disguised as a beggar. In his disguise he provokes ridicule from Shvn-Field, a sea captain; Almanach, a doctor; and Fitton, an emissary court. They equate his poverty with debased and also alien status:

> *SHV.* How the rouge stinks, worse than a Fishmonger[s] sleeues!
> *FIT.* Or Curriers hands!
> *SHV.* And such a perboil'd visage!
> *FIT.* His face lookes like a Diers apron, iust!
> . . . .
> *ALM.* I wonder what religion hee's of!
> *FIT.* No certaine *species* sure. A kinde of mule!
> That's halfe an *Ethnicke,* halfe a *Christian!*
> (VI:II.iv.49-57)

Later these same "jeerers" admonish the prodigal Peni-boy Junior for his familiarity with the canter:

> *FIT.* We say, we wonder not, your man o'*Law,*
> Should be so gracious wi'you; but how it comes,
> This Rouge, this *Canter!*
> *P.Iv.* O, good words.
> *FIT.* A fellow
> That speakes no language—

ALM. But what gingling *Gipsies,*
And *Pedlers* trade in—
FIT. And no honest *Christian*
Can vnderstand. (VI:IV.i.49-54)

In these scenes, debased characters lambaste Canter with slurs that insult London tradesmen for their investment in their trades. Fishmongers smell like the fish they sell, curriers' hands are redolent of the horses they brush; the canter's face is wrinkled and stained as a dyer's apron. And his immersion in his canting trade has removed the canter from the English "species"—his poverty makes him a pagan ("ethnicke" here denotes "heathen, outside the Christian sphere."), and he speaks a foreign, pagan language (*OED* "ethnic"). Meanwhile, Jonson's audience must have contained both fishmongers, gentry, doctors, sea captains, and even those who might belong to an uncertain religious species. How were they to identify themselves within this spectacle of alterity? Even though the speakers in these scenes are mistaken, just as Tattle is mistaken about Pocahontas, their jeers force distinct groups (and groups that were attempting to distinguish themselves from each other) into contending with their relations. If a canter is like a fishmonger, how "ethnicke" is a fishmonger? Who is a gypsy; who a Christian? While *The Staple's* audience watched an Englishman defined by his relationship with undifferentiated Indian treasure, they were asked to define themselves and their own Englishness in relation to the other in the next group of seats. Colonial transformation was not over there but everywhere in London's theaters.

Chapter 4

Colonial Transformations in Court and City Entertainments

In the early seventeenth century, London elected a new Lord Mayor annually from the exalted ranks of the twelve great livery companies, and that mayor's reign began with a lavish spectacle that lauded his achievements and the city's worth.[1] Those companies and the merchants associated with them funded expansion into every corner of England's emerging global market, and their yearly Lord Mayor's pageants celebrated their heroes in the context of their colonial endeavors—endeavors that were becoming more and more central to the city's economic success. Transformative spectacles in themselves, since they accomplished a transformation in status for the new mayor, Lord Mayors' pageants both acknowledged the Company's and London merchants' roles in England's colonial efforts in Ireland, Virginia, Bermuda, and the East Indies, and they were, I will argue, motors of those efforts. Actively mobilizing London's mayors and prominent guild members as explorer-colonists, these annual public entertainments satisfied guild members' visions of themselves as world-dominators, and they produced expectations of dominance that would fuel continuing mercantile expansion. In addition, these pageants represented the city as whitened space, able to bring its enlightenment to what they depicted as England's dark colonial edges.

Jacobean and Caroline court masques also celebrated an imperial England they created on stage. Although masques and pageants clearly helped to form the imperial England they displayed, suprisingly little attention has been paid to them as colonial texts.[2] This chapter will argue that masques and pageants were central to England's emerging imperial identity. Like so many other literary texts this book has already examined, pageants and masques depend on and instantiate coordinated oppressions. In these entertainments, blackness, an undifferentiated Indianness, and uncivilized Irish-ness all become signs of a precolonized, pre-English dominated world that will be saved by the Companies' and the English multitudes' efforts in both the new Atlantic world and the East.[3] Masques and pageants use undiffer-entiated Indians as well as coordinated oppressions to establish the Com-panies' and the crown's dreams of imperial whiteness and imperial power. Imperial references in pageants and masques may take us far from the Atlantic world, even to the moon in the case of one of Jonson's masques, but their own slippages indicate that the audiences for pageants and masques wished to understand England's identity as formed against an undifferentiated field of otherness, often denoted by blackness.[4] When Jacobean court masques are read in relationship to one another, as I do with *The Irish Masque at Court* and the antimasque[5] *For the Honor of Wales*, their different colonial-political agendas become visible; however even these masques postulate unequivocal submission to a colonial domination that was nowhere near completely established except in the fantasy of the masque.

Ignoring England's actual situation in Eastern trade, masques and Lord Mayor's pageants speak to merchants' and court attempts to link all of Eng-land's colonial efforts in the public imagination. Combining an undifferen-tiated East with an undifferentiated West, masques and pageants display a completely imaginary imperial domination that was, however, unfortu-nately proleptic. English trade in the East Indies boomed just before the enormous growth of commerce with Virginia and the West Indies (includ-ing Bermuda), and the first half of the seventeenth century saw both the eastern and western arenas become centrally significant for England. The East India Company was founded in 1599, ten years before the accidental landing in Bermuda that initiated English West Indian colonization. But just as the Virginia colonization effort faced resistance, and Ireland had yet to bow to England, East Indian trade was precarious. Merchants founded their fortunes on spices found in the East, but they did not rule any land or people in the East. However, many Lord Mayor's pageants envision lav-ish East Indian treasure, and those pageants associate that bounty with profit from Atlantic world efforts; they also reveal that the same philosophy of colonialism, based on the superior transformative power of Christian con-version, supported English efforts throughout the world. In the new Atlantic world, particularly in Virginia and Ireland, England was beginning to enjoy an on-the-ground dominance that sporadically generated hard-won wealth; in the Eastern world, English merchants saw a well-guarded field of riches: a lucrative field to which they had only uneasy access. But both masques and pageants imagined a world with England at the center, a world in which principalities that were in reality independent and powerful trading partners bowed their heads to English power. Pageants and masques deliberately ignore colonial and trade realities, instead celebrating a temporally distant British empire as an already developed entity. The undifferentiated Indian on display in masques and pageants speaks to English desires articulated in the earliest English colonial propaganda, the desires satirized by Ben Jonson in *Eastward Hoe:* that these various others—Irish people, the Anglo-Irish, North American Indians, South American Indians, Africans, East Indians—could magically combine into colonial wealth freely given by the willingly dominated. The undifferentiated natives these entertainments display submit happily and completely to their domination in the same way that Spenser's beloved in the *Amoretti* submits to her lover's mastery (chapter 1).

Although pageants and masques played for different audiences, they served similar functions for those audiences. Both types of entertainments display native antiselves in order to encourage colonial investment and to ensure their audiences' identity as purified, whitened colonists. The London multitudes never sat in King James's banqueting house, and masques sometimes display conflicts between city and court; but as they educate their audiences as colonial rulers, both types of entertainments display antiselves and use spectacles of their transformation. In this chapter I mark the differences between masques and pageants. However, English efforts in the new Atlantic world, in India, and in Africa,[6] depended on investment from courtiers as well as merchants; indeed, fortunes made in the colonial arena could blur the boundaries between these social groups and move English people from one group to another, and these boundaries were already blurred by economic realities. Donne notes in his sermon for the Virginia Company that "as *Merchants* growe vp into worshipfull Families, and worshipfull Families let fall branches amongst *Merchants* againe, so for this particular Plantation, you may consider *Citie* and *Countrey* to bee one body" (F). In addition, as Robert Brenner's crucial book, *Merchants and Revolution,* argues, despite periodic tensions between the city and the court, each group depended on the other—the crown for financial support from the merchants, and merchants for trade protection regulated by the crown.[7] Court and city entertainments reveal the same interdependence of merchants and crown that we have seen in the discussion of naming in Virginia in chapter 2. The consonance between the colonial spectacles commissioned by both groups demonstrates their joint interest in England's imperial expansion, even where the spectacles can reveal rifts in that alliance.

Unlike the plays treated in chapter 3, which persistently point to the faultlines running through English imperial efforts, court masques and city triumphs generally imagined a unified, applauded colonial mission which en-lightened—to play on a pageant pun—their audience. The willful disregard for difference in many of these entertainments enabled a collective fantasy. Like Spenser's little love poems, but on a mass level, these spectacles assured their audiences of the success and propriety of colonial and economic expansion. Also like those poems, these spectacles celebrated the willingness of native populations to become fair English people. As my introduction suggests, recent postcolonial literary and anthropological criticism strives to expose the uniqueness of each colonial encounter and to recover the resistances to colonial domination that traditional history has erased; in Nicholas Thomas's words, to recognize that colonizing projects' "coherence . . . was prejudiced both by internal contradictions and the intransigence and resistance of the colonized" (3).[8] For example, Andrew Murphy's book on English images of Ireland laments that in historical and critical work generally, "a sense of difference is not sustained, nor is any acknowledgement made of the crucial distinction separating the Irish on the one hand from the North American Indians, Caribbeans, and South American indigenes on the other" (1999, 4). In this book, I have attempted to keep those differences in constant view. But early modern court and city entertainments strive for a field of disempowered, undifferentiated otherness that would ensure a powerful imperial identity for England. Lord Mayor's pageants and court masques, as well as projecting colonial desires, helped to produce the illusion of an undifferentiated other that contemporary critics are working so hard to overcome. They also helped to produce the enduring stereotypes of American Indians that we are still encountering in tourist attractions, as my epilogue will show, and in everyday American life.

In the early seventeenth century, the London livery companies underwrote the East India Company (Ramen 141). They also contributed lavishly to efforts to colonize Virginia, and they subsidized English expansion into Ireland. A list of Adventurers to Virginia compiled circa 1618 shows a total of 2,085 pounds, 16 shillings, and 8 pence adventured in the names of London companies; the largest company contribution was the Grocers' 487 pounds ten shillings and most of the twelve great companies adventured substantial sums. Smaller companies too, like the Girdlers and the Leathersellers, contributed impressively to the Virginia effort according to this list (*Records* III: 80-90). A 1620 list of sums paid to the Virginia Company's treasurer confirms all of the London livery companies' contributions. It appears the companies, unlike many of the individual adventurers, did not default on their promises. Individual "elite City merchants" funded the Virginia efforts as well (Brenner 89). Sir Thomas Smith, one of the men named on Norwood's Bermuda map was a director of the Virginia Company (see my introduction); He was also a governor of the Sommers Island company, as well as a prestigious London merchant. Prominent guild members also personally invested in Jacobean efforts in Ireland. Sir William Cockayne, London's Lord Mayor from 1619 to 1620, and according to Brenner "one of London's greatest merchant princes" made a fortune "furnishing supplies for the Irish wars" (Brenner 137, Middleton 1953 vi).[9]

The London livery companies also invested as a group in Irish colonization. In 1609, the twelve companies agreed to a plan to colonize six escheated counties in Ulster. The plantation was named Londonderry and the land divided among the Companies. This scheme was conceived as a city effort; the companies would remake Ulster in London's image, not the image shown in *Eastward Hoe* or *Bartholmew Fair*, but the pageants' purified image of industry and building. Sir Thomas Phillips's 1622 survey of the plantation titles the first plan, "A general plat of the lands belonging to the City of London as they are divided and set out to the 12 Companies as they do butt and bound each upon other" (*Londonderry* 148; see figure 4.1). Londonderry was to be wholly controlled by the companies; the plan shows land bordering the Maine Sea variously assigned to the "Clothworkers, Marchantaylers" and other liveries (Frontispiece, figure 4.1).

Phillips's report accuses the companies of neglecting to equip the plantation properly, especially with English tenants, but he also complains bitterly that the companies are exploiting the land for their own purposes. He states that the London companies "have made intolerable spoil of his Majesty's woods in merchandizing them and converting them to their own use" (21). Phillips also contended to Charles I that the city "intended to abuse him and the Service otherwise they would not have omitted the Principal Point thereof which concerned the safety of the Country, the fundamental ground being the AVOIDING OF NATIVES and Planting wholly with British" (5). Phillips's series of complaints against the city for its neglect of proper attention to the efforts he views as fundamental to English success in Ireland again indicates that there was no one singular uncontested English colonial effort, just as his pained acknowledgment that

Figure 4.1—Sir Thomas Phillips, *A Gennerall Plat of the lands Belonginge to the Cittie of London*. Reproduced with permission of the Lambeth Palace Library.

the city was wasting Ulster's woods shows the city's investment in Ireland. While Jacobean court masques stressed allegiance to James, Lord Mayor's pageants focused on profit. The London companies pursued profit in Ulster through every avenue, selling off Irish forests and extorting high rents from their Irish tenants rather than importing English settlers.[10] The type of colonial effort that Phillips deplored—the exploitation of resources for commercial gain—was exactly the work encouraged in Lord Mayor's pageants. Those pageants see the Atlantic and the Eastern world as vast fields for English profit; in terms of their colonial representations, these entertainments are first and foremost mercantile spectacles, even when they attach a religious content to their displays of native subjects. They encourage the city to waste Ireland's woods, to possess undifferentiated Indian riches, and to bring its economic light to native blackness throughout the world.

Unlike Jonson's plays, which divide England into savage and civil populations, but like England's (other) colonial texts, city entertainments subscribe to the idea that England has attained the pinnacle of social progress. One early Jacobean pageant, Anthony Munday's 1605 *The Trivmphs of revnited Britania,* performed for the Merchant-Taylors Company in honor of Sir Leonard Holliday's elevation to Lord Mayor, begins with stories of Britain's origins. Brute addresses Britannia, declaiming,

> Thou that before my honord victorie,
> Wert as a base and oregrowne wildernes,
> Peopled with men of incivility,
> Huge and stearne Gyants, keeping company
> With savage monsters, thus was *Albion* then,
> Till I first furnisht thee with civill men. (8)

This speech's central comparisons—between "civill men" and a "base" land, "incivility" and "savage monsters"—ideologically subtend both pageants and court masques. They are the same comparisons that inform the lottery broadside and other documents that I examined in chapter 3. As a group, Lord Mayor's pageants concern themselves with demonstrating the civility and civil benefits of the merchants they celebrate; and in those pageants putative colonial subjects worship that civility. These pageants elevate the city as the home of English civility, and the pageants' antiselves recognize their own comparative savageness and desire their transformation into the London companies' colonial subjects.

In Thomas Middleton's 1619 *The Trivmphs of Loue and Antiquity,* funded by the Skinners for Sir William Cockayne's mayoralty, Orpheus honors the King when he says that he sees

> The seuerall Countries, in those faces, plaine,
> All owing Fealty to one Soueraigne,
> The Noble *English*, the faire Thriving *Scot*,
> Plaine hearted *Welch*, the *French man* bold and hot,
> The ciuilly instructed *Irish man*,
> And that kind Sauvage, the *Virginian*,

all assembled to congratulate him (B3v). In this imperial vision, England's vassals join the inherently noble Englishmen in saluting their king, this despite the actual complete lack of fealty to England in France and England's tenuous hold on both Ireland and Virginia. The Irish, according to Orpheus, have been brought to civility by the English, and along with the "kind savages" in Virginia they naturally recognize their proper allegiance. Although Orpheus salutes the king, the pageant finally shows the world of mercantile commerce, signified by Cockayne's international success, as supporting the noble English. Like Munday's 1605 pageant, Middleton's 1619 pageant assumes that the English are noble in comparison to other nations, and it connects that nobility explicitly with the submission it inspires from the Irish and the Virginians. Collectively, London's Lord Mayor's pageants declare England and the city imperial dominators of the Atlantic and Eastern worlds, and they laud that empire constituted in commerce.

In these entertainments, England owes its successful empire to the city's mercantile enterprise. Middleton's 1622 *The Triumphs of Honor and Vertue*, a pageant for the installation of the grocer mayor Peter Proby, along with a triumph displaying "the Continent of India," included a display of

> the out-parts of the *Globe* shewing the Worlds Type, in Countries, Seas and Shipping, whereon is depicted or drawne, Ships that haue bene fortunate to this Kingdome, by their happy and successefull Voyages; as also that prosperous Plantation in the Colonie of *Virginia*, and the *Bermudaes*, with all good wishes to the Gouernors, Traders and Aduenturers vnto those Christianly Reformed Islands. (C2)

This pageant displays the entire world (as the English knew it) as England's imperial margins, which are ethnocentrically denoted the globe's "out-parts" and are connected to England by its fortunate ships. Two hundred and sixty-two years before the establishment of Greenwich mean time, this

pageant declares London the center of the world.[11] Grocers were increasingly dependent on imports of goods like tobacco and sugar, demanded by
the London market.[12] In this pageant, London is the center of a network
of "fourtunate" ships. Like England's projected domination of its putative
Eastern margins, the nation owes its "prosperity" in Virginia and Bermuda
to the enterprise of shipping and investment.

Likewise, Thomas Heywood's 1639 pageant, *Londini Status Pacatus: or
Londons Peaceable Estate*, for the Draper Henry Garway, celebrates the city's
commerce in "al Countries, Christian or Heathen, discoveries, plantations,
(as in *Ireland, Virginia, Bromoothos*, or *Summers Islands, St. Christophers, New
England, Harber-grace* in *New-found Land* &c.) . . . nor *Rome* it selfe the
Metropolis of the *Roman* Empire, could in her most flourishing estate and
Potency, (though she Tyranniz'd over the whole World,) in the least compare with *London*" (126). Garway, along with Cockayne, was one of the
five men "who among them controlled the positions of governor, deputy
governer, and treasurer of the East India Company throughout the 1630s"
(Brenner 78). In his pageant, the Atlantic world becomes a mainstay of a
mercantile empire that surpasses the Roman empire. Although England
hardly rivaled the Roman empire in 1639, by 1640 "the rise of commerce
with the Americas" based on the imports of "tobacco and sugar produced
on farms and plantations . . . was already being strongly felt" (Brenner 33).
Brenner notes that "In the years between 1622 and 1638, tobacco imports
from the American colonies to England leaped from about sixty-one
thousand pounds to two million pounds a year, providing the basis for a
new and increasingly important line of mercantile activity" (113). While
London was not quite Ancient Rome, in this pageant's unfortunately proleptic imperial vision, commerce enables domination. Of course, while
both the commerce and the social domination it entailed would become
a reality, it was the domination that supported commerce. The slave trade
enriched many merchants (Brenner 163-65), and slave labor made the
tobacco and sugar plantations vastly profitable. Without the uncompen-
sated labor of the "uncivilized subjects" displayed in England's entertainments, England could not have achieved the empire these pageants declare
already constituted.

Lord Mayor's pageants acknowledge the existence of those subjects but
declare that they are rightfully and even gratefully dominated. Heywood
devotes his 1633 pageant for Clothworker Ralph Freeman's mayoral inauguration to a celebration of merchants, "by whose Adventure and Industry
unknowne Countries have beene discovered, Friendship with forreigne
Princes contracted, barbarous Nations to humane gentlenesse and courte-

sie reduced" (56). He praises the London merchants collectively, "whose Factors in both *Indies* lye, / The East and West," and the entertainment's margin notes that "Virginia, New England, the Bromoothos *and* St. Christof. *are parts of the* West-Indies" (62). In this vision, merchants have transformed the known world in an English or English-allied image; their presence, their "ly[ing]" in the two Indies will insure English interests there, interests that may include the "reduction" of native peoples into gentleness. "Gentlenesse" here is given the adjective "humane," implying that before contact with English merchants the nations in questions behaved subhumanly. According to the *OED*, before 1700, "humane" was "a common earlier spelling of human." While in the eighteenth century the word moved to mean something like "gentle or kindly in demeanour or action; civil, courteous, friendly, obliging,"[13] in Heywood's pageant it clearly retains its use as marking the way people should behave if they are to be regarded as people. This definition of proper human behavior assumes that the unwillingness of any "barbarous nation" to enter into economic bonds with English merchants, or to allow the English to exploit the land or resources "belonging" to that nation, indicates that nation's pariah status; and in the pageant's circular logic, that unwillingness also affords the English a license to subdue that nation.

Of course, none of these pageants dramatizes native unwillingness; instead each represents the transformation of native populations as a welcomed, glorified conversion from devalued darkness into apotheosized mercantile light. In Middleton's 1613 *The Triumphs of Truth,* performed at Sir Thomas Middleton's[14] confirmation as Lord Mayor, a "strange Ship" appears with four passengers, "a King of the Moores, his Queene, and two Attendants of their owne colour" (B4v). The King speculates that the multitudes, "these white people," are staring because they have never seen "a King so blacke before" (B4v). He understands their amazement but quickly justifies his appearance, for by the grace English merchants have bestowed on him he has lost his essential blackness:

> I must confesse many wilde thoughts may rise,
>
> . . . .
>
> At my so strange arriuall, in a Land
> Where true Religion and her Temples stand:
> I being a Moore, then in Opinions lightnesse
> As far from Sanctity as my Face from whitenesse;
>
> . . . .
>
> How euer Darknesse dwels vpon my Face,

Truth in my soule sets vp the Light of Grace;

. . . .

My Queene and People all, at one time wun,
By the Religious Conuersation
Of English Merchants, Factors, Trauailers,
Whose *Truth* did with our Spirits hold Commerce
As their affaires with vs, following their path
Wee all were brought to the true *Christian Faith*. (C)

In the King's discourse, "commerce" becomes a highly valued transformative force, its enlightening benefits automatically accompanying business with English merchants. Christian truth saves (whitens) those whose "complexions"—signs in this entertainment of their previous ignorant and damned sun-worship—would have condemned them to eternal error. Error appears as a character in the triumph, lamenting that "[her] Sweete-fac'st Deuils forsooke [her]" (Cv). In this pageant, Londoners' suspicion of black faces is justified, since the Moors' blackness was a sign of their error, but that suspicion does not take into account the wonderful possibility of mercantile transformation. The Moors' blackness serves as a sign of this Christian truth's power, which can whiten even those whose degenerate nature is printed on their skin.

Middleton's 1622 *Triumphs of Honor and Vertue* uses a black Queen "representing *India*" to claim that the English gift of spiritual light vastly exceeds East Indian crops in value.[15] Black skin is again the marker of un-Christian damnation that can be symbolically whitened; and Africans and Indians are both visibly marked by their blackness, since their spiritual distance from English whiteness makes them essentially the same. The Queen of *Honor and Vertue*, like the King in *Truth*, posits a spiritual whiteness that redeems her visible blackness:

Draw neere this Blacke; is but my Natiue Dye,
. . . you'le then find
A change in the complexion of the Mind;
I'me Beauteous in my Blacknesse . . .
. . . through my best part runnes
A Spring of liuing Waters, cleere and true,
Found first by *Knowledge*, which came first by you,
By you, and your examples; *Blest Commerce*,
That by Exchange settles such happinesse,
Of Gummes and fragrant Spices, I confesse

My Climate Heauen do's with Aboundance blesse,
And those you haue from me; but what are they
Compar'd with Odours whose Sent ne're decay,
And those I haue from you, Plants of your Youth,
The Sauour of eternall Life, sweet Truth,
Exceeding all the odoriferous Sent,
That from the Beds of Spices euer went. (B2)

English commerce becomes a gift to "black" India, and by extension, in these pageants' logic, to all of England's black, barbarous fringes. Accompanying their displays of native wealth, these triumphs display native subjects desiring English "conversation" and devaluing their own resources. In the Queen's speech, the East Indian spices, which were clearly the mercantile desideratum, become decaying worldly pleasures, much less precious than the truth English merchants can offer India. In his sermon for the Virginia Company twelve years earlier, William Crashaw argued that the Company would give the Indians "such things as they want and neede, and are infinitely more excellent then all wee take from them: and that is 1. Ciuilitite for their bodies, 2. *Christianitie* for their soules" (D4). Endorsing Crashaw's English logic, *Honor and Vertue*'s black Queen declares, "All wealth consists in Christian Holynesse, / To which caelestiall knowledge I was led; / By English Merchants first enlightened" (B2). The Queen's pun on "enlightened" contains *Honor*'s version of "truth"—a keyword meaning English (Protestant) Christianity in all of these pageants; English commerce is essentially an enlightenment, bringing light and (explicitly) whiteness to the empire's dark corners. As she repeats the phrase "by you" in her long speech—"Found first by *Knowledge,* which came first by you / By you, and your examples"—she draws the audience of merchants and Londoners into the project of colonial transformation. In these pageants' logic, Londoners are synonymous with "white people" and therefore can be required to bring their light to these black people who so value their own conversions.

Many pageants use Indian and African subjects (whose boundaries are often blurred) as exotic objects for the merchants' visual delectation. Like the African faces in cameos and portraits that Kim Hall examines, these subjects in pageants function as "curiosities who represented the riches that could be obtained" by merchants' activities (Hall 1995 211). Munday's *Camp-Bell,* a 1609 pageant for Sir Thomas Campbell of the Ironmongers, featured a watershow of "A Whale with a Blackamore in his mouth," along with "A flyeninge dragon and vnicorne" and "a bell feild carried in A Chariott and drawne by ij Estriges" (32). In *Monuments of Honor,* the 1624 pageant by John Webster for John Gore, a Merchant-Taylor, a lion and a camel, the two animals on the Merchant-Taylors' coat of arms, parade; "on the Camell rides a *Turke*, such as vse to Trauaile with Carauans, and one the Lyon a Moore or wild *Numidian*" (B4). The third show in Thomas Dekker's 1629 pageant *Londons Tempe or the Feild of Happines*, also for the Ironmongers, in honor of James Campbell, "is an Estridge, cut out of timber to the life, biting a horse-shoe. On this Bird rides an Indian boy, holding in one hand a long *Tobacco* pipe, in the other a dart" (B2). In these three pageants, the African or Indian other appears with an exotic animal, signifying the native subject's wildness and his association with the riches of the East. In addition, Dekker's Indian boy is associated with tobacco, making him at once a figure of East Indian exoticism and also a representative of Virginian and West Indian profit.

Munday's 1611 pageant for the Goldsmiths' Sir James Pemberton, *Chruso-thriambos,* asks its city audience to imagine "sundry Ships, Frigots, and Gallies . . . returned home . . . with no meane quantity of Indian Gold." In the show, the audience watched English "gallants" acting out sea fights, each with "his *Indian* Page attending on him, laden with Ingots of Golde and Silver" (A3v). Here the pages function, as Hall argues attendants do in English portraits, to accentuate the gallants' whiteness; they also function as figures for the treasure they carry. Five years later, Munday wrote a pageant for the Fishmongers' John Leman, *Chrysanaleia,* in which "the King of Moores" rides on a golden Leopard, "hurling gold and siluer euery way about him," followed by six "tributarie Kings" laden with gold and silver ingots, each one carrying a dart (Bv). The show is intended, Munday writes, to indicate the amity between the fishmongers and the goldsmiths. Munday explains the King's "Indian treasure liberally is throwne. / To make his bounteous heart the better knowne" (C2v). Again, the black figures represent the treasure they carry and willingly bestow on their whitened English audience. In addition, the display of Moors and riches represents not Indian miners, the actual laborers who produced treasure, but the English traders' and craftsmen's (desired) control over Indian resources.

The "black Queene" of Middleton's *Honor and Vertue* enters that triumph looking like these inarticulate displayed antiselves. As the Lord Mayor arrived in Paul's churchyard for that 1622 pageant, he was attended by a triumph,

> bearing the Title of the *Continent of India,* A triumph replenished with all manner of Spice-plants, and trees bearing Odour . . . [and] A blacke Personage representing *India,* call'd for her odours and riches, the Queene of

Merchandize, challenging the most eminent Seate, aduanceth her selfe vpon
a bed of Spices, attended by *Indians* in Antique habits. (Bv)

At the same time as the pageants' pious statements assert that an unworldly,
theological end motivates the companies' colonial efforts, those assertions
appear within lavish spectacles of treasure that undermine their altruistic
sentiments; in fact, the combined spectacle reveals that the theology is a
means to a market end. These spectacles piously claim that English colonial efforts throughout the world are merely intended to increase God's
glory. So the gorgeously displayed Queene of Merchandize ends up
degrading merchandise in comparison with the kingdom of God. They
simultaneously appeal to a mercantile paradise, with profit to be found in
every illuminated and converted territory. The apparent contradiction here
is a sign of the uneasy but finally obsequious settlement English Protes-
tantism was making with emerging capitalism. Donne's sermon registers
the unease associated with that settlement in the parenthetical phrases in
his exhortation to the Company to pursue spiritual gain: "(leaue out the
consideration of profit for a time) (for that, and Religion may well consist
together,)" (D4). Of course, I am not claiming that there were not merchants who devoutly believed that their primary mission was to "save"
these Indians. Nor am I claiming that no one in seventeenth-century London could decode these spectacles. Someone was certainly laughing at the
idea that these native populations desired their conversion (probably the
people in London who were forcibly converted themselves or who had to
hide their various dissents). But these contradictory spectacles clearly satisfied their London audiences or they would not have been performed year
after year. They satisfied and helped to establish a particular dominant
understanding of the imperial market that would remain triumphant for
many years to come.

For the time being, these pageants enabled their merchant audiences and
the London multitudes to believe in their own purified motives as well as
their identities within a vast mercantile metropolitan empire. Pageant audiences, who were at least as diverse as the denizens of Jonson's city comedies,
could celebrate their own desired transformations into vastly wealthy white
English subjects at the same time they celebrated the whitening of their
imperial subjects, a whitening approved and desired by their God. *Honor and
Vertue* ends, appropriately, with a speech by Honor to the new Mayor:

So euery good and vnderstanding Spirit
Makes but Vse onely of this Life, t'inherit

An euerlasting Liuing; making Frends
Of *Mammons* Heapes, got by vnrighteous Ends,
Which happy Thou standst free from, the more white
Sits Honor on thee, and the Cost more bright
Thy Noble Brotherhood this Day bestowes. (C3)

Middleton's twisted syntax congratulates Peter Proby for his whiteness, his ability to keep his identity as grocer and mayor free of the pollution of a fortune "got by vnrighteous Ends." Such praise, of course, depends on a shared definition of "righteous ends." While these pageants work hard to achieve that definition, it is unlikely that such a definition would have been agreed upon by the Virginia Indians the grocers' 487 pounds helped to displace, by the former owners of the escheated Ulster Grocers' plat, or by the black slaves beginning to be incorporated into England's "empire."

... Whilst you sit high,
And lead by them, behold your *Britaine* fly
Beyond the line, when what the seas before
Did bound, shall to the sky then stretch his shore.
(*Prince Henry's Barriers* 435-38)[16]

While Lord Mayor's pageants applaud economic triumph, court masques focus on political domination. But like their city counterparts, masques display colonial subjects eagerly embracing domination regardless of whether they are Irish, East Indian, North American, or Welsh. They also gloss over the actual resistance to the empire they celebrate. Merlin's speech to Henry, King James, and Queen Anne about the nation's destiny concludes the speeches at Ben Jonson's 1610 court entertainment, *Prince Henry's Barriers;* it declares England's empire already made. This masque is largely an epic oration that creates in words "a strength of empire fixt / Coterminate with heauen" (340-41). One of Merlin's speech's and the masque's particular projects as they present Prince Henry as the new Prince of Wales becomes clear a few lines later when James's "great victory" is pronounced: "*Ireland* that more in title, then in fact / Before was conquer'd, is his *Lawrels* act" (347-48). In fact, as we have seen, Ireland's colonization by England under James and his predecessor Elizabeth was far from complete; the list of masquers in the Oxford edition of Jonson's works attests to that. Many of the prominent dancers in James's masques were veterans of Essex's disastrous (for the English) 1599 campaign in Ireland. Sir William Constable, Sir Richard Houghton, and Sir Carey Reynolds, all knighted by Essex in 1599 in the Irish wars, tilted together in the barriers of the masque *Hymenaei* in 1606. By 1610, England had nominally completed the conquest of Ireland, but James was by no means secure of his rule there or of his putative subjects' loyalty. *Prince Henry's Barriers* and, even more clearly, Jonson's 1613 *The Irish Masque at Court* enact imaginary triumphs that counteract the actual previous defeats of English efforts to dominate Ireland and that deny the widespread Irish resentment of the conquest that had been accomplished.[17] The masque audience sees Englishmen, previously defeated by the Irish in Essex's war, and made nobles in that effort, participate in magnificent courtly entertainments in which they exult over the English empire in Ireland made real, successful, and complete on stage. As the court dances, the process of that projected subjection is staged.

Celebrated four years after the founding of England's first successful Virginia colony, *Prince Henry's Barriers* declares an incipient imperial England. Like the Lord Mayors and merchant groups funding pageants and like the spectacles they funded, the Jacobean court masques' audiences linked the Irish and American imperial arenas. Courtiers who were veterans of the Irish wars, supporters of Ireland's colonization, and also prominent supporters of the emerging colonization of America (a number of people belonged to all three of these groups), danced in and attended James's masques. If we compare a list of "the Names of Adventurers with their severall sums adventured" that the Virginia Company attached to its 1620 *Declaration of the State of the Colonie and Affaires of Virginia* with the names of the dancers and tilters in James's masques, we find support for the Company running way over 1,000 pounds; if we add to that total the adventurers who were parents and siblings of the masquers, we find many thousands of pounds of support for the Virginia enterprise, support that was crucial to sustaining colonial efforts. Some of the courtiers dancing and in attendance at masques became colonists in America. Viscount Doncaster, son of the first Earl of Carlisle, who danced in *Love's Triumph,* in 1639 "established his hereditary right to the island of Barbados, then called the Carlisle Islands" (v. X, 431). In 1627 the crown had granted Doncaster's father "the proprietorship of all the Carribbean Isles" (Brenner 127). Today, Barbados still has a bay that bears his name: Carlisle Bay. The Earl of Southampton, who directed the Virginia Company from 1619 to 1624, tilted in *Prince Henry's Barriers*. Most of the dancers in Thomas Campion's 1614 masque appear on the 1618 list of Virginia company adventurers; the

Earl of Pembroke contributed 400 pounds; Dorset, Salisbury, Montgomery, Lord Walden and Lord North, all dancers in that masque, together supplied over 550 pounds.

Like the London public who watched Lord Mayor's pageants, these dancers in court masques and the court audience watched idealized colonial relations as they helped to construct them. Bruce Avery suggests that Spenser's *View of the Present State of Ireland* sees Ireland in a British frame and so accomplishes, not without ambivalence, the same task as the efforts of British cartographers to inscribe borders on the Irish land:[18] "Such a view of Ireland, as framed territory where the English may be planted and so redefine that territory as 'English,' is essential for the policy both Eudoxus and Irenius want to pursue. After the initial cultivation, the Irish might be replanted and soon enough come to draw their identity from this redefined land, and call themselves English" (272-73). Sir Thomas Phillips, who surveyed Ulster for the crown in 1622, shared Spenser's dream, but as his discontent indicates, the dream of an English frame was more fantasy than reality in the early seventeenth century. But in the masque form, ambivalence and reality disappear in the fantasies that delighted their court audiences.

Jonson's 1613 *The Irish Masque at Court*, which James enjoyed so much that he saw it performed twice in a week, enacts the framing of the Irish as English subjects. Although the masque implicitly suggests that the Irish have not bowed easily to James, it assures James of their absolute subjection. The *Irish Masque at Court*'s antimasquers, who speak a "rude" Irish dialect, continually invoke the king, calling his name, "Yamish," six times in 144 lines and proclaiming his title over them, "king," fourteen times in the same space; they also call him "ty majesty" four times.[19] Although James is called by name and title in other masques, no other masque so stresses his name. The incessant repetition is incantatory: it has the effect of a charm in which words create a material reality. The Irish bow to James; they love James; they are James's loyal subjects. The repetition inscribes a relationship that the actual world situation, as referenced in the masque—and even the identities of the masquers themselves—calls into question. One of the antimasque characters, Dennish, identifies himself as "borne in te English payle" (54). The need for and existence of a pale, the area under English control, evinces the tenuousness of English colonial rule in Ireland by paradoxically highlighting the area out of English control. Further evidence of this tenuousness is the fact that the masque must represent these Irish as begging the king's pardon: "Be not angry vit te honesh men, for te few rebelsh & knauesh" (117-18), and the repeated assurances that these men are "ty good shubshects of Ireland, an't pleash ty mayesty" (51-52). But the masque ultimately contains and denies any hint of rebellion or dissatisfaction.

An earlier entertainment for the king was less optimistic (from an English perspective) about the situation in Ireland than was *The Irish Masque.* In 1604, one year after the collapse of Tyrone's Ulster rebellion and just nine years before *The Irish Masque,* at an entertainment for James and Anne at Sir William Cornwalleis's house, Jonson had the king and queen greeted by household gods with this verse:

> Welcome, monarch of this Isle,
>   Europes enuie, and her merror;
> Great in each part of thy stile:
>   *Englands* wish, and *Scotlands* blisse,
>   Both *France,* and *Irelands* terror.[20] (1616 Eeee2)

That entertainment acknowledges Irish resistance at the same time that it declares it ineffective; but the entertainment also suggests that the relationship between Ireland and England is violent and debasing for the Irish. In the entertainment, James is so powerful that Ireland cringes before him. In reality the crown was still conciliating Tyrone, indicating England's tenuous control over Ulster. In 1603, "after almost a century of forceful management, Ireland was controlled rather than pacified" (Moody et al. 140). *The Irish Masque* miraculously contains the resistance Jonson's 1604 entertainment acknowledges, transforming that resistance to the work of a few knaves. Ireland herself, the masque asserts, is populated almost wholly with loving Jacobean subjects.

Fifty lines after the first appearance of the word "shubshects" in the masque the word appears again, now repeated three times in three lines in a simultaneous address by four speakers to James:

> Don. Tey be honesht men.
> Pat. Ant goot men: tine owne shubshects.
> Der. Tou hasht very goot shubshects in Ireland.
> Den. A great goot many, o' great goot shubshects.
> Don. Tat loue ty mayesty heartily.
> Den. Ant vil runne t'rough fire, ant vater for tee, ouer te bog, ant te Banncke, be te graish o'got, and graish o'king.
> Der. By got tey vil fight for tee, king YAMISH, ant for my mistresh tere.
> (100-109)

These men insist that the Irish are good and that their goodness stems from their complete subjection to James and his causes. The masque's vehement insistence on this utter and grateful subjection belies as it also reveals the actual situation in Ulster and the rest of Ireland. When the escheated Ulster lands were apportioned in 1610, the granters attempted to pacify the area by granting land to particular Irishmen perceived as loyal; however, "few of the favored Irish received grants of the land which they actually occupied; none received as much as they believed themselves entitled to. They had every reason to remain resentful and unreconciled and their discontent merged with that of the majority, who had received nothing, to generate a hostility that endangered the success of the project" (Moody et al. 202). When James tried to call an Irish parliament in 1613, the year the masque was performed, his difficulties reflected Old English and Irish hostility towards his government.[21] Negating those difficulties, *The Irish Masque* frames the Irish as fully "subjected" and identified with James and England—far from fighting against the English, they will serve as English soldiers.

Not only does the masque enact the cultural fantasy of the Irish voluntarily subjecting themselves to James in fulfillment of *Prince Henry's Barriers'* promise of a conquered Ireland, it also counteracts an active English fear about the status of the English in Ireland at the time. As we saw in chapter 1, Spenser's *View* decries the "Old English in Ireland" who "have degendered from their ancient dignities and are now grown as Irish as O'Hanlan's breech." The *View* adduces the examples of Lord Breminham who has "now waxen the most savage Irish" and Mortimer "now become the most barbarous of them all" (66). Spenser argues that these men have most horrifyingly even allowed themselves to be called by Irish names. In contrast, in the masque we see Irishmen who call themselves Englishmen.

Though *The Irish Masque*'s antimasque represents its Irishmen as unproblematically James's subjects, the masque finally banishes these "rude" Irishmen because they are not quite suitable to address the king on their nation's behalf. Jonson replaces them with a "civill gentleman of the nation," and, most interestingly, given Jonson's own self-proclaimed status as poet of the English nation, by an Irish "Bard." This Irish poet, the masque tells us, has previously sung James's pacification of Ireland (the king's name appears again, now in "proper" English). The bard's prophecy, as related by the "gentleman," promised the "redemption" of Ireland, now personified as a woman in need:[22]

If she would loue his counsels as his lawes,
Her head from seruitude, her feete from fall,

Her fame from barbarisme, her state from want,
And in her all the fruits of blessing plant.

(162-65)

In contrast with the Irish bard in Spenser's *View,* who glorifies the "most licentious of life most bold and lawless in his doings, the most dangerous and desparate in all parts of disobedience and rebellious disposition," this bard sings the blessings of subjection (73).[23] We should pause over this image of an Ireland pregnant with James's blessing, not only because it represents colonial subjection as a redemption from servitude, but also because it invokes two ubiquitous colonial tropes: sexual excess and the barbarism also found in Spenser's characterization of the Irish as well as in the Jonson plays I examined in chapter 3. In early modern England, the word "fame," especially when used in relation to a woman, signified sexual reputation. Barbarism derives from a Greek word originally signifying those who did not speak Greek. In the gentleman's discourse the fallen woman who is Ireland will be cleansed by Englishness: the English will civilize her; they will make her speak English, not babble in barbarity; and the English will feed her, although it was actually the English wars that brought "want" to Irish lands.

The literal dance in *The Irish Masque* of a colonial relation enabled courtiers to see themselves—even those men who had participated in brutal campaigns in Ireland—as benevolent rulers. It provided, therefore, a justification as well as an enactment of this particular colonial relation and a model for this type of relation elsewhere. As the Irishmen in the masque will thank and serve the colonizers, so the Indians who share Irish cultural characteristics will bless the King and become grateful subjects. It is fascinating that until very recently most modern critics have seen nothing in this masque but insignificant "comic" business.[24] In contrast, the masque's contemporary audience could not ignore its implications. Chamberlain writes to Carleton: "the loftie maskers were so well liked at court the last weeke that they were appointed to performe yt again on Monday, yet theyre deuise (w*ch* was a mimicall imitation of the Irish) was not so pleasing to many, w*ch* thinke yt no time (as the case stands) to exasperat that nation by making yt ridiculous" (I, 498). Clearly the Irish still represented a danger to the English. Indeed, as Moryson reports in his 1613 journal of Ireland, the Irish had lately "growne so wanton, so incensed, and so high in the instep, as they had of late mutinously broken of a Parliament, called for the publike good and reformation of the kingdome, and from that time continued to make many clamourous complaints against the English Governours" (II, Ddd4). Although James later placated the Catholic opposition in the Irish parliament, he dismissed them at court in April 1614: "my sentence is that you have carried yourselves tumultuously, and that your proceedings have been rude, disorderly, and worthy of severe punishment" (quoted in Moody et al. 215). His sentence bears an uncanny relationship to the gentleman's response to one of the Irish antimasquer's request that James "shee anoter daunsh, ant be not veary." The gentleman responds, "He may be of your rudenesse. Hold your tongues. / And let your courser manners seeke some place, / Fit for their wildnesse. This is none, be gone" (149-52). If the masque offended the Irish and thereby exacerbated the troubles, it also trained the court in "natural" colonial dominance.

Just as *The Irish Masque at Court* educated the court about its colonial exploits in other parts of the new Atlantic world, masque representations of Wales, another of England's contiguous colonial territories, instructed the court in its colonization efforts in Ireland and America. On January 6, 1618, Jonson's masque, *Pleasure Reconciled to Virtue*, played for James's court. The king disliked the masque, complaining that there was not enough dancing.[25] Jonson wrote a new antimasque for *Pleasure*, which was performed with the substituted antimasque on February 17. Like *The Irish Masque, For the Honor of Wales*, the antimasque that took the place of *Pleasure*'s original antimasque of pygmies, provided a model for colonization outside contiguous areas.[26] While James had found the original production tedious, it is reported that he was much better pleased with the new antimasque. *For the Honor of Wales* opens with a display of a Welsh mountain, before which stand three Welsh gentlemen: Griffith, Jenkin, and Evan. Evan presents two Welsh women as substitutes for the pygmies in the earlier version: "yow tauke of their *Pigmees* too, here is a *Pigmees* of *Wales* now; set forth another *Pigmees* by him!" (201-202). The women emerge first to music speaking in Welsh; later they come on again to dance for the King after a dance of goats. Phyllis Rackin argues that in Shakespeare's *1 Henry IV*, "the country of the Others, a world of witchcraft and magic, of mysterious music, and also of unspeakable atrocity that horrifies the English imagination, Wales is defined in terms very much like those that define the woman" (1990, 170).[27] As in that play, in the masque Wales is the home of feminine alterity. *Wales* substitutes monstrous Welsh women for monstrous pygmies—both innocuous and "amusing" in Jonson's masques.

The particular relationship between England and Wales depicted in the masque starts with the title, which refers, of course, to honoring Charles, Prince of Wales, as well as to the honor of "the Welse nation." In that pun lies the heart of this particular colonial dance. When the masque played for

James, the actual conquest of Wales was to a large extent in the past, having been pursued most actively in the reign of Edward I, 1272-1307. R. R. Davies argues for that conquest as the start of "an England-centered empire; it was constructed by a king who, however international his reputation and connections, deliberately cultivated a heightened sense of English nationhood and did so in the service of his wars in Wales and Scotland." At the time of that conquest, England's tactics looked much like Jacobean methods in Ireland; they "ranged from eviction of the native population at sword-point to negotiated exchange"; and they comprised "military conquest and establishment of alien political rule" (Davies 14, 10, 12). The first English Prince of Wales was Edward I's son; Edward created him Prince in 1301 and gave him conquered Wales. Even 300 years later, the celebration of an English Prince of Wales marked that conquest. Anthony Munday's *London's Love, To the Royal Prince Henrie*, a 1610 entertainment on the Thames celebrating Henry's return from Richmond, makes that point almost directly, announcing: "For our Chronicles and Recordes doe name but eleuen, that (since the Conquest) were Princes of Wales" (B). *For the Honor of Wales* makes literal the possessive preposition in the Prince *of* Wales. The antimasque's Welshmen are angry at the poet who in the original antimasque has placed Charles "in an outlandis Mountaine; when hee is knowne, his Highnesse *has* as goodly Mountaines, and as tawll a Hills *of his own*" (58-60, my emphases). And with their literal attention to the active verb and the genitive phrase, Jenkin and Evan proceed to list Welsh mountains belonging to the English Prince.

For the Honor of Wales purports to dance the praise of a Wales that has served England well; it constructs Wales as a free state rendering obedience and tribute willingly: "though the Nation bee sayd to be unconquer'd, and most loving liberty, yet it was never mutinous (and please your Majestie;) but stout, valiant, courteous, hospitable, temperate, ingenious, capable of all good Arts, most lovingly constant" (399-403). The colonial relation enacted in *Wales,* unlike the relationship between James and Ireland danced in *The Irish Masque,* glosses over subjection. *Wales* asserts that rebellion is a thing of the past—indeed, that it did not exist, that the Welsh have always been "most lovingly constant" to the English crown. *For the Honor of Wales* displays a semblance of a peaceful international relation with a vassal nation. *Wales's* praise of a subjected nation that bows its head willingly, happily, and valiantly could also be read as a lesson to those still-resistant Irishmen implicitly displayed in *The Irish Masque.*

But like *The Irish Masque, The Honor of Wales* shows the traces of the rebellion it denies. Jenkin and Evan invoke a list of Welsh heroes in the place of the hero—Hercules—of the earlier antimasque. Their candidates include "Lluellin" and "Owen Glendower," both champions in the Welsh struggle against the English. Davies calls Gln Dwr's revolt against English rulers "a classic example of an anti-colonial rebellion" (23). Reduced in the masque to innocuous mythological Welsh heroes able to replace Hercules in these rustics' eyes, stripped of their anti-English valence, Lluellen and Gln Dwr move from Welsh national politics to English imperial myth.

If we compare the way the two masques move from those representatives of Ireland and Wales who speak in dialect to those who speak Jonson's English, we can see the substantial difference in Jonson's and the court's conceptualizations of these nations and their places in England's prospective Empire. At the transitional linguistic moment[28] in *The Irish Masque,* the Gentleman says: "He may be [weary] of your rudenesse. Hold your tongues. / And let your courser manners seeke some place, / Fit for your wildnesse. This is none, be gone" (150-52). At the transitional moment in *For the Honor of Wales,* Griffith tells his audience of the Welsh rustics' tribute to James: "Very homely done it is, I am well assur'd, if not very rudely: But it is hop'd your Madestee will not interpret the honour, merits, love, and affection of so noble a portion of your people, by the povertie of these who have so imperfectly uttered it" (384-88). The difference between the derogatory qualifiers—two nouns and an adjective in the former, and three adverbs and a noun in the latter—is the distinction between the dismissive colonial relation of the English and the Irish, and the "international" relation of the English and the Welsh, as depicted in the two masques. The Irish antimasquers embody "rudenesse" and "wildnesse," their manners are "courser," and they must be abruptly banished. The Welshmen's discourse was spoken "homely" and "rudely"; they are poor and they speak "imperfectly." The Irish are interchangeable with wildness and rudeness—England must either domesticate them or dismiss them. The Welsh can be excused and improved.

The Welsh dialect in *For the Honor of Wales* deviates less from Jonson's customary English than does the Irish dialect in *The Irish Masque.* The first line of the earlier masque reads "For chreeshes sayk, phair ish te king?" The first line of *Wales* is "Cossin, I know what belongs to this place symwhat petter then yow."[29] The Oxford editors of Jonson's masques call the language of *Irish* "Anglo-Irish jargon" (X, 542). The only relevant dictionary definition of jargon is "any talk or writing that one does not understand." Although the spelling of the first masque looks less like "standard" English than that of the second, *The Irish Masque*'s language is not "jargon." The audience is not supposed to misunderstand the language of *The Irish Masque*, rather, it is supposed to understand it in two different ways. The audience should understand its content and laugh at its form. Each word provides both a link with its signified, conveying meaning, and a slippage from its signified, meant to provoke laughter; therefore each word signifies doubly: as a "bad" appropriation of a comprehensible English word and as a joke on the speaker's ineptitude with the English language.

The language in both masques is, then, dialect: "a variety of a language that is distinguished from other varieties of the same language by features of phonology, grammar and vocabulary, and by its use by a group of speakers who are set off from others geographically or socially."[30] In both the masques the markers of dialect are almost purely phonological. Stephen Orgel attributes part of the success of *Wales* "to the crudeness of its humor—dialect jokes," he says, "were always good for a laugh" (1968 145). Orgel reads the "real" joke of this masque as its commentary on the previous masque's failure; he calls *Wales*, therefore, "a Jacobean in-group joke" and concludes that these masques "seem to be about masques themselves" (146). I would argue, in contrast, that the real joke of the masque is in the dialect, and that a further analysis of this "joke" reveals the masque's colonial ambitions.

As we have seen in chapter 3, in the opening discussion of Bryan the Irish footman in *The Second Part of the Honest Whore*, dialect is always about relations of power between language groups, its comedy "good for a laugh" only from the perspective of the dominant group for whose eyes it is written. Raymond Williams argues in *Keywords* that

> it is indeed in the stabilization of a "national" language, and then within that centralizing process of a "standard," that wholly native, authentic and long-standing variations became designated as culturally subordinate. The language, seen neutrally, exists as this body of variations. But within the process of cultural domination, what is projected is not only a selected authoritative version, from which all other variations can be judged to be inferior or actually incorrect, but also a virtually metaphysical notion of the language as existing in other than its actual variations. There is not only *standard English* and then dialects; there is also, by this projection, a singular *English* and then *dialects of English*. (105-106)

Williams describes this process as having taken place in England in the late seventeenth and the eighteenth centuries. But these early-seventeenth-century masques clearly evince the distinction between "English" and "dialects of English." Both masques end with a tribute to the king in "English" and both condemn the previous speakers for improper speech, as we have seen above.

The "comic" business in these masques is very clearly about more than the masques themselves. It is about the making of language groups and the construction of England's contiguous empire. Jonson's contemporary audience seems to have understood this. Two letter writers at the time make no mention of *Wales* as a comment on the earlier failed antimasque of *Pleasure*. They do mention, however, both the king's reaction to the Welsh men, and a possible Welsh reaction to the masque: Sir Gerard Herbert wrote to Carleton on February 22, 1618, "It was much better liked then twelueth night; by reason of the newe Conceites & ante maskes & pleasant merry speeches made to the kinge, by such as Counterfeyted welsmen, & wisht the kinges Comynge into Wales." And Nathaniel Brent wrote also to Carleton on February 21, "the princes maske was shewwd againe at Court on Tuesday night with som few additions of Goats and welshe speeches sufficient to make an English man laugh and a welsh-man cholerique." (quoted in *Works*, X, 576-77). These contemporary observers' responses suggest that this masque was not equally humorous for all potential audience members.

Eric Cheyfitz suggests that "at the farthest reaches of the frontier forged by the figure of center/periphery, the reaches of the far-fetched, the English of the lowest class (those who hold virtually no property) [or those, like the Irish and the Welsh, whose property has been taken away] can become a single foreign tongue in the perception of those in power" (99). Perhaps that is how we can read English speculation that the Native Americans they were encountering spoke some form of Welsh. Explorers could at times hear, in languages they had never heard before, languages that they knew the sound of but could not understand. They could plan to make vassals of those whose discourse resembled that of an already familiar vassal nation. The colonial relation at work in *For the Honor of Wales* was present for a short time in colonial Virginia as Captain John Smith's early narratives, which postulate relations with the "king" Powhatan and the "princess" Pocahontas, suggest. It did not take long for that model of international relations to become a vision of subordination that more resembled what we have seen in *The Irish Masque*. These court masques, which Jonson published in his 1616 *Works*, enabled masquers and court audiences to feel a natural superiority over both the "rude" and "wild" Irish and the "homely" and "imperfect" Welsh, who possessed a natural inferiority that could be transferred to other subordinated populations.

As well as ridiculing Irish and Welsh antimasquers, masques, like pageants, displayed Indian and African subjects in the service of installing a dominant English whiteness. When masquers danced their measures in these entertainments, they became symbols of whiteness in contrast to denigrated antimasque nativeness. Like English sonnet sequences (Kim Hall), and like Lord Mayor's pageants, Jacobean and Caroline masques see whiteness as allied with goodness, blackness with evil and error; and as in those other texts, these masques repeatedly map that abstract Manichaean color scheme onto skin color. That mapping is remarkably explicit in Middleton's 1619 entertainment by the gentlemen of the inner temple for "many worthy ladies"; *The Inner Temple Masque or Masque of Heros* displays a spectacle of embodied days: "The three *Good Dayes,* attyred all in white garments, fitting close to their bodies" (B4v); "The three *Bad Dayes* all in black Garments, their Faces blacke . . . [and] *The Indifferent Dayes.* In Garments halfe white, halfe blacke, their Faces seamed with that party Colour" (C). Later in the masque, Harmony sings to the masquers:

> Moue on, Moue, be still the same,
> You Beauteous Sonnes of Brightnesse,
> You adde to Honour Spirit and Flame,
> To Vertue, Grace, and Whitenesse. (C2v)

Depending on how one reads the last two lines of this stanza, as the masquers dance, they either add their spirit and flame to an abstract honor and their grace and whiteness to virtue, or their dance creates additional honor, spirit, flame, virtue, grace, and whiteness. Either reading allies whiteness with virtue and grace, and gives the bright masquers the ability to embody and create its essential goodness.

In this formative period for racial difference, in this foundational period of empire, we see the crucible of whiteness, that category that will become so powerful that by the twentieth century it is unspoken (Dyer). Unlike late-twentieth-century representations that constantly evoke whiteness' power without naming it, that see it as normative, Jacobean and Caroline masques and entertainments continually name the category in their displays of colonial others, explicitly installing whiteness as normative by contrasting it to native strangeness. Hall brilliantly demonstrates this process working in Jonson's first two masques commissioned by Anne, *The Masque of Blackness* and *The Masque of Beauty* (128–141).[31] As she argues, "it is no accident that this first court masque is both an elucidation of the nature of blackness and a celebration of empire" (1995, 133). That masque, as she notes, stresses England's name, Albion, and so equates England with snowy whiteness.

Masques made England equivalent to whiteness; they also equated native otherness with darkness. Inigo Jones and George Chapman's *The Memorable Maske,* which celebrated James and Anne's daughter Elizabeth's marriage, opened with a march through London and culminated at court on February 15, 1613. In the march, two triumphal cars carry six musicians "attir'd like Virginean Priests . . . [with] strange Hoods of feathers . . . and on their heads turbants, stucke with seuerall colour'd feathers, spotted with wings of Flies, of extraordinary bignesse; like those of their countrie" (A4v). As in the city pageants, "Indians" are associated with exotic fauna, and decked with feathers:

> Then rode the chiefe Maskers, in Indian habits . . . richly embroidered with golden Sunnes, & about euery Sunne ran a traile of gold, imitating Indian worke . . . betwixt euery pane of embroidery, went a rowe of white Estridge feathers . . . and about their neckes Ruffes of feathers, spangled with pearle and siluer. On their heads high sprig'd feathers, compast in Coronets, like the Virginian Princes they presented. (B)

The text calls them "altogether estrangefull, and *Indian* like" (Bv). Unlike the masquers in *Blacknesse* who wore blackface, these masquers wear the more customary masks, but they are distinctly dark: "Their vizards of oliue collour; but pleasingly visag'd: their hayre, blacke & large" (Bv). They ride on horses, each "sided" by "two Moores, attir'd like *Indian* slaues" (A2v). In the spectacle at court the maskers are "triumphantly seated" in "a rich and refulgent Mine of golde" (C).

The masque postulates an undifferentiated Indianness: the masquers are called Virginian Princes, but their gold embroidery and ostrich feathers connect them with the East Indies, and the riches of the gold mine in the court spectacle link them to both the Spanish South American colonies and the East. In the masque's fiction they have traveled to England, the unchanging center of the world, on a "rich Island lying in the South-sea, called *Poeana*" (D2), which is "yet in command of the Virginian continent" (D2v). Their Indianness is equated with strangeness: to be Indian is to be a stranger, to be estranged. And it is also visually linked with darkness. The "Virginian Princes" have olive skin and long black hair, and they ride with Moores. The black slaves that attend the masquers work "as figures for the exotic or foreign" (Hall 227). But unlike the black slaves in English aristocrats' portraits, these figures associate the masquers with darkness rather than emphasizing a masterful whiteness. The masque postulates a hierarchy of difference—"Virginian Priests" and "Moores, . . . like *Indian* slaues"— but the hierarchy works within a frame in which all these Indian and African populations are subordinated to what is being established as the normative whiteness of James and the English court.

At the entertainment's climax, Honor's priest Eunomia asks the princes to subject themselves to James, England's sun:

> Virginian Princes, ye must now renounce
> Your superstitious worship of these Sunnes,
> Subiect to cloudy darknings and descents,
> And of your sweet deuotions, turne the euents
> To this our Britan *Pheobus*, whose bright skie
> (Enlightend with a Christian Piety)
> Is neuer subiect to black Errors night,
> And hath already offer'd heauens true light,
> To your darke Region; which acknowledge now;
> Descend, and to him all your homage vow. (D4v)

This speech repeatedly links its Indians with tropes of darkness, and James, and by extension England, with pure light: the Indians worship a sun "subiect to cloudy darknings" and "black Errors night"; they live in a "darke Region." In contrast, James is a sun that will never darken because his light derives from "Christian Piety"; only he can carry the pure light of heaven to Virginia. The masquers descend from their rock and dance out their enlightenment. The masque continues its paean to whiteness in the song of love and beauty dedicated to Elizabeth and her bridegroom. In the song, Love and Bewty were born to the couple Panthea and Eros:

> Bright *Panthea* borne to *Pan*,
> Of the Noblest Race of Man,
> Her white hand to *Eros* giuing,
> With a kisse, ion'd Heauen to Earth
> And begot so faire a birth,
> As yet neuer grac't the liuing. (F4v)

Here the discourse of white female beauty joins the celebration of England's enlightened state to postulate a whiteness essential to "the golden world" the masque creates (G).

As Jerzy Limon notes, the march through London of all of the masque's participants, including the Virginian Priests, and the Chief Maskers attired like "Indians," was an integral piece of this entertainment (142-43). This aspect of the masque along with its display of Indian gold demonstrates its affinity with the Lord Mayor's pageants I have already discussed. Also as in those entertainments, this masque celebrates the union of profit and honor. In the argument that Chapman attaches to the masque's text, he explains that the masque as well as celebrating the royal marriage, celebrates the redemption of the god Plutus, who has been represented by classical authors as "naturally blind, deformd, and dull witted [but] is here by his loue of Honor, made see, made sightly, made ingenious; made liberall" (a4).[32] Like Lord Mayor's pageants, the masque celebrates the pursuit of unlimited colonial wealth as an honorable endeavor. Chapman and Jones' *Memorable Maske* produced a whitened city as well as a whitened court.

Like *Eastward Hoe*'s Virginia, as D. J. Gordon notes, this masque's notion of a lavish, gold-producing Virginia is anachronistic.[33] As in *Eastward Hoe* also, that notion is conditioned by Raleigh's Guianan dreams, and the masque may well have been intended to promote those dreams. Of course, those dreams were elusive and finally fatal to Raleigh. But the masque's installation of a whiteness defined against Indian, Native American, and African darkness was prescient and has continued to haunt America's, England's, and Europe's collective imaginations.[34]

The *Memorable Maske* was one of the first of a genre of masques that featured antiself Indians. It played about a month before *The Irish Masque at Court*, which appeared on December 29 and January 3, 1613 (old dating) as part of a series of entertainments celebrating the Earl of Somerset's marriage to the Earl of Suffolk's soon-to-be-infamous daughter Frances Howard. Suffolk, England's Lord Chamberlain in 1613, was a prominent Virginia adventurer. Three days after *The Irish Masque*, on January 6, the gentlemen of Gray's Inn presented *The Maske of Flowers* at court in honor of that marriage. *Flowers*'s premise is a commission by the Sun to Winter and Spring, who are required to present two entertainments for the marriage of "two noble Persons, in the principall Island of our vniuersall Empire" (B2). Winter must present a challenge between Silenus (wine) and Kawasha (tobacco) to determine their relative worthiness. Given James's distaste for tobacco, that vice is predictably caricatured. Kawasha appears "riding vpon a Kowle-staffe . . . borne vpon two *Indians* shoulders attired like *Floridans*." He wears a nightcap topped with a chimney, with holes through which his ears stick out "hung with two great Pendants . . . his body and legges of Oliue-colour stuffe, made close like the skinne, bases of

Tobaceo-colour stuffe cut like Tobacco leaues, sprinkled with orcedure,[35] in his hand an *Indian* Bow and Arrowes" (B3). Kawasha is a completely undifferentiated composite of various American Indian cultures: borne by "Floridans," Silenus calls him "this great Potan," probably a reference to Powhatan, the Virginia Indian ruler. Kawasha is almost certainly a version of the Virginian idol Kiwasa described by De Bry in the sixteenth century, a description reprinted in the early-seventeenth-century *Purchas his Pilgrimage* (*A Book of Masques* 154n.1).

In an outrageous displacement, the masque asserts that North American Indians caused the English demand for tobacco. Native Virginia tobacco (*Nicotania rustica*) was too bitter for English tastes developed on South American tobacco, and the crop did not succeed in England until John Rolfe introduced *Nicotania tabacum* from the Spanish colonies. But the masque accuses the Indians, whose land the English were seizing in order to grow their imported commodity, of causing English addiction. Silenus sings, "He is come from a farre Countrey, / To make our Noses a chiminey" (B4), and a later verse, possibly sung by the chorus, reads "*Kawasha* and his Nation / Found out these holy rites" (B4v). The masque blames the Virginia Indians for inducing English smoking, a practice that in reality was sponsored by English merchants looking for a fortune. *Flowers* associates Kawasha both with animals, exotic foreigners, and city denizens. Silenus proposes to return Kawasha "to his Munkeis, / From whence he came" (B3v). Silenus and Kawasha each present an antimasque dance, Kawasha's including a "Bawd," a "Roaring Boy," a "Citizen," a "Mountebancke," and a "Iewesse of Portugall" (C). Unlike the Lord Mayor's pageants, but very like the plays we looked at in chapter 3, *Flowers* associates a vice it names as colonial with the excesses of the city, its bawds and roaring gallants. However, the masque blames Virginia Indians for city excess.

Perhaps because of Suffolk's involvement with Virginia, the entertainments for Lady Howard and the Suffolk-Somerset alliance constantly refer to England's colonies. Thomas Campion's 1614 masque for the alliance, danced by many prominent Virginia adventurers, displays "*America* in a skin coate of the colour of the juyce of Mulberries, on her head large brims of many coloured feathers, and in the midst of it a small Crowne" (B). The feathers identify this figure as an undifferentiated Indian; her coat is colored with another potential Virginia commodity, the mulberry tree. The Virginia Company's propaganda constantly touted the presence of large numbers of mulberry trees in the colony. For years, reports of these trees fed the dream of an English silk industry, a dream King James particularly cherished. Campion's masque, which represents each of the "four parts of the earth" as women, makes Europe "an Empresse, with an Emperiall Crowne on her head," Africa "a Queene of the Moores, with a crown," and puts Asia in "a Persian Ladies habit with a Crowne on her head" (B).[36] America's small crown thus makes her the most dominated "part" of a world subordinated to Europe, and her costume associates her most clearly with natural resources. She is displayed as the site of European investment and colonial dominance. This masque separated Asian people from American Indians but it collapsed all North and South American Indians into one mulberry-colored, feathered Indian.[37]

The undifferentiated Indian appeared again in a number of masques for the Caroline court of the 1630s. Although by this time a considerable amount of additional information about the Virginia Indians had been disseminated in England, this image still clearly served the court's imperial imagination of itself and England. Inigo Jones and Aurelian Townshend's 1631 *Tempe Restord*, like earlier masques and Spenser's *Faerie Queene*, utilizes the Circe story for its transformation theme.[38] *Tempe Restored* has a particularly dehumanizing representation in which Circe watches antimasques consisting of "Indians and barbarians, who naturally are bestiall." The dances include one of "Indians adoring their Pagole" dancing with an "Idoll," a "Hare," "Apes," "Barbarians" and other animals (B). The word "pagole" doesn't appear in the *OED*, although "pagod" does with variants "pagode" and "pagothe." Travel accounts from the East Indies and China use the word "pagode" to designate an idol (*OED*), and it seems like a reasonable (if offensive) conjecture for "pagole" in this masque. Inigo Jones and William Davenant coproduced another Caroline court masque in 1634 called *The Temple of Love*. The masque's temple is ornamented with "Indian trophees: on the one side upon a basement sate a naked Indian on a whitish Elephant, his legges shortning towards the necke of the beast, his tire and bases of severall coloured feathers, representing the Indian monarchy: On the other side an Asiatique in the habit of an Indian borderer, riding on a Camell; his Turbant and Coat differing from that of the Turkes, figured for the Asian monarchy" (A3). Although the white elephant visually associates this naked Indian with an East Indian monarchy, his nakedness and his costume's feathers smack of the undifferentiated American Indian. The masque makes no mention of skin color in this spectacle, which may indicate that it was unremarkable. However, it is distinctly possible that the elephant's whiteness was intended as a visual contrast with the Indian's darkness. These masques not only present undifferentiated Indians, they also claim that there is one "Asiatique." In Davenant's 1635 Caroline masque, *The Triumphs of the Prince D'Amour,* the curtain opens on "a Village consisting of *Alehouses* and *Tobacco shops,* each fronted with a red Lettice, on which blacke Indian Boyes sate bestriding Roles of Tobacco" (B). Once again English vices are signified by Indians whose difference equals blackness. These various representations of the 1630s show the range covered by the undifferentiated Indian. While some court masques distinguished between people from India and people from America, none represented the variety of cultures in the East or the West; none distinguished between the many groups populating the North and South American continents, and others posited a completely undifferentiated Indian associated with barbaric darkness.

Aurelian Townshend and Inigo Jones's 1631 court masque *Albion's Triumph* culminates in a speech by Peace that reveals the dream behind all these displays of undifferentiated Indians. The printed masque opens by explaining that the spectacle will celebrate the king's union with a Queen, ALBA, "whose native Beauties have a great affinity with all Purity and Whitenesse" (A2). Townshend commends the name Albion's propriety, since it signifies England, and England's queen epitomizes whiteness. The masque ends with Peace's address to the gods Neptune, Plutus, Bellona, and Cebele. Peace charges them:

> Neptune to Sea, and let no Sayle,
> Meete ALBIONS Fleete, But make it veile.
> *Bellona* Arme, That Foes may see,
> Their Lillies kept by Lyons be.
> Their fruitfull fields (*Cebele*) make
> Pay Centuple for all they take.
> And let Both Indies (*Pluto*) meete,
> And lay their wealth at ALBA's feete.
> (C4)

These masques dream of undifferentiated Indian wealth freely given as an offering to a land whose whiteness shines in contrast to their native darkness. England is a white lily guarded by fierce lions; the Indies will bow before England and lay their treasure at its feet. Although England's rulers in this masque worship peace, it is a peace sponsored by Indian wealth and backed up by English naval might. Jacobean and Caroline court masques imagined an empire spanning the globe, its subjects worshipping the white beauty these entertainments installed.

So far this chapter has dealt with pageants and masques that in one way or another directly address England's colonial enterprises. But even in masques that do not ostensibly take empire or native antiselves as their subjects, we can see the form's preoccupation with the transforming new Atlantic world. When the 1620 court masque *News From the New World Discovered in the Moon* was published posthumously in Jonson's second folio, the first five words of the title were displayed in larger type than the last four.[39] THE NEW WORLD towers above the moone. At first glance, the reader might believe, then as now, when reading Stephen Orgel's edition of the *Complete Masques,* that she was reading news from the "New World"—a display of a world and its creatures found not in the heavens but on the earth across the ocean. Perhaps she would not have been mistaken. We can read the page headings as an act of textual interpretation, the text interpreting itself—or rather, the exigencies of printing and typesetting interpreting the text. For the masque plays out one English fantasy of good news from a new world. In the masque a poet—recognizably Jonson—carries a report to the court of the inhabitants of a world in the moon. He brings his report by means of poetry, which he calls "the mistress of discovery."

The features of that world provide us with a sketch of a possible English idea of another world, a world in some ways like that of England and in some ways radically other. The antimasque world as an example of speculative fiction is thus very like many explorers' reports. Hakluyt's collections contain numerous reports of lands that resemble England in their government and social hierarchies and yet contain strange wonders and monsters. The English "found" monarchies and courts wherever they traveled; they also "found" cannibals and amazons, witches and devil worship. One could argue, of course, that the masque's creation is less constrained by an external reality, but one would have a difficult time making the case for the relations of explorers and settlers as reporters of an empirical reality, especially for English reports of the early seventeenth century. We find instead much the same degree of mixture of the familiar and the monstrous as we find in the antimasque of *News From the New World Discovered in the Moon.*

In the case of this antimasque, the familiar is political and social organization. The moon is "found to be an Earth inhabited!"

1st Herald. With navigable Seas and Rivers!
2nd Herald. Varietie of Nations, Polities, Lawes!

1st Herald. With Havens in't, Castles and Port-Townes!
2nd Herald. In-land Cities, Boroughes, Hamlets, Faires and Markets!
1st Herald. Hundreds, and Weapontakes! Forrests, Parks, Coney-ground, Meadow-pasture, what not?
2nd Herald. But differing from ours.
(128-36)

This is a comprehensible world, in macroscopic structure much like Europe, its smaller subdivisions English; a world both accessible and "navigable," with "Ladies," "Knights, and Squires," "servants, and Coaches."[40] As in Jonson's plays, the "New World" of the masque, like the "New Worlds" reported by travelers to curious English and Europeans, provides an opportunity to critique the existing social structure. The masque digs at the expanding news industry—a critique that Jonson took up and elaborated in *The Staple of News*. For its audience's joy, the masque postulates a world in which lawyers have lost their voices, a world devoid of tailors and the self-love engendered and nourished by the exigencies of fashion.[41]

However, although English travelers would likely recognize their surroundings and the social hierarchy on this moon, the moon also has "phantasticall creatures" to delight those travelers and the masque audience. Its inhabitants speak "onely by signes and gestures" (a fantasy of a people without a language); they travel in clouds, and the world has a curious island, "the Isle of the Epicoenes." This island becomes the antimasque's focal point, its creatures providing the antimasque dance and display. Its inhabitants, the Epicoenes, are creatures of indeterminate gender: "under one Article both kindes are signified." They cannot be distinguished by dress—the main marker of hierarchy in English society at the time. The Epicoenes seem to be the overdetermined locus for much cultural contestation at the time. In one condensed figure, these monsters represent the struggle over the boy actor, the active debate over sex roles,[42] as well as the inability of the English to classify the men and women encountered across the Atlantic in traditional gender roles. English explorers often found the women of the Americas doing "men's" work, and they also found Indian dress codes radically unfamiliar. Captain John Smith tells his readers that the Powhatan "women and children doe the rest of the worke. They make mats, baskets, pots, morters, pound their corne, make their bread, prepare their victuals, plant their corne, gather their corne, beare all kind of burdens, and such like" (II, 116). Unable to read the protocols of Powhatan dress, Smith sees all women who wear jewelry and sit in proximity to the "king" as concubines. All of these difficulties for the English—the controversy over women, the boy actor, and the reading of overseas gender—cannot be separated; Jonson locates them safely in the role of the masque's monsters. The Epicoenes reproduce by laying eggs from which come the Volatees: "a race of Creatures like men, but are indeed a sort of Fowle, in part covered with feathers . . . that hop from Island to Island" (286-89). As we have seen above, feathers often mark masque players as some sort of "Indian."[43]

However, *News From the New World* does not represent any particular group; instead it offers the opportunity for free play in the colonial field—the reported monsters do not have to exist; the English discover only their own shiny beautiful Englishmen. The masque after the antimasque of the dancing Volatees reveals these creatures as having had no reality; they were only the poet's inventions created to delight the King: "hitherto we have moved rather to your delight than your beliefe." Led by Prince Charles as the figure of Truth, the courtiers descend from a height above the moon to replace the created monsters, the Volatees, and to praise their ruler. A herald tells the King that the Volatees were fictitious news; he should now "expect a more noble discovery worthie of your eare": the courtiers and their praise. As the poet's constructions with no separate ontological status, the Volatees resemble the human specimens brought over from voyages and displayed to the court and the public—people like the woman the English called Pocahontas, who will be a primary subject of chapter 5.

That is, Pocahontas, who was displayed in the audience of another Jonson masque, *The Vision of Delight*, was certainly a real human being; but her persona as presented to the court was wholly a creation of the Virginia Company's strategists. Named, clothed, and baptized by her handlers, she was dubbed the Indian princess despite that fact that inheritance in her culture was matrilineal. As Kirkpatrick Sale points out in *The Conquest of Paradise*, she would never have inherited what the English called (also in error) the "Powhatan Kingdom" (279). She was introduced at court as Lady Rebecca, and her portrait was engraved in full English costume. And according to Peter Hulme, when an oil painting was made from that engraving, she even began "to lose her AmerIndian features" (169). The Englishmen who saw her in London called her "the Virginian woman." The records of the parish church of St. George, Gravesend name her "Rebecca Wrothe" "a Virginia Lady." That record is wholly an English creation; for her name is now English, her last name her husband's, and her designated origin the land the English named after their queen. The record provides a fitting testament to the erasure of the Indian identity of the woman kidnapped by Samuel Argall in the English colony and to her integration into English cultural expectations.

In 1620, *News From the New World* played coterminously, and perhaps not coincidentally, with actual attempts to make the Indians on Virginia's coast disappear. By the 1620s, the Virginia Company had developed an insatiable need for land to plant tobacco and to pay off the original joint-stock settlers. In 1619, eleven ships landed in Virginia carrying 1,216 people. But the English in Virginia did not accomplish genocide as easily as Jonson's masque accomplishes the erasure of the Volatees. The Native Americans refused either to conform to a peaceful colonial relation of masters and slaves or to cede their land when they were no longer necessary to the settlement. In fact, these Algonquian Indians fiercely resisted being replaced by a constituency that would worship James. From the 1606 landing of 144 colonists on, these Indians attempted unsuccessfully to discourage the English settlement.

As the Introduction discusses, the 1622 "massacre" of the English settlers was said by Smith to have happened largely because the English had not separated themselves properly from the Indians. Integration with the native population meant that the English settlements could not be fortified against Indian attack. Smith's account of that rebellion laments that the English "bettered . . . and safely sheltered and defended" "the Salvages" and left "their houses genereally open" to them. The Indians were "friendly fed at their tables, and lodged in their bed-chambers, which made their way plaine to effect their intents" (II 293-94). The "massacre," then, in the English documents surrounding it, largely resulted from the Indians having mixed with the English. Actually, as I suggest in my introduction, the English colonization of Virginia was threatened from the beginning by the appeal that the Indian lifestyle held for colonists. *English* mixture with and attrition to the Indians was anathema to the project of subduing, subjecting, and ultimately destroying these people. Therefore the English documents gloss over that threat and even read it in reverse: they want to be like the English, but English people do not want to be like them.[44]

Jonson's most popular masque, the 1621 *Gypsies Metamorphos'd*,[45] can be interpreted as a programmatic response to this threat. When critics have read this masque in other than formal terms, they have placed it in the intricacies of Jacobean court politics, specifically as a gloss on James's relationship with his favorite, George Villiers, the Marquess of Buckingham.[46] In addition, Dale B. J. Randall's book on the masque, *Jonson's Gypsies Unmasked,* provides a good overview of the social position of gypsies and of English attitudes towards gypsies in the early seventeenth century. Randall's book reads the masque as an allegory in which Jonson warned James of the gypsylike character of Buckingham and his family. Martin Butler's brilliant article on the masque contests Randall's reading, arguing instead that Jonson's masque provides a forum in which to cleanse Buckingham; Butler also reads the masque in relation to Jonson's position on contemporary continental politics. Since England's domestic and continental politics were intimately related to its colonial ambitions, the masque is also a vehicle for colonial desires. In addition, this masque plays out a particular colonial logic in which antiselves are incorporated and tamed. A complex extended show, the masque contains flattering and advisory palm readings of James, Charles, and various courtiers; the ballad of Cock-lorell that became Jonson's most published piece of poetry; and numerous dances and songs; it was Jonson's longest and most frequently performed masque.[47] The conceit of the courtiers as gypsies frames these various entertainments. Besides providing a vehicle for exploring James's relations with his favorites, that conceit enables the masque to play with the boundaries of English identity. The masque shows first the appeal of a wildness, visually denoted by a different skin color, and then the incorporation of that wildness into Englishness. This colonial problematic remains a feature of late-twentieth-century colonial relations as American sports teams gain power from association with the "wildness" of "Braves" and "Redskins."[48]

Like other court masques, *The Gypsies Metamorphos'd* assumes that England is an imperial power, and it also maps goodness and purity onto whiteness and links dark skin with evil and danger. In addition, because the masque played at a time when James was considering an alliance with Spain, *Gypsies* alludes to Spain's colonial empire. After the gypsies' Captain tells James's fortune, another gypsy reads his son Charles's palm. This gypsy lauds Charles's prospective marriage to the Spanish infanta:

> Shee is sister of a Starre,
> One the noblest nowe that are,
>   Bright *Hesper*,
> Whome the *Indians* in the East
> *Phosphore* call, and in the west
> Hight vesper. (369-74)

According to the gypsy, both east and west Indians see Philip III as the brightest star, the morning star of the east and the evening star of the west. Again, as in other masques and pageants, Indians are undifferentiated worshippers, but here they worship Spain, England's traditional enemy. This masque illustrates the dynamic I discussed in chapter 3, in which the Spanish are at once competitors but also models for colonial exploits. In *The*

Gypsies Metamorphos'd, the Spanish threat, which was seen as so powerful in the Protestant colony of Bermuda (see chapter 2) and as a threat to English domination in Ireland, becomes neutralized by the prospect of James's son's marriage to the infanta. Once again, we can see that colonial references in English literature reveal different English agendas as well as similar English imperial ambitions.

In the gypsy's reading of Charles's palm, the alliance with the infanta and Charles's own prospective kingship promise English dominance. The gypsy reads a future in which Charles, "by being long the ayde / Of the *Empire*, [will] make afraide / ill neighbours" (390-92). This claim in 1621 is reminiscent of Jonson's suggestion in the 1604 entertainment at Cornwalleis's house that James is "Both *France*, and *Irelands* terror." In *Gypsies*, England is declared an imperial power which has the capacity to terrify any neighboring nation that might be ill-disposed to its empire. Presumably, the alliance with Spain will only increase England's own imperial might.

In *Gypsies*, as in so many other court and city entertainments, that imperial might, which is seen as England's true identity, is connected with white skin. However, *The Gypsies Metamorphos'd* momentarily maps darkness onto Englishness, if only as a way to contain the threat of Englishmen "going native." The masque is laced with references to the gypsies' dark skin. When the foolish Jack Cockrell admires the gypsies in the antimasque section, he calls them "the finest oliue-coloured sprites" (735). Likewise, in the beginning section of the masque, the principal gypsy, the Iackman, played at Burley by Buckingham, declares that the gypsies have knacks "that will delight you, / Slightes of hand that will invite you / To indure our tawney[49] faces, / And not cause you *cut your laces*" (133-36). The gypsies' tricks and dexterity will overcome the shock of their darkness, a darkness that might itself otherwise overcome the court's ladies (the cutting of the laces that bound women's undergarments was a customary way to revive a lady who was faint or had fainted). When the gypsies return to the stage to mingle with the rustic foolish men of the antimasque, the gypsy Patrico tells them, "Although wee looke tawnie, / Wee are healthie and brawnie" (820-21). Once again, the gypsies' darkness is marked in the masque and seen as a potentially frightening or upsetting feature. When Patrico invites the gypsy performer of the ballad of Cock-Lorell to sing his song, he says, "Come in, my long sharke, / With thy face browne[50] and darke" (1054-55). Patrico's rhyme links the gypsy's darkness to his criminal behavior. While *The Gypsies Metamorphos'd* does not speak directly about English whiteness, its continual references to the gypsies' dark skin, and its implications that that darkness will frighten both court ladies and country bumpkins, create a general English whiteness by contrast.

Of course the masque complicates this picture in exactly the ways that so many texts already examined in this book complicate the picture of the unified English colonist. Because the people playing these dark gypsies are courtiers, and their tricks and antics are so appealing, the masque presents dark wildness as, at least momentarily, very attractive. As we have seen in discussions of Henry Spelman in this book's introduction, and of Spenser and other Irish colonist's fears about colonial transformation in chapter 1 and chapter 2, the attractiveness of native life and native populations was one of the key problems facing England's imperial efforts. Randall's book on the masque shows that gypsies in England and Europe provoked the same worry from authorities. Randall notes that there was great English fear of, as well as documented cases of, Englishmen "becoming" gypsies.

Jonson's masque handles that threat by incorporating gypsy wildness into whitened English identity and producing that wildness as theater. In the masque, the gypsies appear; they read the court's palms; the rustics come on stage; the gypsies read their palms and pick their pockets. When the rustics complain, their belongings are restored, and they admire the gypsies so for this "magic," that one of them, Puppy, wishes to become a gypsy. The gypsy Patrico explains the trick of being a gypsy—one must be lazy, acquire a talent for theft, and one must darken one's complexion. He explicates the wonders that a gypsy can perform, and as he lists those wonders, he produces the king and the prince for the admiration of the rustics. Then the gypsies are discovered as courtiers; Jonson's stage direction reads only "the gypsies changed." The masque shows, finally, as its title indicates, not Englishmen "going gypsy" but gypsies "going English." Like English colonial documents, the masque depicts those who would desire to desert to the gypsies as base men. In the transformation of the gypsies into courtiers, then, the masque shows not only that attractive wildness is only theater produced by talented courtiers; it also shows that what base men really admire most is the glory of the court.

Again as in Jonson's earlier masques, the alien and the base are naturalized and thereby neutralized, but the naturalization is a different process, a different colonial relation. Most fascinating is both the amount of mixing between the gypsies and the English that occurs in the masque before the metamorphosis—which Randall points to and says cannot be erased—and the clear effort at the separation of the two groups after the "change." The rustic characters find the gypsies and their life intensely appealing. Puppy declares: "I haue a terrible grudging now vpon mee to be one of your Companie. . . . I would binde my selfe to him, bodie & soule" (1141–44). And Cockrell seconds him: "wee'll be all his followers" (1157–58). At the moment the masque represents this desire it initiates the "change," and the gypsies' greatest wonder becomes not their lawless life but the English court. The gypsies are not only transformed to courtiers, the mechanisms of their transformations both into gypsies and back into courtiers are exposed. This emphasis marks the crucial difference between the movement toward different forms of national relations in *The Irish Masque* and *Wales* and the naturalization of the gypsies in the later masque. In *Gypsies,* the court audience saw not a subjected nation, not a separate culture, but a wildness that could be tamed, harnessed, and though not completely erased, made completely English.

At the end of the masque Patrio declares that the "truth" that he told before the metamorphoses will out:

> The *Gipsies* were here,
> Like Lords to appeare,
> With such theire Attenders
> As you thought offenders,
> Who nowe become newe men,
> You'll knowe 'em for true men
> . . . .
> Eache Clowne here in sight,
> Before day light
> Shall proue a good knight;
> And your lasses, pages
> Worthy theire wages.
> (1263-69, 1314-18)

The masque exposes the mechanisms of the theater including the boy actors—the pages who have played lasses—and the arts of costume and makeup. Even before the change, Patrico tells Cockerell that if he wants to be a gypsy he can act like one, and he can "change [his] Complexion / With the noble Confection / Of wallnuttes and hoggs geace" (1212-14). Not only is the gypsies' darkness exposed as makeup, that makeup is also broken down into its homely components.

The epilogue that appears in all but one extant copy of the masque reinforces the message that gypsy wildness is a product of English theater, and it especially stresses the English production of the makeup that darkened the courtiers' skin:

> But least it proue like wonder to the sight,
> To see a *Gipsie* (as an *Æthiop*) white,
> Knowe, that what dide our faces was an oyntmen[t]
> Made and laid on by Mr *woolfs* appointment,
> The Courtes *Lycanthropos:* yet without spelles,
> By a meere Barbor, and no magicke elles,
> It was fetcht of with water and a ball;
> And, to our transformation, this was all
> . . . .
> The power of *poesie* can neuer faile her,
> Assisted by a *Barbor* and a *Taylor.* (1479-92).

This epilogue asserts that real gypsies, and by implication other real dark people, can never be whitened, for their darkness is as indelible as that of the proverbial Ethiope. These gypsies, however, were only gypsies by virtue of their makeup, and the removal of that makeup required no magical transformation, only English water and soap.[51] There are no gypsies and no rustics, only theater, which can make wildness appear and disappear. By the time of the masque, the gypsies were already legislated out of England.[52] On England's newest colonial front, where the gypsylike wildness and appeal of the Native American population was still a threat, native subjects would be murdered, only to be reenacted in the imaginations, the games, the theater, and the sporting events of the colonists.

The transformation from wildness into Englishness that *Gypsies* enacts should already be familiar from the Virginia Company documents and the plays that I examined in chapter 3. In masques, that transformation was actively celebrated and danced by the court and courtiers who would make decisions that would advance England's colonial efforts in Virginia. In the Jacobean masque prominent patrons of England's North American expansion danced their new identities as colonists. Orgel's seminal book on Jacobean court masques, *The Illusion of Power,* argues finally that Jonson's masques created visions of their power for the king and court. However, Orgel suggests, the English revolution destroyed those illusions, when the English people refused to accede to the dream of absolute monarchy. The imagined world in which James I and then Charles I ruled as absolute monarchs collapsed, and the masque form ultimately disappeared. But if the masque fashioned the king and court as colonizers, its power did not fail. The dreams it shared with its less exalted city counterparts continued to motivate English colonial efforts. And if Charles on the scaffold is a sign for Orgel of the failure of *one* of the masque's illusions of power—the Stuarts' images of themselves as absolute monarchs—colonial Virginia, the ever problematic Northern Ireland, colonial India, and even present day America are signs of the masque's success.

Chapter 5

"A Virginia Maske"

I n *The Staple of News,* Ben Jonson's character, Peni-boy Canter, refers to "the historian" Captain John Smith for his information about "Pokahontas." Smith was the early Virginia colonist who produced the most copious texts about the colony, including the *Generall Historie of Virginia, New England and the Sommer Islands,* divided into six books. He is particularly noted for his "ethnographic" material about the Powhatan Indians, material that, as the epilogue will show, remains authoritative to this day. In his masque, Jonson designed his displays of "New World" feathered creatures, the Volatees, to please and flatter the court; this chapter will show that following his lead, Smith also displayed singing and dancing Indians in his texts to promote his own interests with the Virginia Company and the court. Just as Jonson looked to Smith's texts for information about Virginia, Smith used Jonson's printed masques to understand the desires of the court, an audience Smith clearly wanted to reach. In this chapter, I look closely at a specific incident of "singing and dancing" Indians in Smith's text. The 1623 broadside prospectus for Captain John Smith's *Generall Historie of Virginia* advertises that in book III one can read "how Pokahontas entertained him with a maske." In the 1624 *Generall Historie,* Smith rewrote an account of a Powhatan Indian women's dance, which had originally appeared in print in 1612, as a formal "Virginia Maske." Seven years after Ben Jonson published his masques in his 1616 folio *The Workes of Beniamin Jonson,* Smith's prospectus invited the English reading public to witness in print a "Maske" by the "salvage" inhabitants of the newly explored and settled territory of Virginia.

The "Virginia Maske" is a short piece of text that crystallizes two processes of colonial transformation in the new Atlantic world. As we have seen in the discussion of Bermuda in chapter 2, the American colonies became a site for English dreamers who saw colonization as a chance to transform their social status and material conditions. Smith's self-assertion, seen brilliantly in the "Maske," is a bid for that type of colonial transformation. Smith's folio *Generall Historie*, like Jonson's folio *Workes*, was central to his project of establishing his authority. When Smith wrote his "maske," he created himself as author just as Jonson wrote himself as author in his textualizations of the masques. Ben Jonson's *Workes* served as a model for Smith's assertion of authority in a printed masque; the masque form enabled him to assert that authority in relation to his king as well as to the Indians. In his "Virginia Maske," Smith transforms himself, as well, into the authoritative audience, identifying himself with both his king and his king's principal writer, Jonson.

The "Virginia Maske" is also a witness to a related process of colonial transformation. For in order to represent himself as authoritative and to represent the Powhatan Indians as masquers, Smith had to transform a native dance textually into an English dance form alien to the dancers it supposedly described. By figuring the Indian dance as a savage version of an English masque, Smith evacuated its Indian meanings and represented the Indian women performing it as entertainers without power in their culture. Smith's rewriting of a dance by Powhatan women as a type of English masque co-opts the experience of American Indians for the benefit of the white writer. Smith's rewritten "Maske" fashioned the Powhatan people, for the reader, into an English monarchy; like Jonson's English masques at court, that "Maske" constructed a king or an ideal of kingship, an audience for that ideal, and an author. It placed Smith in the king's position at the center of attention. Smith's "Maske" also presented the dancing women as English fantasies of witches and concubines rather than as economically and politically significant people. To transform himself into a powerful authority, Smith transformed the culture he depicted into a recognizable culture operating on an English model. Thus the two colonial transformations—Smith's self-assertion and his disempowerment of Powhatan women—are intimately related.

Those transformations would prove fatal to a historical understanding of Powhatan culture in its own terms; they enabled a reading of Powhatan culture that would erase its material realities from history, and they created Smith as the ultimate reader of Powhatan culture. By paying close attention to Smith's "maske" I hope to contribute to what Alvin M. Josephy Jr. calls for: an effort "to retell the history of America and the Western hemisphere, restructuring it more wholly and accurately to include the participation of the Indians and the roles and contributions of their cultures and traditions that have been misunderstood, traduced and ignored" (7). Ironically, Smith's own texts implicitly reveal a Powhatan culture in which women were politically and economically central, a culture Smith misreads in his transformation of the dance. This chapter looks at ways of rereading Smith's masque to reveal both what it said to its English reading audience and what other meanings it may have erased for that audience and for later history.

As a number of critics have suggested, *The Workes of Beniamin Jonson* was as much a paean to Ben Jonson's identity as author in his time and to his authority for posterity as it was a collection of writings.[1] And, I would argue, the masques in the folio have a special relationship to that claim for authority.[2] Jonson's folio culminates in the masques, preceded by a title page that reads "MASQVES AT COVRT." The Author B.I. (see figure 5.1). Even more than the title pages of the plays in the Folio, that title page produces a singular author and authority. For although the play title pages attribute the plays to the "Author B.I.," they each also place above that ascription the name of the company that performed the play. And each play's text ends with a page that gives the name of the acting company again, the names of the principal players, and the year the play was first acted, along with a note of license from the Master of the Revels. Thus, while the title pages of the plays in the folio postulate a singular author, they also undercut that authority. The final pages of the plays, which directly face the title pages of the next play, assert a multiple authority—divided between the "author B.I.," the Master of the Revels, and the acting company whose names appear in capital letters. Despite all the additions and emendations that Jonson makes to the play-texts to make them his own in the Folio, the title pages declare a plural authority. In contrast, the title page of the masques faces the final page of the "Entertaynments," a page of text whose last line is *The Author* B.I. When we turn to the masques in the folio, we find a striking double claim for singular authority.

Of course, Jonson's running quarrel with Inigo Jones over the authorship of the masques, and the fact that the outcome of that quarrel would determine Jonson's chances for royal patronage, gave the poet an added impetus to assert an indivisible authority on the masques' title page. Jonson

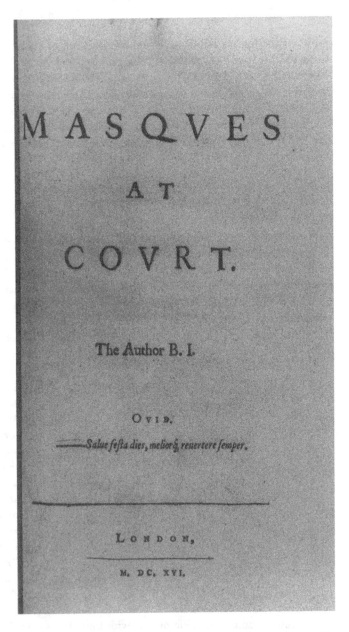

Figure 5.1—The internal title page. Ben Jonson, *Masques at Court* in Ben Jonson. *The Workes of Beniamin Jonson*. London, 1616. Reproduced with permission of the Folger Shakespeare Library.

was looking for the king's sanction and patronage, and he was also allying himself in his masque writing with the authority of the king. Don Wayne argues that

> The place of the author is finally privileged in opposition to that of the theater audience by an identification of his own judgement with the "power to judge' of the king. In this way, the place of the king, the highest earthly place, functions as more than just that of another audience of the play; it is the place of final authority, and as such it is the locus of the voice of the Father, the voice that *author-izes* . . . the writer's voice even as it limits its scope. (1982, 118)

Wayne makes this comment in reference to the induction to *Bartholomew Fair,* but one can make an even stronger case for this view of Jonson and authorship with regard to the masques, all of which played for the king, and all of which strongly identify the king and the author; in the fictions of Jonson's masques, the king has the transformative power to move the plots—to affect the action on the stage.[3] In these fictions, Jonson allies his singular authority with the king's.

Like Jonson's production of his folio *Workes,* Smith's production of his folio *Generall Historie* was a bid for authorship and authority. And like Jonson's masques, Smith's "maske" had a special relationship to his claim of authority. The "entertainment" Smith writes emerges only slowly in his texts as a royal entertainment, a "Virginia Maske." In the *Proceedings of the Virginia Company,* the second part of the 1612 text *A Map of Virginia,* the textual gloss on the incident reads: "the women's entertainment at Werawocomoco." The reader sees Captain John Smith "presented with [an] anticke." The text describes the dancing as a "maskarado" (I, 235-36). According to the *OED,* an "anticke" at the time could signify any pageant of grotesques. In his essay "On Masques and Triumphs," Francis Bacon includes "antics" in his list of the antimasquers he had seen. But antics were not only present in masques proper. The word denoted any appearance of grotesquery or mixture; it seems to have emerged from discoveries of classical statuary that mixed representations of beasts and humans. The word "maskarado" referred to a costumed ball or anything like a costumed ball, entertainments, of course, related to the masque as Jonson and Inigo Jones, Samuel Daniel, and others developed it at court, but not the same as a masque in 1612. Especially in Jonson's hands, the masque had evolved as a form separate and distinct from its roots in masquerade. As Joseph Loewenstein says, "the form itself was an accomplishment: Inigo Jones and Jonson

imbued the loose conventions that had accumulated in Tudor royal entertainments . . . with recognizable generic coherence" (1985, 169). In the *Proceedings,* Smith describes a dance of grotesques, a dance more kin to an
interval of dancing in a play than to a full-fledged Jacobean masque.
That loosely categorized dance takes place in print in a loosely organized collaborative text. The *Proceedings* appears textually as just that, a
compilation of the proceedings of the company, consisting of numerous
sections, each signed by many colonists. The title page reads:

> with the discourses, Orations. and relations of the Salvages, and the accidents
> that befell them in all their Journies and discoveries, Taken Faithfully as They
> were written out of the writings of Doctor Russel Tho. Studley Anas Tod-
> kill Jeffra Abot Richard Wiefin Will. Phettiplace Nathaniel Powell Richard
> Pots. And the relations of divers other diligent observers there present then,
> and now many of them in England. By W.S. (Smith I, 199)

A letter to the reader from one T. Abbay introduces the text, announcing
that the *Proceedings* was not written by "mercenary contemplators . . . neither am I the author, for they are many, whose particular discourses are
signed by their names. This solid treatise, first was compiled by Richard
Pots" (I 201). John Smith claims neither authorship nor even collection of
the *Proceedings,* although that text appears as the second section of the *Map,*
the title page of which declares: "written by Captaine Smith." Three other
colonists, Richard Wiffin, William Phettiplace, and Anas Todkill, sign the
part of the *Proceedings* containing the dance.

The *Generall Historie* is a more massive compilation than the *Proceedings,*
more on the order of Samuel Purchas's or Richard Hakluyt's collections
than a proceedings; but unlike Purchas's or Hakluyt's collections, which
advance a general notion of the English as colonizers,[4] Smith's collection
looks like an effort to promote himself and his own idea of the colonial
project. As Philip Barbour notes in his edition of Smith's complete works,
the printer issued the prospectus for the *Generall Historie* at the same time
that James I was being petitioned to investigate the tangled affairs of the
Virginia Company (II 5). That investigation led to a takeover by the crown
in 1624. By the time of the prospectus, Smith had been in England for
years trying to get patronage for further adventures in North America. The
Historie, then, is part of that effort at a time when this prospect looked
brighter for Smith, as the company that had rejected him was failing. The
text declares Smith the proper interpreter of all the events leading to settlement and of settlement and colonization.

Although largely taken from other men's works, the *Generall Historie* nevertheless testifies to Smith and to his authority. He maintains some multiplicity of authorship in this text; the section of the *Proceedings* that contains the entertainment even gets an additional signatory, Jeffery Abot, when Smith includes it in the *Generall Historie*. But Smith claims the whole firmly on the title page: "By Captaine John Smith." Unlike the prefatory matter of the *Proceedings,* the prefatory matter of the *Generall Historie* repeatedly names Smith as author. His broadside prospectus declares that the text to be published was "Discovered, observed, or collected by Captaine Io. SMITH sometime Governour of Virginia, and Admirall of New England" (II 7). The *Historie* has a preface by Smith and also prefatory matter that includes nine dedicatory verses to him by name; the first of ten altogether is addressed to "the Author." And although he will leave others' signatures in the texts, Smith, unlike Hakluyt and Purchas,[5] often alters and adapts stories, especially those in which he figures strongly as a character.

His textual changes contribute to the project of constituting Smith's authority. When Smith retells "the women's entertainment at Werawocomoco" in the *Generall Historie* he turns it into a "maske." The text's gloss now reads "a Virginia Maske"; that "Maske" gets an additional character in the royal Pocahontas, a new introduction, and its frame in the narrative acquires novel details. The dance is now a formal entertainment, a "maske," and it carries with it all the cultural weight of the masque in Jacobean England. The most important addition in making that "maske" is Pocahontas, whom Smith had previously developed in his narratives as a significant character. In his texts, she becomes a purified Indian; more refined than the Indians around her, she carries the cachet of royalty. "The very nomparell of [Powhatan's] kingdome," her sympathies lie, of course, with the English (I 274). In the later version, Pocahontas is both the producer and part of the spectacle of the entertainment, and her presence in those roles marks the entertainment as a royal masque.

The *Proceedings* introduces that entertainment as follows: "Powhatan being 30 myles of[f], who, presently was sent for: in the meane time his women entertained Smith in this manner" (I 235). In the *Generall Historie,* however, "Powhatan being 30 myles of, was presently sent for: in the meane time, Pocahontas and her women entertained Captaine Smith in this manner" (II 182). The *Proceedings* calls the dancing women Powhatan's—"*his* women" (my emphasis). Smith's reading public could see the entertainment as merely a dance put on by Powhatan's "concubines," especially since Smith has already identified the women surrounding the king as a sort of harem.[6] In contrast, "Pocahontas and her women" demand to be read, in the text's codes, as a princess and her court. Although that court may also consist of concubines, Smith is now being entertained directly by a princess; that princess takes the role of Queen Anne at the English court— both the presenter of and participant in spectacle, and its privileged audience. In Pocahontas's presence, Smith takes his title "Captaine," so he becomes a more noble spectator worthy of a place at her court.

As the scene opens in the later text, Smith once again interpolates Pocahontas. When the dancing women are about to appear in the *Proceedings,* a great noise frightens the English:"supposing Powhatan with all his power came to surprise them, but the beholders which were many, men, women, and children, satisfied the Captaine there was no such matter" (I 235). Pocahontas does not appear in the earlier text. In the second version, Pocahontas arrives to calm the English fears; Smith's men are startled "supposing Powhatan was come to surprise them. But presently Pocahontas came, willing him to kill her if any hurt were intended" (II 182-83). Here Pocahontas functions as a liminal link between the English and the Indians just as she did in the story that Smith tells of his captivity. She repeats her central role in Smith's narratives as the Indian woman who identifies with, protects, and serves the English colony—the role in which she has entered American history. Once again she allies herself with Smith; but where in that originary captivity scenario her father had the ultimate power over Smith, here she gives Smith authority over her:"willing him to kill her." In Smith's "Virginia maske," the royal presenter, Pocahontas, defers to the Captain.

In both versions, Smith watches alone: " *he* sitting uppon a mat" (my emphasis), watching while thirty naked women, in body paint, wearing horns and skins and carrying arms, dance and sing, and subsequently entreat his love. Smith is thus the primary spectator. In *The Illusion of Power,* Orgel argues that the court masque functions both to display the king's own power to the king and to display the king to the court. We can see this "maske" working in a similar manner for Smith, since the text emphasizes Smith's placement. Although we hear that others watch—the "beholders": men, women, children, presumably Pocahontas, other English—we hear nothing about where those others are standing or sitting. Like the king, Smith is the focus of all eyes. Many of the masques published in Jonson's *Workes* stress the king's placement in the audience. For example, *Love Freed From Ignorance and Folly* opens, "So soone as the Kings Maiestie was set, and in expectation, there was heard a strange Musique of wilde Instruments." Likewise, *The Irish Masque at Court* begins by stating, "The King being set in expectation, out ranne a fellow attir'd like a cittizen." Smith opens his description of the "Maske" as follows: "In a fayre plaine field they made a fire, before which, he sitting on a mat, suddainly amongst the woods was heard such a hydeous noise and shreeking" (I 182). As in Jonson's written masques, Smith's text places and prepares the primary spectator and the spectacle begins.

The text displays the royal Pocahontas presenting an entertainment directed at Smith, and the entertainment's finale again places Smith at the center of attention. At the end of the "maske" the dancing women who belong to Pocahontas's train offer themselves to Smith. They surround him, crying, "Love you not me? Love you not me?" In that action they offer not only their bodies, they offer him the position of the king; they surround him as he has told us they surround Powhatan.

With Orgel's theoretical framework in mind, a further textual interpolation becomes significant, and it indicates once again the significance of naming in the colonial world. Immediately before the text narrates the "Maske," Smith describes English plans and preparations at this point in the settlement. He assures the reader that he keeps his troops in readiness. We read in the *Proceedings* that "the whole company every Satturday exercised in a fielde prepared for that purpose" (I 234). In the *Generall Historie,* we get more detail:

> the whole Company every Satturday exercised, in the plaine by the west Bulwarke, prepared for that purpose, we called Smithfield: where sometimes more then an hundred Salvages would stand in an amazement to behold, how a fyle would batter a tree, where he would make them a marke to shoot at. (II 181)

Here the English and the visual evidence of English might are spectacles for the Indians, who watch a demonstration of martial prowess on a piece of their land claimed by the English and named Smithfield. Smithfield, of course, signifies the market space near London, but it also contains Smith's name. The dedicatory poems to the *Generall Historie* name Smith, and they also play on that name—as a Smith, he is an artificer. His name provides an irresistible metaphor for his actions. Robert Norton's poem declares:

> *Envie* avant. For *Smith,* whose *Anvill* was *Experience,*
> Could Take his *heat,* knew how and when *to Strike* . . . *(II 50)*

In this poem, Smith's experience as a warrior provides him with the tools to shape the colonies, and his given name determines the violence of his work upon them. Given this punning penchant of the text, we can read a play on the name Smith as a sign pointing to its author. Smithfield is both a sign and a synecdoche for Smith; it is he—his "marke"—whom "an hundred Salvages . . . stand in an amazement to behold." In this introduction of a martial Smith-field immediately before the "maske," readers observe Smith as spectacle, just as the court watched James while his masquers danced. The martial display prepares the ground for the "maske"; watch Smith, it says. Readers then watch Smith watch his "Maske" and, finally, they see him surrounded, the focus of the dancers' adulation, made king in the spectacle.

But as a colonist, Smith's relationship to kingship was always anxious. Although both in his position as "maske" writer and in the action of the "Maske," Smith allied his own placement with that of the king, Smith, like Jonson, needed the king's patronage—in Smith's case, the favor of being authorized to colonize. Smith was also still concerned at this moment in the text with promoting an evanescent model of colonialism, an English relationship with Powhatan as vassal monarch. In addition, he was negotiating a space defined in the English mind as part of England, but outside the reach of the body and the eyes of James. Thus Smith's bid for ultimate authority was in at least a triple tension with the authority of "real" kings, both English and Indian. Smith's authority could not be independent of the king's since it depended on Smith's promotion of a particular vision of kingship and of relationship between the English and the Indian monarchs, and since it required the absolute trust of his monarch.

The concern in his texts with kingship—who will rule this "new" area—surfaces elsewhere in Smith's denials of other people's accusations. According to the *Proceedings,* the Virginia company imprisoned him for a time due to "scandalous suggestions" that he would "make himself king" (I 207). Later in that text, in a scornful passage dealing with his rivals for power in the colony, we read that "some propheticall spirit calculated hee had the Salvages in such subjection, hee would have made himselfe a king, by marrying Pocahontas, Powhatan's daughter" (I 274). The text treats these accusations with utter disdain, but its need to record and deny them opens the question of Smith's relation to kingship. The accusations function as Freud tells us negation functions; they affirm as they deny Smith's anxious aspirations.

So it is suggestive that the text elaborates the "maske" immediately preceding an account of the controversial crowning of Powhatan. For a brief time the Virginia Company fantasized that if it gave Powhatan the nominal trappings of English monarchy, he would cooperate fully in the disempowerment of his people. Although from the beginning of his dealings with these Algonquian Indians, Smith had identified Powhatan as their monarch and Pocahontas as a princess, he strongly opposed the formal coronation of Powhatan that the Virginia Company and its representative, Captain Newport, desired; Smith writes that the coronation is a complete waste of the Company's resources. Coming as it did with Newport, the coronation challenged Smith's status. Smith wished to see himself as the anointed representative of the English, as the Englishman who understood these Indians and who could interpret them to his patrons. Had Smith required, or even approved of, that coronation for his political purposes, we could see his formalizing a dance into a "maske" as simply giving the appropriate entertainment at the court of Powhatan, on an occasion of importance, with Smith as ambassadorial observer. In light of his objections to the coronation, however, we need to see this "maske" as a more ambiguous construct. It does celebrate royalty, in the form of Pocahontas, the king's daughter, but it celebrates as well the authority and, finally, the power of a particular colonizer. Smith is clearly the privileged eye; and the end of the "Maske" presents Smith surrounded by his court. In his description of the "Maske," Smith attempts to transform himself into the authoritative ruler of Virginia.

If Smith's "Maske" showed English readers Smith as the privileged audience, what did they see him watch? The answer to that question is obviously women, thirty naked young women; there seem to be nothing but Indian women in this "Maske." Although the observers include men, specifically old men and children, the dancers are all women; and when John Smith adds a player, he adds another woman, Pocahontas. Who are the women that Smith displays? How do they function for Smith and for the projects of his text?

A large body of evidence suggests that most North American Indian tribes were structured nothing like the patriarchal patrilineal European societies they first encountered in the early modern period.[7] Although Powhatan culture was not the matriarchal utopia that some scholars of North American Indian cultures would like to find, it was also distinctly not the same kind of patriarchal society that Smith left behind in England. All Powhatan women had relatively more freedom and power than their English counterparts. Unlike English women, Powhatan women did not face legal and institutional obstacles to full participation in their culture (Rountree 1998). Smith's use of an English entertainment form to structure his account of what he saw these Indian women doing denied his audience a glimpse into a world that was not England, into an alternative sex/gender system.

Smith prominently displays women in the "Maske" as he does in all of his descriptions of Powhatan life. Smith might be marking the importance of women in this society; he and other Virginia colonists cannot escape noting their presence even in gatherings of Indians in power. When the Indians first bring Smith captive to Powhatan,

> On either hand did sit a young wench of 16 or 18 yeares, and along on each side the house, two rowes of men, and behind them as many women, with all their heads and shoulders painted red; many of their heads bedecked with the white down of Birds; but every one with something: and a great chayne of white beads about their neck. (II 150)

Smith later identifies these women as concubines, but the very ubiquity of women in prominent positions in Powhatan social and political space makes that identification questionable.[8] In his 1615 text, *A True Discourse of the Present Estate of Virginia*, Ralph Hamor relates that when he came to Powhatan with a message from Sir Thomas Dale, he

> certified him [his] message was private, to be delivered to himselfe, without the presence of any, save one of his Councellers, by name Pepaschicher, one of my guides, who was acquainted with my businesse he instantly commauded all, both men and women out of the house, his two Queenes only excepted, who upon no occasion whatsoever, may sequester themselves. (40)

Hamor notes that Powhatan dismissed women as well as men; therefore, women were inside Powhatan's house. Also Powhatan either permitted the women at his side to remain or else he did not have the power to dismiss them when considering important business. Smith notes that Powhatan society is matrilineal, although he has enormous difficulty conceptualizing how such a scheme of inheritance might work (I 61). Both observations, the ubiquity of Powhatan women and inheritance based on the women's line, could force him to reassess his own preconceptions of the place of women in the system he confronts. Instead, in the face of contrary evidence, he continually assigns unlikely roles to Powhatan women.

Smith's texts' focus on Indian women is not solely a function of his observation of Powhatan society. One of the projects of his texts is to assert his bravery and competence for his multiple audiences. The texts also had to reassure potential investors in the Virginia colony that the enterprise was safe. Thus Smith might have foregrounded women in order to minimize the threat that the English perceived from Indian men. When Smith tells the reader that the "far greater number [of Indians] are women and children" he is mollifying English fears (I 160). His texts also perform a voyeuristic display of Indian women's sexual availability as a lure to prospective colonists. Hence he comments that Powhatan gives his chief men his women when he tires of them; Smith means the reader to link this titillating comment to Smith's insistence that the Indians consider him a chief. The end of Smith's "Maske," when he becomes a sort of sex object, is another instance of that project of display. But Smith's texts also mark, perhaps despite themselves, the importance of women politically and economically; they and other texts of the Virginia colonization note Indian women in proximity to power in leadership and advisory roles.

Women certainly led some East Coast Indian tribes. Smith noted a woman ruler in his first published text, *A True Relation of such occurences and accidents of note, as hath hapned in Virginia* . . . : "the next day the Queene of Apamatuck kindely intreated us, her people being no lesse contented then the rest" (I 31). Although Smith's own culture had recently had a female monarch, Smith automatically assumes that the Powhatans might be less happy with a woman ruler. But in the world of the people Smith encountered, unlike in his own world, a female ruler was not a last resort, but simply a woman who lead. William Strachey's 1612 narrative mentions both this woman, the "Queene of Appamatuck," and another female commander, "Oholasc Queene of Coiacohhanauke" (64). Cross-cultural anthropological analyses suggest that women's status was high in hunting-gathering and nonintensive agricultural societies and in matrilineal groups. As Smith and the other colonists note, all Powhatan women played important roles in the group's subsistence—they were the primary farmers, and they seem to have had control over food and resources.[9] European colonists, including Smith, consistently claim that Indian women are forced to produce food. The colonists see these women as oppressed and see themselves as saving women from the hardship of production. But the women they encounter do not report oppression, nor do they ask for the saving grace of European sex/gender systems. Instead, along with their control of resources, Powhatan women seem to have had vital political roles in their society.

Over the past five hundred years European colonizers have introduced their forms of patriarchy into the cultures of the colonized as a way to control and "regularize" subaltern behavior.[10] M. Annette Jaimes and Theresa Halsey write eloquently from their perspectives as late-twentieth-century Native Americans of the deformation of relations between the sexes due to colonization of their peoples. They describe a pattern of male dominance imposed by colonial authorities. Rayna Rapp argues that "Colonial and imperial incursions are not simply 'economic.' They reorganize many aspects of social relations and cultural meaning which in turn affect gender relations" (505). And gender relations themselves present a direct target for colonial authorities. Throughout the world, colonizers have reinforced native patriarchal forms and imposed their own. Colonial authorities have often dealt only with the men in a society. For example, under the Canadian 1876 Indian Act, the government accepted children with Indian fathers as Indians with whatever tribal and land rights accrued to this status, while they denied children of Indian mothers rights as Indians, and this in societies that were previously matrilineal. In America, whites forced Indians to convert to Christianity and to live by its patriarchal moral and sexual ideologies, and the Church enforced male dominance in the institution of marriage.

Frequently, colonial authorities have succeeded in enlisting native men in the disempowerment of native women. This has been recognized on a general theoretical level across colonial cultures. Albert Memmi suggests that "the colonized in the throes of assimilation hides his past, his traditions, in fact all his origins which have become ignominious" (122). Many Native Americans have seen assimilation as the only way to escape their poverty and the subjection of their peoples; in Memmi's analysis, "the first ambition of the colonized is to become equal to that splendid model [the colonizer] and to resemble him to the point of disappearing in him"(120).[11] In order to gain power in the white culture, or to retain their lands, North American Indian tribes disenfranchised women and removed them from positions of authority.

As it presents dancing women offering themselves to Smith, far from showing us Powhatan women, the "Virginia Maske" assigns them roles in the English patriarchy; it exemplifies a colonial disempowerment of Indian women. The "Maske" offers Smith's audience Powhatan women as English women, as women whom the English could understand, as women who had social position only in relation to men. "Chiefs, warriors, politicians, physicians, property owners, wives and mothers—only the last two are roles which Indian women generally shared with their white contemporaries" (Mathes 137). Of course, male dominance in England was not seamless. As

Margaret J. M. Ezell argues, "the institutional structures of seventeenth-century England were patriarchal by tradition and law" (161); however, "the patriarch's wife, both in the family and in society, wielded considerable power, whether acknowledged in theory or not. . . . But that power was to a large extent displayed on a private level, not through the public institutions" (163).[12] It is unsurprising that despite Elizabeth's recently ended long successful reign in England, Smith feels the need to note that the Queene of Appamatuck contents her people as well as male Indian rulers satisfy theirs. Smith's "Maske," as I have argued above, displays the power and authority of the English male observing eye and displays that power at the expense of any appreciation of women's community, ritual sodalities, or authority.

Smith provides a lengthy description of the Powhatan women dancing:

> thirtie young women came naked out of the woods, onely covered behind and before with a few greene leaves, their bodies all painted, some of one colour some of another, but all differing, their leader had a fayre payre of Bucks hornes on her head, and an Otters skinne at her girdle, and another at her arme, a quiver of arrowes at her backe, a bow and arrowes in her hand; the next had in her hand a sword, another a club, another a pot-sticke; all horned alike: the rest everyone with their severall devises. These fiends with most hellish shouts and cryes, rushing from amongst the trees, cast themselves in a ring about the fire, singing and dauncing with most excellent ill varietie, oft falling into their infernall passions, and solemnly againe to sing and daunce; having spent neare an houre in this Mascarado, as they entered in like manner they departed. (II 183)

The women's leader ever so gently evokes the English topos of the Amazon—fully armed "with a quiver of arrowes at her back," and "a bow and arrrowes in her hand." These Powhatan women are a version of the Amazons English explorers often "found" on the edge of societies they encountered; Amazons were always located at cultural margins.[13] At the center of the colonial encounter, where this topos can exist precisely as role—as entertainment—and not as threatening reality, Smith's text can evoke it only faintly. As the discussion of Spenser's Radigund in chapter 1 shows, the Amazons of romance and myth took power away from men. Their threat lay exactly in their alterity to European patriarchal structures, which the codes of their female-dominated society inverted. But the women in Smith's "Maske" are not quite Amazons; they have no power; they are naked, and their weapons and horns are costumes. A spectacle of infinite variety, each woman is painted a different color, each carries her own "device." The *OED* informs us that the word "device" was used at the time to signify a heraldic sign, and as a synonym for masque. Smith employs it to designate what the women carry. Their burdens become in effect emblems or props, markers of character. We see an unspecified infinite variety of signs, not heavily armed, threatening women.

Smith's characterization of the dancing women also invokes the topos of the witch.[14] The Powhatan women are collectively called "fiends" who make "hellish" noises. Smith marks them as un-Christian, as directed by the devil. Smith's English audience would have found the spectacle of witches—women who appealed to the devil and who were agents of deviltry—extremely familiar. The English "saw" witches in witch trials, in other travel narratives, in their king's fascination with witches, and often on the English stage, in plays like Shakespeare's *Macbeth,* and in Jonson's published masques.

One masque in particular resembles Smith's "Maske": Jonson's 1609 *Masque of Queenes,* sponsored by Queen Anne and "celebrated from the House of Fame, by the Queene of Great Britaine with her Ladies." Jonson introduces that masque as "a spectacle of strangenesse, producing multiplicity of Gesture." His witches "with a kind of hollow and infernall musique" come forth "all differently attir'd: some with ratts on theyr heads; some, on theyr shoulders; others with oyntment-potts at theyr girdles; All wth spindells, timbrells, rattles or other venificall instruments, making a confused noyse, wth strange gestures" (20-36). The antimasque section of *Queenes* culminates in their dance:

at wch wth a strange and sodayne Musique, they fell into a magicall Daunce, full of praeposterous change, and gesticulation, but most applying to theyr property: who, at theyr meetings, do all thinges contrary to the custome of Men, dauncing, back to back, hip to hip, theyr hands ioyn'd, and making theyr circles backward, to the left hand, wth strange phantastique motions of theyr heads, and bodyes. (344-50)

The witches' antimasque in *The Masque of Queenes* and Smith's "Maske" are strikingly similar, both, as Skiles Howard has called the antimasque, "icon[s] of otherness that inver[t] every aspect of [English] courtly custom" (120). *Queenes* was celebrated by Queen Anne "with her honorable Ladyes," just as Smith's *Virginia* was celebrated by "Pocahontas and her women." The presentation of Smith's "Virginia Maske" by female royalty and those women belonging to the distaff court, therefore, has precedent in England.

Anne commissioned and her ladies performed all of Jonson's first masques for James's court. Smith's decision to have Pocahontas present the "Maske" enhances the "Maske's" resemblances to *Queenes*. Smith's dancers enter to a "hydeous noise and shreking," Jonson's to a "kind of hollow and infernall musique." Both Smith's and Jonson's dancers appear "all differently attir'd." Both carry signs of their deviltry: Jonson's "other veneficall instruments," Smith's "severall devices." Both sing and dance with "excellent ill varietie." The texts ask us to read the variety in both cases as abhorred mixture, as a spectacle that denotes disorder, diversity, antisignificance, and disarray. As I have noted previously, Smith describes the women as grotesques, Bahktinian figures of mixture.

These correspondences between the two masques ultimately suggest that Smith's spectacle is a display not of Indian women, but of English fantasy—the fantasies of the Amazon and the witch that were played out on the English stage by men and boys (although women did act in masques). I am not insisting that Smith used *Queenes* as a model, although he very well might have, as Jonson published *The Masque of Queenes* in quarto in 1609 and Smith did not issue the *Proceedings* until 1612. Whether or not the resemblance is due to direct influence, both masques represent English fantasies of outlandish women; the colonizer represents Indian women as stage Amazons and witches, fearsome women drained by being staged of whatever power they might possess in "reality." James I's *Demonologie* attests to an actual English fear of witches at the highest level of society, however much that fear was itself conditioned by representations. Jonson's masque, however, attests not to the fear of witches but to the mastery of that fear. His witches are purely contained, purely produced, purely theater, as his published masques insist; he includes the name of the choreographer of their "magical dance." Smith, also, produces not actual witches but naked young women in costume entertaining him in a "Maske."

As the example of James's *Demonologie* suggests, what I am calling English fantasies do not exist in a realm unrelated to what we now call empirical observation. Rather, fantasies conditioned reports in colonial texts such as Smith's and were themselves conditioned by observation. Various forms of outlandish women have always existed in the Western tradition, and Smith "saw" some of these outlandish women at the same time that he saw thirty Powhatan women. These thirty women presumably existed and danced. Even my analysis rests on their existence, and the ways in which Smith saw them and transformed them into an antimasque do not obviate their existence. However, Smith's representations do condition how the women were perceived by his audience, and they also conditioned further

English performances of outlandish women. Thus, the English knew of Amazons and witches from various representations. They then saw and heard of outlandish women as they traveled and encountered new people; what they saw enabled both confirmations of earlier representations and refined additional representations.

In the codes of his English fantasy, the women in Smith's "maske" are "fiends," but not threatening fiends; they are spectacle. Through his narration of the dance itself and its framing events, Smith carefully controls any suggestion that these women or this spectacle may endanger the English. Smith's description empties the "infernall passions" of fear and shock value by surrounding them in his text with singing and dancing: "singing and daunceing with most excellent ill varietie, oft falling into their infernall passions, and solemnly againe to sing and daunce." Smith shows the reader choreographed disorder. In the *Proceedings,* at the first sound, the English grab their weapons until the "beholders" reassure them that no one will attack. In the *Generall Historie,* not only does Smith add the spectacle of Pocahontas offering to immolate herself should the Indians contemplate any injury to the English, but the English seize, along with their weapons, "two or three old men by them." The account in the *Generall Historie* thus triply assures the reader that this is, as the glosses specify, a "Maske" and an "entertainment," and that the English are in control of the episode.

Of course, each of these framing devices, and especially their accumulation, points also to the aggressive potential of passionate women with weapons. Thus this passage performs the very delicate operation of producing Indian women as outlandish women and simultaneously producing them as only a performance of outlandish women. Smith's "Virginia Maske" cannot be read simply as a version of Jonson's *Queenes.* In *Queenes,* the antimasque of witches was presumably performed by boys, professional actors presenting an English fantasy. In Smith's "Maske," his "fiends" are really Indian women, and Indian women are really "fiends" and at the same time not fiends, not frightening, merely Indian concubines in costume. This passage offers the reader both the frisson of witnessing savagery and wildness and the security of English colonial control.[15]

To emphasize that what we have seen was costume, Smith's entertainment ends with the reentry of the women as "Nymphes": "having reaccomodated themselves, they solemnly invited him to their lodgings." There they surround him, entreat his love, and feast him. Some of the women "attend" him at the feast, while others sing and dance. Once they take off their skins, horns, and weapons, in the logic of the text the Indian women reveal themselves as wanton women—the identity of Indian women—all in love with Smith, all his attendants in one way or another. As I have already suggested, this is not the first time in Smith's texts that we see a man surrounded by women. Indeed, for a variety of complicated reasons, not the least of which was the actual position of Indian women in Powhatan culture, Smith seldom represented Indian men except as surrounded by Indian women. In Smith's chapter on Powhatan's government, he describes Powhatan's wives: "He hath as many women as he will, whereof when he lieth on his bed, one sitteth at his head, and another at his feet, but when he sitteth, one sitteth on his right hand and another on his left" (II 126). And when Powhatan entertains Smith in state, the king sits with "at his head and feete a handsome young woman: on each side his house sat twentie of his Concubines, their heads and shoulders painted red, with a great chaine of white beads about each of their neckes" (II 155).[16] Smith's "Virginia Maske" culminates with Smith in the king's place, surrounded by the women Smith has interpreted as wanton.

The second movement of Smith's "Maske," or the masque of "Nymphes" after his opening antimasque of "fiends," functions doubly. First, it reinforces the frame's placement of Smith in the position of supreme authority. As Smith's description of himself as the designated spectator of the "Maske" puts him in the position of the English king, so his position in the "Maske" at the center of a bevy of adoring women puts the captain in the position of the Powhatan king. Secondly, his narration creates these women as the "concubines" he says surround the king. It marks them as available women, especially to Smith, but also to his audience. Costumed, the women who entertain Smith are horrible in their diversity; "reaccomodated," they are all one spectacle of accessible femininity. They serve him, all crying, "Love you not me, love you not me." That "me," said in concert, signifies the Indian woman as unindividuated sexuality.

The spectacle of a king surrounded by women appears again in the 1630 *The True Travels, Adventures, and Observations of Captaine John Smith,* a species of early autobiography purporting to be a record of Smith's adventures in "Europe, Asia, Africke, and America." In that text, Smith writes of "Crym-Tartar" chiefs: men, surrounded by women, who have "captived women to breed upon" (III 194), and "wives, of whom they have as many as they will" (III 195). In this description readers saw the stereotypical Turk that Edgar playing Tom O'Bedlam in *King Lear* "in woman out paramour'd" (TLN 1871, III.iv.89). Smith culled his descriptions of the Tartars from Friar William de Rubruqis's *Iterarium,* written in 1253.[17] Smith's "ethnography" of the Powhatan Indians echoes and replays that earlier text's visions of exotic rulers surrounded by willing women as rewritten in the *True Travels;* Rubruquis's women are the ground for Smith's representation of the women who surround him adoringly in the denouement of his "Maske." Hakluyt reprinted Rubruquis and then Purchas in his *Pilgrimage* reprinted Rubruquis again along with Smith's text.

In the *True Travels* we can clearly see a closed circuit of colonialism. A treasure trove of Orientalism,[18] the *True Travels* shows how images of "barbarian" cultures circulate and support each other. These circulations erase the differences in material culture between societies, their undifferentiated images functioning like the images of undifferentiated Indians in Jacobean and Caroline pageants and masques to create a savagery that justifies colonization. An example of this circulation of images, Smith's "Maske" culls its images of Amazons and witches and concubines from masques and other colonial accounts and then offers them as a description of Indian women and their practices.

Generally, Smith's English culture was able to see all women only in relation to men.[19] Women were unmarried daughters, or wives, or widows, or mothers,[20] but legally they had no independent status, although, as Ezell argues, they may have exercised considerable power in those roles. When Smith saw women dancing, he saw women in relation to men, women dressing as male fantasies: whores offering themselves to men, and a princess subjecting herself to a man's desires. The *Generall Historie* carries a dedication to an Englishwoman, the Lady Frances, Duchess of Richmond and Lenox. Ostensibly signs of a privileged woman's power as a patron, that dedication and the portrait of Lady Frances that accompanies it themselves indicate the limitations on the power of even white noblewomen. The last paragraph of Smith's dedication asks Lady Francis [sic] to "accept" his works so that the protection of her name "will inable them to be presented to the Kings royall Majestie, the most admired Prince Charles and the Queene of Bohemia." The dedication names Elizabeth, the king's daughter, only under her husband's claim to a crown; and Frances is a conduit to the king. The paragraph ends with Smith's prayer that his patron "may still continue the renowned of your sexe, the most honored of men, and the highly blessed of God" (II 42). Always confined by, subject to, and judged according to the authority of men, Lady Frances is also defined by men: her titles were her husband's, and her husband's coat of arms hangs over her left side in the portrait. The portrait carries the caption: "The portraiture of the illustreous Princesse Frances Duchess of Richmond and Lenox daugter of Thomas LD. Howard of Bindon Sonne of Thomas Duke of Norfok. whose mother was Elizabeth daughter of Edward Duke of Buckingham." Frances's lineage includes the woman Elizabeth, whose titles do not bear mentioning, only in order to make the link back to Edward, who was doubly descended from royalty.

Lady Frances, one of the most privileged English women, was still ultimately defined in her society by her relations with men. Constrained by his culture's patriarchal ideologies, Smith constitutionally could not see an Indian woman in or out of his "Maske." Although he could and did report Indian women's' doings, he could not see their agency except in his or another man's behalf, nor could he see Indian women's relations among themselves. When Smith observed polygyny practiced among the Powhatan people, he interpreted it as a form of prostitution. However, anthropologists have seen such practices in North American Indian cultures as opportunities for community among women. The Powhatan polygynous household, from Smith's point of view a patriarch's heaven, may actually have been a woman-based household providing for important relations and decision making among women. According to Ann Thrift Nelson's analyses of the predictive factors for female ritual sodalities, the Powhatan culture that Smith encountered was likely home to women's groups with important ritual functions as well as economic power. Although his texts are full of women, a dedicatory poem to the *True Travels* makes women's place in Smith's texts clear. It contains a list of all of the foreign women who aided Smith in his travels:

> *even Earth, Sea, Heaven above,*
> Tragabigzanda, Callamata's *love,*
> *Deare* Pocahontas, Madam Shanoi's *too,*
> *Who did what love with modesty could doe.*
> (III 145)

Women have love to give, men power. When an English man encounters another culture he has political dealings with the men and, potentially, amorous rites with the women. Through their love, women facilitate men's political ends.

Another of the *True Travels*'s dedicatory verses reads that Smith has

> . . . found a common weale
> In faire America where; thou hast wonne
> No lesse renowne amongst their Savage Kings
> Than Turkish warres, that thus thy honour sings.
> (III 148)

In Smith's original epitaph, we read

> How that he subdu'd Kings unto his Yoke,
> And made those Heathen flee, as Wind doth Smoke:
> And made their land, being of so large a Station,
> An Habitation for our Christian Nation.
> (III 390)

These verses write the political project, the colonial project, of Smith's texts as a project among and between men. To gain renown, even in America, Smith must subject kings. The heathen who flee him, I would argue, are also men, the only possible political subjects of kings. Smith's masque-making is one way in which Indian women were disenfranchised. In the final picture that Smith's epitaph paints of his colonial project, Indian women are nowhere in evidence.

The disempowerment of Indian women within the "Virginia Maske" becomes particularly ironic when we realize that the "Maske" was itself in all likelihood a show of power. Since we do not know much of what the women said or sang in the performance, it is difficult to say what or even if power was being celebrated. Either unable or unwilling to comprehend or to relay his comprehension of the Powhatan language in the performance, Captain John Smith characterized that language as "hellish shouts and cryes." We have no way to judge Smith's competence as a translator[21]; however, Smith certainly understood some of the Powhatan language, and a few of the English men and boys he knew who had lived longer with the Indians knew more (see my introduction). Smith inserted a Powhatan word list with English translations at the beginning of *A Map of Virginia,* and he republished that list at the end of the second book of the *Generall Historie.* Strachey's manuscript, to which Smith may have had access, contains a much longer list of vocabulary; Strachey also transcribes and translates an Indian song culled from a victory celebration (1953, 85–86). From Smith's account of the "women's entertainment," we can infer that the performance he witnessed and created as a "Maske" was full of Indian words and gestures. Besides the "shouts and cryes," Smith reports nearly an hour's worth of singing and dancing. Rather than attempt a translation of even one song, as Strachey did, Smith relayed a fiction of incomprehensibility, and thus could insert his own signifying system—a display of devil worship and "infernall passions." This was certainly Smith's narrative choice; he could as easily have displayed his own lack of understanding (if such was the case) and thereby have conveyed that there was something to be understood. His decision to write that the performance contained "cryes" instead of words not only enabled him to generate a fictional meaning (infernal ecstasy); it also erased from English history whatever meanings the dancers were generating for their own use or toward a goal of communication with the English.

Out of all the language in the performance that Smith terms "A Virginia Maske" and "The Women's Entertainment," Smith offers four English words: "Love you not me." This feast of signification after a famine of gibberish, pure "passion," lends these words weight. Apparently, the women speak these words in English, or if not, the text renders the translation invisible or unnecessary. The English language of love and seduction becomes the meaning of Smith's "Maske," as Smith's account reduces all the other language to categories of noise—shouting, crying, singing. This noise has a performative but not a communicative function and is not finally language. His focus on those four words again draws our attention to Smith and especially to Smith in relation to young Indian womanhood.

Smith's account reduces all the other language in the performance to "infernall passions." I do not mean to suggest that the language Smith writes about but does not write, if he had translated it, would have conveyed to Smith's contemporary audience and then to us some pure seventeenth-century Indian meanings to which we could have unobstructed access. I do want to suggest that whatever the women were singing, shouting, saying, or crying is completely lost to us now, even more completely than it would have been had we some version of it that we could barely read due to loss of context, historical difference, and due to the very condition of language that makes pure communication an impossible fantasy.[22] Had they existed, Smith's transcription and translation would always have been many levels away from any early modern Indian meanings. However instead of translating, he chooses to foreground himself as reluctant lover and to give us four words: "Love," conveying the rapidly evolving code of English bourgeois romantic love relations; "not," the negative that implies a choice for Smith—he can accept or reject the love of young Indian women at his discretion; "you" and "me," timeless and culturally unspecific human positionalities. The choice of the language of love over other possible languages in the performance—of resistance, war, display, pride, spirituality, nature, and myriad other possibilities—is an act of "cultural genocide."

I take the term "cultural genocide" from an essay in which Ward Churchill describes the cooptation of late-twentieth-century Native American spiritualism by various New Age hucksters. Using Russell Means's term, "cultural genocide," Churchill systematically dismantles claims that the appropriation of native spiritual traditions for sale to white culture and the representation of native spiritualism in white terms is healthy, or caring, or anything but an act of destruction. The central question for Churchill is not whether or not Lynn Andrews or Carlos Castaneda, for example, have been moved, affected, or changed by their "Indian" experiences, or whether white readers can benefit or become more spiritually aware from reading them. The sincerity of their responses is immaterial. What matters is the material effects of their writings of those responses and experiences on the Indians involved. Writers like these enable perceptions of Indians as exotic, primitive (early versions of a more advanced us), and currently irreversibly endangered; stable ethnohistorical subjects, art objects to be worn and bought; tourist destinations to be appropriated for white use; and specifically not people suffering an ongoing colonial war. The terms of Churchill's analysis apply to Smith's representation four hundred years ago of the language of the dance he witnessed. It was an act of "cultural genocide" to offer the reader noise instead of language, to offer the language of English romantic love as *the* language of this performance. "To speak a language is to take on a world, a culture" (Fanon 1967, 38). To represent these Indian women as speaking the English language of romantic love was to represent them in history as a part of an English culture and simultaneously to dissolve the culture their dance was enacting.

Smith's masque-making erased more than the language of the performance. The Indian women's dance culminated in what Smith relates as a feast. In the *Proceedings'* account we read details of that meal:"the feast was set, consisting of fruit in baskets, fish, and flesh in wooden platters, beans and pease there wanted not (for 20 hogges) nor any Salvage daintie their invention could devise" (I 236). When Smith rewrote the dance as "Maske" he minimized the feast:"the feast was set, consisting of all the Salvage dainties they could devise" (II 183). The excision of the extensive feast is thorough; The *Proceedings* ends the set piece of the dance with "this mirth *and banquet* being ended," while the *Generall Historie* finishes "which mirth being ended" (my emphasis). Although English masques, at least those that took place in James's banqueting hall, could end in feasts, the published descriptions of the masques most often did not include details of the feasting.[23] Thus Smith's removal of the list of foods served to him by the women made his account look more like the accounts he might have read. So we have a possible explanation of the deletion on the basis of its utility for Smith as masque-maker.

Also, when we look at any representation of food in a text of the early-seventeenth-century Virginia colonization, we must also remember that food was the crucial link between Smith's colonizing culture and the cultures of the Indians. Food was the axis along which this particular colonial relation turned.[24] According to Smith and the other Virginia colonists, the Indians were responsible for feeding the English. Smith informs his readers that "we had the Salvages in that decorum (their harvest being newly gathered) that we feared not to get victuals for 500" (II 181). "Decorum" is an interesting choice of word here because it connotes propriety and, in the late sixteenth and early seventeenth century, also notions of rank and order. The *OED* offers two usages that have since dropped out of English: "That which is proper to the character, position, rank, or dignity of a real person," and "orderly condition, orderliness" (*OED* "decorum"). Smith tells his readers both that the Indians should feed the colonists, as is proper to their relative rank, and that the Indians are in a state of order due to English control: "we *had* the Salvages in that decorum" (my emphasis). The provision of food is the condition of this colonial relation and the condition of the Indians' subjugation. Smith's "Maske" offers us a microcosm of Smith's preferred colonial position. Smith sits while Indian women (unthreatening and seductive) serve him food.

But it was also crucial to this colonial enterprise that readers see the Powhatan culture as relatively impoverished. Here those men writing about colonial Virginia had to walk a fine line; they could not depict Indians as starving because potential colonists and patrons had to see Virginia as fruitful. On the other hand, their texts could not show the Indians too well fed and satisfied, otherwise why would they need the English, how could the audience see them as bereft of God, and why would not the English wish to become like them?[25] Strachey tells his readers that his project is to bring the Indians

out of the warme Sun, into Gods Blessing, to bring them from bodely wants, confusion, misery, and these outward anguishes, to the knowledge of a better practize, and ymproving of those benefitts (to a more and ever during advantage and to a civeler use) which god hath given unto them, but envolved and hid in the bowells and womb of their land (to them barren and unprofitable, because unknowne). (1953 24)

Smith tells us that the Powhatan people are improvident, cannot store food and so grow thin in times of scarcity (II 117). In this colonial logic "all the Salvage dainties"—a meal possibly impoverished, "Salvage," made without invention—are more palatable than a veritable feast of fruit, fish, flesh, beans, and peas.

The genocidal act here is more than the elision of the elements of the feast in order to create the perfect orderly but impoverished Indian, although the elision evokes that act. Smith erases the details of the food in the *Generall Historie* and indeed provides few particulars in the relatively rich description in the *Proceedings*. Both accounts obscure from history the importance of food to the Powhatans: especially to the Powhatan women largely responsible for its farming, gathering and preparation. We learn from Smith's account of a dance that ends in a feast only what we already know from Smith about the importance of food to the colonists, about the place of the Indians in the English schema for survival and domination. We learn nothing about the meanings that the provision of food to the English had for the Indians. We learn nothing about the place of food in all the particularities of its cultivation, gathering, preparation, and consumption in Powhatan culture. The dance using the skins and horns of animals, bows and arrows—possibly weapons of war, but also very possibly tools to obtain food—might well have been directly related to the food served and to the manner of serving and preparation. Even from Smith's description in the *Proceedings* we learn nothing about how the food was cooked, how it was arranged on the platters, the types of fruit, beans, fish, and flesh, how the animals were killed and prepared. All of these details would have had cultural significance and may have been linked to other elements of the dance.

Those elements—skins and arrows, body paint and leaves, as well as the dancers' movements—that Smith regards as props and costumes and aspects of English theatrical practice, were also more than likely charged with significances outside of Smith's understanding of theatrical performance. Although real insights and important interpretations can be derived from analyses of performance as ritual, Smith's writing of Powhatan performance in terms of English theater was designed instead to erase the essential differences between the two cultures.[26] Smith notes elements of what I will call ritual—activity marked off from daily living, activity that calls attention to itself as somehow special—but he interprets it as a ritualized practice in his own culture: theater. As Catherine Bell suggests, however, "basic to ritualization is the inherent significance it derives from its interplay and contrast with other practices. From this viewpoint there would be little content to any attempt to generate a cross-cultural or universal meaning to ritual" (90). Smith's writing of the Indians' dance as his "Maske" presents, then, the same problem presented by current cross-cultural studies of performance as ritual. Although current theorists may be more self-conscious, both flatten cultural difference.

The representation of what might have been a central religious practice as a performance of an English fantasy erases Powhatan religion; and we must not forget that the fantasy that enabled the English persecution of "witches," translated, would enable the persecution of these Indians and their forced conversions. To take the tools that may have been used in these women's labor as props in a diabolical performance is to empty them of value and to erase any relationships and meanings formed in that labor. Of course English theater was not devoid of meaning; there is no entertainment without ritual significance. English antitheatricalists complained that paid actors wearing the clothes of people of higher rank transgressed order and place and that performance could cause disturbing social change. Such was a possible meaning and threat of theatrical practice in England. But these women were not English players using the props of English theater; they were handling the tools of their own daily lives, presenting their culture. Smith's genocidal move was to read what for lack of a better word I will call their ritual as his theater, their body paint as ornamentation for his eyes, their variety of dress, signifying probably a feast of meanings, as his "spectacle of strangeness." Smith read himself as audience and them as theater where they may well have been playing with him and thus producing something together that he could not understand.

The reader of this chapter will have noticed that in my discussions of the particular erasures that Smith's "Maske" effects I have continually invoked conditional expressions. When I attempt to write about the dance as experienced by these Indians, my rhetoric is always in the subjunctive: "possible," "more than likely," "might have been," "may have been," "probably." Such expressions point to the success of Smith's transformative masque-making and to its genocidal effects on Powhatan history and especially Powhatan women's history. The masque's historical success is demonstrated in one late-twentieth-century historian's passing account in an otherwise critical article: "Women were offered to important visitors like Smith, who was mortified when he and his party were treated to passionate dancing and overtures from thirty barely clad girls" (Quitt 237). But we have other options when reading Smith—our major source for the material of this encounter—than reading his "Maske" in his terms. As I end this chapter I want to move fully into the conditional. What could these women have been dancing?

Earlier in his adventures, before the "Maske" of Powhatan women, Captain John Smith describes an attack by "Salvages": "Sixtie or seaventie of them, some blacke, some red, some white, some party-coloured, came in a square order, singing and dauncing." These Indians, armed with "Clubs, Targets, Bowes and Arrowes" charged the English and were rebuffed by superior firepower (II 144). Smith's narrative, then, describes costumes and equipment similar to those in the "Maske" used in an actual act of resistance. Smith's descriptions of the Indians' triumph at his capture includes many of the same elements (II 147-48). The crucial difference between these incidents and his "Maske" seems to lie in the gender of the dancers. Once the text reveals that naked women have made the "hydeous noise," we see only innocuous titillation. But Smith's "Maske" could well have been a demonstration of women's power in preparation for a fight or a hunt. It could also have been a celebration of a successful fight or hunt. Smith may have been a convenient audience, the designated audience as he believed, or merely part of the larger group of watchers necessary or not necessary to the ritual. The dance may have been intended to convey a warning, or the women could have been proposing a joint action with the English. They might even have been celebrating Smith as an Indian warrior. The point is that none of these possibilities is less likely and most are more likely than Smith's "Maske." The Powhatans would have had no use for the masque form, a product of English court culture, and it would have looked as strange to their eyes as their dance looked to Smith's. Could Smith have looked through their eyes, the conquest of colonial Virginia might have been slower or even quite different, and we who are searching for an Indian-centered interpretation of the colonial past might not have been forced to use our imaginations.

Our current attention to the nuances of Smith's "Maske" can help us to see Smith's "ethnography" as the cant of conquest. And the lens we can develop by rereading Smith's "Maske" is useful today for understanding the tradition it initiated of making Indians into entertainment. In a twelve-page fashion spread in the October 1991 issue of *Self* magazine, titled "Visions of Style: Rediscovering the Beauty of Native American spirit," the editorial text declares that "Ancient Navajo tradition tells us that to 'walk in beauty' means to live in harmony with the Earth. Ever in search of ways to make peace with the self, Americans are rediscovering the Native American way at spas, in makeup and clothing, in food, in fitness classes, in specialized vacations." In the spread and commentary, *Self* interprets Native American ceremonies, designs, and philosophies as spectacle for a non-Indian audience and as new and fun versions of an "American" quest to remake the "self." Smith's "Virginia Maske" is an early precursor of magazine articles such as this one, which interprets Native American traditions as "style" choices for the non-Native reader with disposable income. Articles such as this, and the products they tout perpetuate the cultural genocide initiated by early writers like Smith. We cannot really change what Smith's masque-making did to the Indians he described, although we can certainly question Smith's authority. But we can bear witness to the tendencies of our contemporaries to write the ongoing colonial struggle in North America as entertainment and spiritual renaissance for the non-Indian audience. And we can refuse to be entertained.

Epilogue

Late-Twentieth-Century Transformations: Pocahontas and Captain John Smith in Late-Twentieth-Century Jamestown

> The shape of the thinkable future depends on how the past
> is portrayed and on how its relations to the present are
> depicted. (Tony Bennett, *The Birth of the Museum* 162)

In 1999, the undifferentiated Indian of the seventeenth-century masque
lives on in the tourist gift shop at the Jamestown Settlement museum near
Williamsburg, Virginia. This museum gift shop distinguishes between Indians from India and the Americas, but it does not sell products only from or
about Virginia Indians. Enter this shop and you will find books, trinkets,
and children's toys about Indians as one generic type and Indians from all
over North America. The Jamestown Settlement museum sells rings made
by the Eastern Cherokee, books and tapes about Indians of the Southwest,
books about Indians of the Lower Mississippi, and the ubiquitous dream
catchers, products designed to get Americans in touch with "their" Indian
natures and heritage.[1] Such products might not (and obviously do not)
cause a tourist to pause, since, after all, this museum tells a story about Indians. But the assumptions behind this gift shop display becomes apparent
when its Indian offerings are compared with its offerings about the "other
side" of the story. When the shop is dealing with English colonial material, it does not sell material about the French colonization of America or
about Spanish efforts in the Caribbean, or even about English activity in
the New England colonies that became important to England soon after
the Virginia settlement was underway. Were one to ask the staff why those
other materials are not available in the gift shop the answer would be quite
clear—those are not the stories this museum is about. This museum is

specifically about the first few decades of English settlement at Jamestown.[2] But the assumption behind its gift shops' Indian displays is that of an updated, undifferentiated Indian.

Despite the evident good intentions of the curators at both this museum and the Colonial National Historical Park, the late-twentieth-century tourist attractions/history museums that celebrate the Jamestown experience repeat a number of the tragic problems surrounding the stories they tell; their gift shops unproblematically produce the undifferentiated Indian, and their exhibits unproblematically reproduce English stories and mythologies. What I mean here by "good intentions" is the visible work these museums do (especially the Settlement museum) to tell other sides of the story, to include the Native American point of view in the story, to introduce the African experience into the story, to regret both the history of slavery and the destruction of Native American lifeways, and even to argue with their own central mission: the celebration of settlement in Virginia. These good intentions can be seen in the Settlement museum's acknowledgment of the inaccurate romanticization of Pocahontas: the room focused on her makes no bones about her age at the time of her kidnapping; it does not treat that kidnapping as romance, and it acknowledges the history of her story—its construction as romance.

Those good intentions are especially evident in the excellent work of the costumed interpreters at the recreated Powhatan village, at least three of whom self identified as Native American[3] when I visited the site in the summer of 1998. Those interpreters were careful to explain to tourist groups and individual visitors that the site was created from English accounts; for the most part they do not pretend that the site tells the truth about Powhatan Indian life in 1607. The interpreters also stressed that there were multiple perspectives on the history the museum is supposed to tell. For example, on one day of my July 1998 visit, an older gentleman asked a female interpreter how the Virginia Indians had gotten to Virginia. She replied that some scientists believed these groups had crossed a land bridge from Asia but that there were other theories, and that if you asked a Native Virginian he or she would probably say they had always been there. When the tourist pressed her, she stressed that there were different stories about Indian origins in America and that they had different weight depending on one's perspective. This same interpreter carefully inserted into her storytelling the fact that there were still many Native Americans living in Virginia and elsewhere in the United States, a fact that most white Americans are trained almost from birth to deny.

But the Settlement's exhibit structure continually undermines these good intentions, disabling efforts to question the celebratory story of development the museum tells. That exhibit structure was largely determined in the circumstances surrounding the Jamestown Settlement museum's founding, and subsequent well-intentioned attempts to tone down the celebration have not eliminated the museum's underlying Eurocentric perspective. The museum and the three living history displays—the Powhatan village, the recreated English fort and village, and the replicas of the ships *Discovery, Godspeed,* and *Susan Constant*—that presently constitute the entertainment called Jamestown Settlement opened in 1957 to celebrate the 350th anniversary of the English landing at the point they named Cape Henry, the landing that resulted in the first permanent English settlement in North America. When the site opened for the public in 1957, it was named Jamestown Festival Park. As I hope to have shown in chapter 2, naming is always significant in the colonial world, and the name "Jamestown Festival Park" shows as much about the cultural world of late-fifties America as the name "Cape Henry" does about the patronage network central to the early settlement. For a social history that attempts to see the events involved in that landing and settlement from a bottom-up perspective, the entertainment's 1998 problems are clear from the entertainment's unproblematic recounting of its own history. Certainly at least one of the major groups of people whom the story is about, the "Powhatan Indians," might be little inclined to celebrate that landing and, indeed, their initial military response indicates their reluctance. Rather than seeing it as an occasion for "festival," these people might mark the event with mourning. The celebration of this 350th anniversary, like the troublesome celebrations of Columbus's landing,[4] is a celebration by the victors of what many on the other side see as a devastation to be deplored.

The new social history has challenged the celebratory atmosphere of the museum; and at least partly in response to that challenge, in 1990 the museum was overhauled, new galleries were added, and the site's name was changed to "Jamestown Settlement." The 1990 name change, from "Festival Park" with its happy, laudatory tone, to "Settlement," with its air of undeniable historical fact, indicates the designers' awareness of the name and mission's potential offensiveness. But the problems even with the seemingly innocent name, "Jamestown Settlement," are one sign of the entertainment's possibly inevitable failure as a site of balanced social history. For the site in 1607, or even in 1610—the approximate time of the colonial settlement that the recreated fort represents—was only "Jamestown Settlement" in the eyes and documents of the English living there. For the Indians, the site was a piece of Tsencommah invaded by some Englishmen and boys who might either be absorbed into their native culture or driven back to England. The English perspective indicated by the name "Jamestown Settlement" pervades the Settlement museum's exhibits.

For example, the curators have tried to include a bigger picture of the Atlantic world into their English gallery by including a corner exhibit in the English Gallery titled "Ireland: a Colony Close to Home." But the exhibit tells the story of English colonization in Ireland from the perspective of English gains and losses, even though, in the case of the Irish, a developed alternative historiographical tradition from the early modern period is available. When I was walking through the exhibit, I overheard a conversation that points to the problems with the exhibit's perspective. A white man in his late forties or early fifties called his well-groomed teenage son over to the display: "Come look at this—the English attitudes to the Irish." The young man came over, took a look, and responded, "I don't know, some of them, they still throw their trash out of the window." His father replied, "They're slow coming into the modern world, I guess." Of course, no museum can control its patrons' reactions, but exhibits set up the parameters that condition patrons' responses. In this case, the museum's patrons could see clearly that history means English people, "our" ancestors, who moved into the modern world, in relationship to "their" ancestors, who unfortunately did not.

The difficulty with even the name "Jamestown Settlement" is one sign of the museum's intractable problems, problems that may well be inherent in the tourist attraction's form. Richard Handler and Eric Gable's brilliant analysis of nearby Colonial Williamsburg finds a central and crucial difficulty with the genre, living history museum. They convincingly argue that "Mimetic realism . . . deadens the historical sensibility of the public. It teaches people not to question historian's stories, not to imagine other, alternative histories, but to accept an embodied tableau as the really real" (224). The denial of alternative stories that Handler and Gable find at Colonial Williamsburg, which depicts eighteenth-century Williamsburg, is even more unsettling at the Jamestown sites since these sites tell a much earlier story with fewer historical sources. The museum depends on the most prominent of those few sources, Captain John Smith's narratives, to tell a coherent story of settlement against great odds. Even the archaeological evidence used to construct the sites at Jamestown is so governed by those few sources and their assumptions that the archaeology, especially that featured at the national park site, repeats those stories, instantiating them scientifically as the "really real."

The Jamestown sites are late-twentieth-century colonial transformations. That is, they are places in which tourists are transformed into Americans or foreigners with a particular attitude toward the colonial American past. They are also sites in which that past is transformed for current consumption. But the sites are transformations that, despite their good intentions, actively resist depicting the feedback entailed in the complex processes of colonial transformation, the transformations of dominant cultures that this book has repeatedly demonstrated. Although the Jamestown Settlement brochure declares that "Colonizing Virginia meant overcoming problems, adaptation and change," the museum does not depict the English attrition to the Indians, and it does not address the problems the English had keeping the Indians away from English weapons. Many of the stories told in the Virginia Company records question the museums' basic narratives. Interestingly, these late-twentieth-century storytelling machines, the Jamestown Settlement and the Colonial National Park Site, like the 1993 Disney film *Pocahontas,* are more likely than earlier stories to be influenced by alternative, postcolonial understandings of the stories they tell. Yet, ironically, they can obscure these alternative visions more than do the sixteenth- and seventeenth-century texts this book has explored.

Although these museums' curators have clearly been influenced by their late-twentieth-century milieu, their displays show that they share some of the deep assumptions that governed English storytelling four centuries ago. These assumptions include the mythology of the single hero(ine), the idea that the world is organized in binary ways, the vision of history as progress, and the related idea that technology equals complexity and that both are good. The Jamestown sites and the early English sources share these assumptions despite the fact that the museums are completely creatures of their own time, with foundational ideas (such as the unquestioned assumption that individual merit and not rank should determine success) that most sixteenth- and seventeenth-century English people would have found completely alien, if not revolting. Thus the museums are essentially organized like the historical sources they use, but the museums must misinterpret or selectively interpret the sources in order to approve them and to tell the stories they wish to tell. That crux forms the late-twentieth-century Jamestown entertainments that teach the American public about England's seventeenth-century takeover of Virginia.

The indoor museum at the Jamestown Settlement contains three main galleries, the "English Gallery," the "Powhatan Gallery," and the "Jamestown Gallery," each subdivided into smaller exhibits. Visitors encounter the English Gallery first, although they may also view the educational film the museum provides and then visit the galleries. Thus the physical layout identifies visitors with English colonists. The gallery's first subsection announces "England Moves to Colonize" and "The Stage Is Set." As visitors walk through the indoor museum, they walk with the agenda of English colonists in mind, following their trajectory, encountering the rest of the museum as the English would have encountered it. However, such a trajectory is self-evident only if the English perspective is assumed. Imagine a mythical "museum of the earth" story about an alien invasion of earth that began with visitors experiencing the alien world and the alien's needs. Unlike that never-to-be-realized alien-centered vision, in its physical layout the Jamestown Settlement performs a similar operation to the court masque dance, physically imprinting a story on its desired audience and creating that audience as white Americans whose history is English history.

Both the Jamestown Settlement museum and the nearby Colonial National Historical Park offer an educational film to orient visitors. The Jamestown Settlement's educational film tells the story of the English settlement's early years fairly—that is, it does not depict the English as particularly righteous people encountering savages—but it is narrated by an "English colonist" and it shares his perspective without questioning his story. The film clearly asks the audience to sympathize with its narrator, who relates his own and the settlement's tribulations, and who gets in what is only the first of the museum's gibes at "gentlemen" who refused to work at Jamestown but were finally forced to acknowledge that only labor would save the day. The Jamestown Settlement museum film is vastly better than the film offered at the Colonial National Historical Park. That film, which is mostly a narration over a static view of a large-scale painting, includes lines like "a few red men watched" the English landing. The Colonial National Historical Park's film recounts English tribulations in Virginia as steps backwards, and the film's narrator states, "In 1622 the step backward wore soft moccasins." Indians in the film are thus red, moccasined impediments to white Virginian success. While the National Park's film is blatantly biased in favor of the English side of the story, the Settlement museum's film acknowledges the Indians as more than stereotyped red people in moccasins. But the narrator's point of view is still English, although it is critical of English nobility.

This narrator's anti-nobility and anti-gentry perspective is one sign that the Jamestown Settlement museum could have been designed by Captain John Smith, whose narratives frequently castigate useless gentlemen. Smith's narratives get pride of place in the museum; excerpts appear throughout, lending a sense of direct historicity to the texts below the quotations, exhibit labels whose matter mostly derives from Smith's narratives. This bizarre circularity would be apparent only to a reader familiar with Smith's texts; and it has the odd effect, if one is not familiar with those texts, of lending particular historical weight to the labels themselves, which relate the historical facts that, not surprisingly, confirm Smith's seventeenth-century voice. The museum's wall devoted to Smith—the only chronicler so honored—acknowledges Smith's critics.[5] One line of text above his picture states, "Many contemporaries accused John Smith of being 'an ambityous, unworthy and vayneglorious fellowe,' but history has made him a hero." But the museum nowhere acknowledges its own almost unbelievable complicity in that history of heroism. Instead, Smith's words stand for the truth about the English Settlement and, most egregiously, about the Powhatan Indians. In the Powhatan Gallery, Smith's texts hold a prominent pride of place. So Smith, an Englishman who was undeniably economically and ideologically interested, becomes the disinterested empirical observer[6] who introduces the late-twentieth-century American (and foreign) tourist into the world of the Powhatan Indians in 1607.

Perhaps the Jamestown Settlement museum loves Smith so much because it mistakenly sees in him the initiation of its own beloved foundational assumption, that the English temporarily failed in Jamestown because of their reliance on soon-to-be outdated and morally wrong ideas about rank.[7] The Jamestown Settlement museum (mis)reads Smith as a twentieth-century capitalist who believed in individual merit and the rise and survival of the fittest, instead of as a man born into a world of rank and privilege who was attempting to attain rank and privilege for himself. Smith's stories about useless gentlemen were not attempts to disrupt the system of rank and privilege; rather they worked within that system, attempting to displace those gentlemen who didn't act like proper gentlemen, and attempting to burnish Smith's own condition. Smith was certainly no Henry Paine (whom we read about in chapter 2) intent on overturning hierarchy, but a realistic appraisal of Smith's motives in his time would disrupt the museum's hagiography. It might also force the museum to contend with the truth: that there is not one even slightly unmotivated story to be found about the early Indian and English experience at Jamestown.

The museum seems to need heroes, and because of that need it avoids complexity where the historical sources offer not heroes but problems, confusions, and mixed motives. Perhaps the search for heroes is inherent to the genre of tourist attraction/living history museum—what would Disneyland be without its larger-than-life characters? But such hagiography prevents the public from making its own decisions about the sources and about the people involved in the invasion/settlement. At the least, the installation of heroes seriously distorts history. For example, the Jamestown Settlement museum is involved in the cult of Queen Elizabeth I,[8] an odd focus for a museum about the early seventeenth century. Elizabeth, who happily supported lucrative pirate enterprises, was much less enthusiastic about less immediately profitable American ventures—witness Sir Walter Raleigh's continual unsuccessful attempts to get her to fund his South American ventures. And the queen died in 1603, four years before the *Discovery, Godspeed,* and *Susan Constant* landed in Virginia. Yet a large statue of Elizabeth greets museum visitors, and the queen is mentioned frequently in labels. King James I gets much less play, although he was actively involved in the settlement's affairs. Since the museum wants to tell a quantitatively verifiable story about tobacco's role in the colony's eventual expansion, and since James was opposed to tobacco cultivation and smoking, perhaps he can't be made a hero. But the museum's neglect of his multiple roles in England's Virginia ventures in favor of the more charismatic Elizabeth repeats seventeenth-century mythmaking at the expense of a more accurate and critical history. In July 1998, the visitor who walked into the outdoor gift shop at the Settlement museum's entrance was greeted by an endcap display rack offering trinket boxes with Elizabeth's picture on them next to boxes displaying Pocahontas's picture, the conjunction in all probability leading to a distorting temporal confusion, as Elizabeth died before any English person had met Pocahontas.

While the museum's hagiographical impulses determine its use of Captain John Smith and the Virgin Queen, the museum's Powhatan Gallery displays are governed by progressivism—the idea that technological complexity equals valorized progress. Assuming that our imaginary typical visitor, interpellated as a seventeenth-century English person by the English Gallery, has passed through that gallery, he or she now enters the Powhatan Gallery through a short hall on which John Smith's words about these Indians are inscribed. The museum's brochure assures its visitors that "Objects in the [Powhatan] gallery teach an appreciation of an efficient and complex culture"; but this attempt at inculcating cultural relativism again betrays the museum's foundational assumptions. The brochure, of course, doesn't feel the need to assure visitors of English culture's efficiency and complexity, nor does it challenge the idea that efficiency and complexity are values any culture would be happy to represent. In fact Western "complexity" becomes the scale by which the Powhatans are judged. As our imaginary visitor enters the gallery, he or she sees a wall with a long semicircular display case that depicts human events in North America compared to human events in the rest of the world from 400,000 B.C. to the time of the Powhatan culture the English encountered. In this display case, the visitor looks at labels that read, "pottery making began in the Middle East ca. 7000 B.C." and "the first cities 'rose' there in the period from 4000-2500 B.C." The display informs our visitor that the Southeastern woodland Indians (in implicit comparison) were first cultivating crops and building permanent villages, after the "Golden Age of Greece" and during the period that the great wall of China was built (ca. 215 B.C.) and the first Roman Empire was in place, from 27 B.C. to 476 A.D. The display case assumes a belief that all societies progress toward increased complexity and that this progression is a necessary good. Included in that governing assumption is the idea that the woodland Indians the English encountered in Virginia, the Indians the entire gallery subsumes under the name Powhatan, were clearly progressing on the same trajectory as the other civilizations in the Western and Eastern world. They were just a little slower.

Thus, while the display case is obviously intended to show commonalities between cultures, its insistent narrative drive toward "complexity" ascribes a childlike status to Powhatan culture. The case "shows" that while the people of the Middle East, the "cradle" of Western civilization, had achieved agriculture, pottery making and cities by the year 2500 B.C., the Powhatan Indians had attained these cultural heights only by about 1215 A.D. By that time, Homer had composed *The Iliad* and the Magna Carta had been signed. Such a comparative trajectory automatically privileges development, never questioning the human cost of the events and monuments it celebrates. For example, the Great Wall of China and the Egyptian pyramids, both marked on this display case's history of civilization, are presented as massive accomplishments rather than as documents of the toil and suffering of the human beings forced to build them.

Under this progressivist ideology, Powhatan culture in the Powhatan Gallery is necessarily an early manifestation of the greatness that the human race in general is constantly striving to achieve. The Indian culture, as the museum unwittingly presents it, is both early, childish, and pristine, absolutely separated from the much more complex English culture presented in the previous exhibits which also happen to be much more physically colorful. Thus the Powhatan culture's loss becomes a loss to a white audience of a simpler way of life, closer to nature—the kind of simplicity that American culture perennially calls for in the face of the pressures of modern life. This nostalgic appreciation of another culture as the childhood of one's own culture is one of the dangers that Kenneth Hudson points to when he states, "in today's world, to emphasize 'traditional culture' is not, in my view, a particularly responsible or constructive thing to do, however attractive it may be from the point of view of showmanship" (460). Hudson suggests that privileging the past over the present enables museum visitors to lament a loss rather than to engage constructively in the contemporary problems of the people on display. The Settlement museum's Powhatan Gallery makes Hudson's argument powerfully, since it does not encourage visitors to think about the descendants of the people it displays as Indians in the United States, whose historical claims to land and to justice might be considered today.[9] The white viewer is left to mourn a loss rather than to consider public action to remedy the problems created in the settlement the museum depicts.

Both the mythology of heroism and a progressivist ideology also determine the materials available at the Colonial National Historical Park site. The text on the brochure and map a visitor receives upon entering this site does not even consider the place of Indians in its narrative of progress towards a laudable present. Under its silhouette photograph of revolutionary soldiers (obviously costumed men from a historical recreation) and above the map that shows Jamestown Island in relationship to Williamsburg, Busch Gardens, and the Yorktown Victory Center, the brochure has inset a description of the park sites by the most famous twentieth-century interpreter of National Parks, Freeman Tilden. Like the Settlement museum displays, Tilden's text assumes that human nature tends toward individualism, toward disregarding systems of rank, and toward heroism. Captain John Smith is again a figure for the future in Tilden's text:

> But leaders like John Smith and John Rolfe had finally infused a spirit of order in the town. And what happened is perhaps inevitable in colonialism. Outlanders cling to old ways, but they develop new ways. They retain loyalties, but the loyalties are no longer blind. Self-reliance and the struggle for existence take effect. They look like the same men, they talk like the same men, but they are stubbornly themselves, and want the fact acknowledged. It is a period of coming of age.

Tilden's story is a story of colonial transformation, but unlike the documents from the early seventeenth century that tell stories of transformation, Tilden's story does not register the multiple possibilities and dangers to English identity inherent in colonial transformation. Instead, his story is a story of progress; the English experience in Virginia is the childhood of the great individualist twentieth-century America. Colonialism, in Tilden's text, is the story of the *men* who leave their home in the mother country to face the challenge of growing up into "themselves" in the outland. Tilden's inset text assumes an American tourist identity that celebrates and was formed in all the events at the colonial sites. He states: "It may be that the ultimate value of history comes through the mere contact with historic preservations. . . . In a place of great historical importance, the visitor subtly becomes part of that history. . . . We find in history the explanation of why we are *we*, why I am *I*, why you are *you*." Tilden's emphasis on a *we*, an *I*, and a *you*, at the end of a text that erases Indians and Africans from American history, creates a white identity that is the end result of historical progress.

Although the Jamestown Settlement museum incorporates Indians and Africans into its story of the American "we," its Powhatan gallery displays share the narrative of progress with the more retrograde National Park site. And any narrative of progress is ultimately a story that celebrates what has happened as inevitable. To many white Americans, trained to celebrate America and the American way, Tilden's interpretation and the narrative of progress that stands behind both the Colonial National Historical Park and the Jamestown Settlement museum would seem unremarkable. But should anyone celebrate the purposeful destruction of cultures just because that destruction helped to form the culture we live in today? Such a logic, as Churchill points out, would lead to German celebration of the Nazi era, a prospect that many Americans would view with horror (Churchill 1996).

People from two very different historical moments and places, late-twentieth-century Virginia and early modern England, share the foundational assumptions that stories have single heroes, that history is a story of progress, and that human society (happily) tends towards technological complexity. They also both organized the world in clear binaries, like the early modern ideas of savagism and Englishness, binaries that colonial transformation continually threatens to disrupt. The Jamestown Settlement museum and the Colonial National Historical Park both depend on a binary thinking that determines their finally destructive visions of history. That binary is evident in the design of the Settlement museum's galleries and its outdoor living history sites. The indoor museum at the Settlement has a Powhatan Gallery and an English gallery, as if the story of settlement started with separation. But it clearly did not, since the sources we have for the story are all sources that depict the cultures during and after encounter. The Settlement's outdoor museum has a Powhatan Village and an English Fort completely separated from one another. During my visits, the English Fort never had costumed Indians walking around in it, and the Powhatan

Village was likewise devoid of costumed colonial English people. This living history museum therefore denies the transformations going on in both cultures as they encountered each other, transformations pointed to even in the John Smith narratives the museum depends on for much of its information. All of the Virginia colonial narratives show that the English fort always contained Indians and that Indian villages had English inhabitants and visitors. And, as this book has shown, sometimes the boundary lines between these identities blurred completely.

The Colonial National Historical Park, which has sponsored archaeological work on the early period of settlement, can claim a scientific basis for its depiction of early colonial culture.[10] But the archaeology at the sites is governed by both the English stories from the settlement period and by the binary thinking that determines both the tenor of those stories and the shape of these twentieth-century tourist attractions. Archaeologists, after all, don't approach sites without doing their historical homework, and thus far that homework has been a reading of the primary sources of the encounter as if they told the coherent story of settlement that Tilden's interpretation represents. Therefore the artifacts on display found at the digs are sorted by cultural origins: guns and metal implements belong to English settlers, and items of Indian manufacture belong to Indians. This displayed archaeology denies the mixture of cultures that I discussed in this book's introduction. I find it tragic that science can confirm what so many English people actively struggled to present: a story of a whitened English identity and a backward Indian culture.

This book has attempted to show that the new Atlantic world created during the English literary renaissance was intimately connected to that renaissance. That literary renaissance dealt continually with the colonial transformations forming that world. When we teach that renaissance, when we study it, and when we re-create it in productions, in children's stories, in films, we are actively either telling the stories of colonial transformations or omitting them in a conscious or unconscious effort to obscure those connections. Along with telling the stories of some of those connections, I have shown how the transformations entailed in that new Atlantic world have contributed to forming that world as we know it today. That historical process is real even though it was not inevitable. It is up to us in the American twenty-first century to tell other stories.[11]

Notes

Introduction

1. Quotations from Shakespeare are from the Hinman folio, which, though it is a composite text, at least represents a text of its time. For ease of reference I have also included citations to *The Norton Shakespeare*. Throughout this book I have tried to cite either original texts or old-spelling editions wherever possible.

2. See Cavanagh 1993, Bradshaw, Carroll, Hadfield, Highley, and Maley.

3. See Brown 32.

4. On the question of place and topicality in Shakespeare, see Marcus 1988, especially chapter 4, "London."

5. See Ralph Lane's letter to Sidney from the Roanoke colony. Hall 1995 describes Sidney's imperial interests, 73-74; see also Fuller 17, 19, 146, 178-79n.13.

6. Bruce McLeod's introduction also attempts to counter this argument by showing "the interaction between literary culture and the developing world of Britain's first empire" (7), although most of the texts McLeod reads were written in the Restoration and early eighteenth century. See also Lim's introduction.

7. Within the boundaries of the contemporary state of Virginia there were probably at least thirty-six Algonquian, Iroquoian, and Siouan groups. Algonquian: Accohanoc, Accomac, Appomattoc, Arrohattoc, Chesapeake, Chickahominy, Chiskiack, Cuttatawomen, Kecoughtan, Mattaponi, Moraughtacund, Nansemond, Nantaughtacund, Onawmanient, Pamunky, Paspahegh, Piankatank, Pissaseck, Patawomeke, Powhatan, Quiyoughco-hanock, Secacawoni, Tauxenent, Warrasqueoc, Weanoc, Wicocomoco, Youghtanund; Iroquoian: Cherokee, Nottoway, Meherrin; Siouan: Mana-hoac, Monacan, Saponi, Tutelo, Mohetan, Occaneechi. For a breakdown of groups with approximate numbers, see McCary. Also see Rountree 1989,

especially 9–12. Of course all of these names and numbers were gathered out of English sources and are therefore clearly not definitive.

8. Even to call the Virginia colonists English is somewhat of a problem since in pursuit of oft-mentioned dreams that the Virginia colony would produce all the goods that England previously had to import from the continent and Asia, the English adventurers in London recruited and sent French vignerons, Dutch ironworkers, glass-blowers, and builders, and other foreign nationals to the colony. See for one example the 1620 Broadside "A Note of the Shipping, Men, and Provisions, sent and Prouided for Virginia: "Foure Dutch-men from *Hambrough*, to erect Sawing-Mills . . . Eight French *Vignerons*, procured from *Languedock*, who are very skilfull also in breeding of *Silke*-wornes, and making *Silke*" (*Records* III 240). These men were also part of the Virginia colony, although it seems from Captain John Smith's repeated xenophobic comments about them that many English did not accept them as English, and they may not have identified themselves with the English colony (see Smith II 209, 226).

9. "Colonizer" is also a late word in English. The *OED* finds its first usage in 1781. Of the pantheon of related words—"colony," "colonize," "colonial," "colonialism," "colonialist," "colonist," and "colonization"—only "colony" and "colonize" are sixteenth- and seventeenth-century terms respectively. William Strachey struggles for a term like the modern colonist as he addresses rules of conduct to "euery Colonell, Gouernour Captaine, and other Officer . . . as they are Souldiers, but happily not yet as they are, or may be Coloni, members of a Colony" (1612 F3). The usages the *OED* finds for "colony," from a 1548-49 English discourse on Scotland to Richard Eden's translation of Peter Martyr's account of the Spanish conquests in South America, to Samuel Purchas, to Davies' discourse on Ireland, accurately delineate a genealogy of English imperialism. Although they are anachronistic, then, I will use the terms as extrapolations from their progenitor "colony."

10. See also Stuart Hall 247.

11. Because none of the general names used for the people living in North and South American before Europeans arrived is satisfactory, I will use the equally unsatisfactory "Indian," which is often used by activists in the United States. See Sayre's introduction for a good discussion of the problems involved. My second chapter also addresses the theoretical difficulties involved.

12. While the scholarly work on the encounter places more than twenty of these groups under the rule of Powhatan, the chief of one of these groups, and therefore calls them the Powhatan Indians, even this consolidation seems questionable on at least two grounds: 1) It relies solely on English accounts that are both contradictory on this point and interested in making it. The English were searching for one supreme leader on the model of their

own country and of the Spanish American empires. They found Powhatan relatively early and established trading relations with him. But if his "empire" was as powerful and coherent as the scholarship generally claims, how did the Chickahominy, who lived so near the Powhatans retain their independence? Spelman's account of his residence with Powhatan and then his leaving Powhatan for the king of Patowemeke also makes the claim for Powhatan's power over neighboring groups suspect. 2) The coherence of groups that the English "found," if it existed at all when they arrived in 1609, may well have been formed in relation to their arrival as an outside enemy; or, if it was formed prior to the influx of English settlers early in the seventeenth century, it could well have been formed in relation to earlier Spanish missionary efforts, to Spanish revenge expeditions after those missionaries were killed (see Lewis and Loomie), to the abortive Roanoke efforts in the 1580s, and/or to the inevitable disease that must have accompanied at least some of these European encounters.

In the early 1620s Powhatan's brother Opechancanough organized a rebellion against the settlements that included a number of Indian groups, and English indiscriminate retaliation enlisted groups that had not participated in that rebellion into subsequent resistance efforts. See Rountree 1989 for a rationale for calling these groups the Powhatan Indians.

13. Recent ethnohistorical and anthropological work has tried to be attentive to differences among the Indians in Virginia. However that work has been stymied by its (forced) reliance on English sources. Therefore even this excellent work generalizes the Indians finally into one "Powhatan" culture with one set of cultural characteristics. For an example of this sensitive work, see Gleach.

14. On this question see the active debate in postcolonial studies, which includes debate over the term "postcolonial," in the 1995 special issue of *Ariel* 26:1, in Chambers and Curti, and in Aschcroft, Griffiths, and Tiffin. *Colonial Transformations* contributes to an ongoing project that aspires to keep power in sight while acknowledging complexity. Catherine Hall describes this project as the effort to remember "empires differently . . . [with] the recognition of inter-connection and inter-dependence, albeit structured through power, rather than a notion of hierarchy with the 'centre' firmly in place and the peripheries' marginalized" (66-67). *The Postcolonial Studies Reader's* editors assert that "Theorizing this complex 'intimacy' [the 'peculiar intimacy of colonizer and colonized'] without giving away the fact of persisting and historic *inequalities* within those relations and structures is perhaps *the* major focus of contemporary post-colonial theory" (86).

15. Likewise, England's Irish colonial policy was severely criticized in the seventeenth century. See Highley 151.

16. While certain texts from 1622 advocate total extermination, the English

writings and policies were quite divided and contradictory on the question of policy toward the Indians in the wake of the attack. In August 1622, the London Virginia Council wrote: "we must advise you to roote out from being any longer a people, so cursed a nation, vngratefull to all benefittes, and vncapable of all goodnesse," but even they advocated saving the children's lives in favor of what Churchill terms "cultural genocide" (*Records* III 672). Five months later John Martin, a colonist, wrote that the colonists needed the Indians to control the natural world around the settlements (*Records* III 705-706).

17. Note that the colonists' hunger can't excuse their destruction. Although this course of action doesn't seem to have been seen as a viable option, they could always have gone back to England forever.

18. One example from the literature on Virginia is Ralph Hamor's account of Pocahontas's kidnapping, where another Indian, Iapazeus, and his wife betray her for "a small Copper kettle, and som other les valuable toies so highly by him esteemed, that doubtlesse he would haue betraied his owne father for them" (5).

19. This is the 1622 quarto reading. The folio reads "base Iudean." Most modern editions famously choose the quarto here; the shifting possibilities illustrate my point below about representations of differences and hatreds taking their energies from one another.

20. See for example William Capps's letter to Doctor Thomas Wynston (*Records* IV 37-39) and the more famous Edward Waterhouse text that Hulme bases his argument on.

21. On the consonance between low-ranked English people and the "meaner sort" of Indians in the eyes of their social superiors, see Kupperman's essential book. My analysis is indebted to hers although we differ sharply on the issue of English accuracy in reports about the Indians.

22. Another attempt in the same year to silence detractors is the twelfth of William Strachey's laws for the colony published in London in 1612:

> No manner of person whatsoeuer, shall dare to detract, slaunder, calumniate, or vtter vnseemely and vnfitting speeches, either against his Maiesties Honourable Councell for this Colony, resident in England, or against the Committies, Assistants vnto the said Councell, or against the zealous indeauors, & intentions of the whole body of Aduenturers for this pious and Christian Plantation, or against any publique booke, or bookes, which by their mature aduise, and graue wisedomes, shall be thought fit, to be set foorth and publisht, for the aduancement of the good of this Colony, and the felicity thereof vpon paine for the first time so offending, to bee whipt three seuerall times, and vpon his knees to acknowledge his offence, and to aske forgiuenesse vpon the Saboth day in the assem-

bly of the congregation, and for the second time so offending to be condemned to the Gally for three yeares, and for the third time so offending to be punished with death.

From the calumnious history that followed this law's publication, it clearly deterred few if any people angry with the Company and its policies. Indeed, had the law been pursued, a number of the Company's noble adventurers would have been hung for their remarks during the internecine struggles that prompted the Company's demise.

23. Two important recent books, Gesa MacKenthun's *Metaphors of Dispossession* (on links between Spanish and English colonial discourses) and Gordon M. Sayre's *Les Sauvages Américains: Representations of Native Americans in French and English Colonial Literature*, place England's American expansion in relation to continental colonial enterprises.

24. Andrew Hadfield, 1998, discusses Richard Hakluyt's similar concerns (104).

25. See Craven's *Dissolution* and Ransome's cogent introduction to Ferrar.

26. Such advancements were unusual but not impossible. On November 28, 1618, John Chamberlaine wrote to Sir Dudley Carleton (who in 1610 had hoped to go to Ireland as Secretary) that "one Captain Yardley a meane fellow by way of provision goes as governor [of Virginia], and to grace him the more the King knighted him this weeke at Newmarket; which hath set him up so high that he flaunts yt up and downe the streets in extraordinarie braverie, with fowrteeen or fifteen fayre liveries after him" (II 188). Chamberlain's parodic picture of Yardley's braving exposes the threat such advancements posed to the structured hierarchical court world.

27. See Strachey 1612, for example.

28. This claim could be made for English and U.S. cultures throughout their histories on the grounds that cultures must always enculturate their children, but it applies especially well to early modern English culture before the rise of individualism and cults of childhood.

29. See also laws against Englishmen traveling and going to Indian settlements (*Records* I 171; Strachey 1612 B4,Gv). The English had the same problem with their troops in Ireland deserting to the Gaelic enemy forces. Highley speculates that Gaelic lifeways may have seemed preferable to poor Englishmen (90-91). On this phenomenon throughout colonial America, see Axtell 1985, 302-27. See also Morgan 81.

30. At the age of thirteen, Thomas Savage had been given to Powhatan by Captain Newport and Captain John Smith.

31. See Cheyfitz and Murray for vital work on translation in colonial enterprises.

32. See J. H. Lefroy's nineteenth-century comment that the early narratives of Bermuda's settlement "enable us to account with certainty for the introduction of many species which are now naturalised, but which might, in the absence of information, have been taken for native" (I, 54).

33. In her brilliant 1988 attack on English colonialism and its successor, tourism, *A Small Place*, Kincaid explains that Antigua's sunny appeal to tourists stems from that drought, and that "since you are a tourist, the thought of what it might be like for someone who had to live day in, day out in a place that suffers constantly from drought, and so has to watch carefully every drop of fresh water used (while at the same time surrounded by a sea and on an ocean—the Caribbean Sea on one side, the Atlantic Ocean on the other), must never cross your mind" (4).

34. See Hulme and Mackenthun.

35. Cronon's magnificent history of the ecological changes wrought in New England by English colonization offers an indispensable model for understanding the effects of such colonial efforts. Similar studies of the other regions in North America, including Virginia, are badly needed. Cronon makes the important point that all human land use changes the land. Indian practices of burning forests to clear land for farming clearly changed the land on which they dwelled, but English livestock, industry, and ploughing were vastly more transformative and their effects more irrevocable.

36. Joseph Roach notes that "the traditional Mohawk word for the Dutch, *Kristoni*, "metalworkers," complemented their term describing Europeans in general, *Asseroni*, or "ax makers" (122, after Richter 75). Chapter 2 investigates the implications of such namings.

37. Just as livestock had a transformative effect on the land and people in England's colonial margins, so Virginia and Bermuda's only successful cash crop, tobacco, which required vast cleared fields, irrevocably transformed those lands' ecosystems. And tobacco, although it goes up in smoke when lit, materially transformed English culture as well as North American land and cultures. Tobacco made money, transferring the flow of English capital to Spain to an internal colonial capital exchange, and it also changed English physiologies and cultural habits due to its addictive and transformative nature. Tobacco was also immensely profitable, and colonial records document the search for profit through this commodity far more than they record its ephemeral quality. The late-twentieth-century U.S. and state government attempts to regulate tobacco, and the tobacco industry's willingness to settle lawsuits for billions of dollars, speak to the vast and myriad material effects of English addiction to smoke. Compare Knapp's interesting discussion, chapter 4.

38. Unlike a number of recent Renaissance scholars who read colonial documents, I did not come to these texts through Greenblatt's work. That interest was kindled by my graduate work with a scholar of early American literature, Myra Jehlen. Reading that literature, I realized that it was as much or more a part of the early modern English culture that I primarily studied as it was a part of something that could be called American literature. But Greenblatt's work has clearly enabled my efforts as it has so many others' by

making my project conceivable in the field described by the term English Renaissance literary studies.

39. Apropos this book's focus on transformation, the first English poetry we know of produced in Virginia was Sir Edwin Sandys' brother George's translation of fourteen books of Ovid's *Metamorphoses*, from the 1623 edition of which this introduction takes its epigraph. For other productive work breaking those barriers down see Greene 1994 and 1995, Bartels, Highley, and Montrose.

40. Although my major concern is with English territorial expansion in the Atlantic world, that expansion was also linked to commercial expansion in the East Indies and Africa. In chapter 4, I examine some of those links. Forthcoming work by John Michael Archer and Walter Cohen examines the vital importance of that commercial expansion to the early modern English cultural world. See also Bartels 1997 and Banerjee.

41. And the process worked the other way as well, as soldiers and other adventurers reached literary heights in their descriptions of colonial exploits. See Haile's introduction to Strachey's *True Reportory*. The text that is attributed to Nathaniel Butler (see chapter 2) is also magnificently literary, in a traditional sense, as is Captain John Smith's *Generall Historie*.

42. See Abbot Emerson Smith 8.

43. In the coming years, the English would begin bringing African slaves to other islands in the West Indies. Brenner notes that in 1626 two merchants sent sixty slaves to their plantation on St. Kitts (127). The Bermuda settlement became an initiation point for English colonial efforts in the West Indies as a whole. See Brenner 161-66.

44. For a metropolitan adventurer's implication in the earliest colonial slaveholding, see the *Historye*'s account of Sir Edwin Sandys's dispute over who owned a group of Bermuda slaves (242-44). That dispute was one of the issues in the bitter fight for control of the Virginia Company that lead to its demise.

45. On the trope of "washing the Ethiop white," see Hall, the essays in Hendricks and Parker; also see McClintock; also see Dyer on later constructions of whiteness in relation to washing and purity.

Notes to Chapter 1

1. All citations to Spenser's shorter poetry are taken from the Yale edition.

2. Spenser frequently notes Ireland's seductive physical beauty; see the description of Ireland's rivers IIII.xi.43 and Arlo-hill in the Mutabilitie canto VI.

3. See Shannon Miller for an interpretation of the Bower of Bliss episode that points to "England's idleness in pursuing New World projects" (46).

4. See Orgel 1996 and Breitenberg for analyses of this fear in early modern English culture.

5. See also Bernard; MacArthur offers an insightful review of the criticism; see Hunter for an astute commentary on the numerological criticism.

6. For earlier considerations of the poet as planter, see Jenkins 1932, 1937, 1938 and Henley.

7. Rambuss's important study offers another way to situate Spenser's poetic career within his daily life. See Miller 75-82 for a reading that places *Colin Clouts Come Home Againe* in an Irish and "New World" context.

8. But see Wells, Bates for important exceptions. See also Thompson for an interesting analysis that reads the poems as recording the transition from Protestant martyrdom under Mary to Elizabethan Protestant triumph.

9. See Jones and Stallybrass for accounts of the gendering and sexualizing of Ireland.

10. See Hulme's fascinating discussion of this poem. Also see Montrose on Raleigh's *Discoverie of Guiana* (1991).

11. See Mackenthun's analysis of Morton's feminizing discourse.

12. See also Robert Dunlop.

13. Michael MacCarthy-Morrogh argues that "ultimately the 1588 commission's role was insignificant. It did not provide an everlasting decree concerning these land disputes; it did not put a stop to the claims and petitions; it failed to halt the growing decrease of the plantation's area" (100). Nonetheless, if "by 1611 about one-third of the whole plantation area had been returned to the local inhabitants," the new settlers still retained two thirds of the escheated land (106).

14. See *The Spenser Encyclopedia* which notes that "three extant letters to Richard . . . indicate her gratitude for his generosity" (109). Canny suggests that Boyle's godchildren were well-treated: "Such christenings were seemingly of importance to him since he recorded them in his diary. Furthermore, he followed the careers of his godchildren more closely than those of their siblings" (1982, 45).

15. Rambuss and Waller both offer helpful accounts of the poet's career difficulties.

16. I am not suggesting that other Elizabethan sonnet sequences were wholly removed from England's colonial life and courtly or public world. See Marotti and Hall for expositions of these connections. Certainly *Astrophel and Stella* can be read in relation to Sidney's and Sidney's father's Irish interests (Canny 1976, 66). Also see Fuller 18-19 on Sidney's colonial interests. Language relating to tyranny and sovereignty pervades sonnet sequences; see especially Samuel Daniel's *Delia*. Paradoxically, it is *Amoretti*'s "happy ending"—the lover's triumph—that has led both to critics celebrating the sequence as lauding mutual love and also to my understanding of its colonial interests and consequences. Assessments of the poems as celebrating "true love" must, however, disregard the telling ambiguities in the final sonnets.

17. Of course this link is not confined to the early modern period. The story

is classical, told by Herodotus. And effeminization (although its meaning and mode has changed) is still a way of dealing with enemy and conquered nations.

18. In book two, canto three of the *FQ*, Belphoebe, although she seems "heavenly because of her "stately portance" (21) explicitly rejects the "pompe of proud estate" and it is Bragadocchio who is compared to a "fearfull fowle" who "prune[s]" proudly and "renewes her natiue pride" (36).

19. See Prescott for a fascinating comprehensive analysis of Spenser's possible sources and contexts for this poem.

20. Compare Ralegh's triumphant description of his entry into Guiana's heart:

> On both sides of this riuer, we passed the most beautifull countrie that euer mine eies beheld: and whereas all that we had seen before was nothing but woods, prickles, bushed and thornes, heere we beheld plaines of twenty miles in length, the grasse short and greene, and in diuers parts groues of trees by themselues, as if they had been by al the art and labour in the world so made of purpose: and stil as we rowed, the Deer came downe feeding by the waters side, as if they had beene vsed to a keepers call. (57)

21. See Canny's discussion of Boyle's integration into Irish culture (1982, 126–27).

22. Settlers in the next wave were not quite as generous about their predecessors' innocence. Fynes Moryson writes of the rebellion that drove Spenser out of Ireland:

> And to speake truth, Munster undertakers . . . were in great part cause of this defection, and of their own fatall miseries. For whereas they should have built Castles, and brought over Colonies of English, and have admitted no Irish Tenant, but onely English, these and like covenants were in no part performed by them. Of whom the men of best qualitie never came over, but made profit of the land; others brought no more English then their owne Families, and all entertained Irish servants and tenants, which were now the first to betray them. (II, 219)

23. See also Moryson, who following Camden writes that the Italians and Spaniards "yeelded themselves, in the yeare 1583, and were put to the sword, as the necessitie of the State, and their manner of invading the land, was said to require" (II, 171-72).

24. See Ellis 282 and Haynes-McCoy 108.

25. Tadhg Dall Ó Huiginn also uses this tale in a praise poem to Conn O'Donnell. His maiden Ireland has been enchanted into a "great, forbidding she-

dragon" (Knott 1922, XXIII 3). Ó Huiginn's enchanted maiden is a less active suitor than Ó hEoghusa's, but she takes an active part in awaiting her destiny of being rescued by the true Irish King, taking a vow not to appear for anyone but him (4). Both poems participate in a complex colonial transformation, since the source story for both poets appears to be a 1475 Irish translation of Mandeville (see Knott XXIII 191 note to stanza 19).

26. See Canny's discussion of English official response to Gaelic poets (1988, 20-22).

27. See also Cairns and Richards, chapter 3.

28. See Balzano and Herr for the impact of this mythology on Irish women and Irish culture generally. Herr notes that "The Mother Ireland Myth does show remarkable resilience even in the face of ongoing critique" (24n. 46). Also see Keane on the Gaelic origins and twentieth century manifestations of the myth of Ireland as a devouring mother. Nandy describes a similar dynamic where colonial consciousness which constructed "femininity-in-masculinity . . . as the final negation of a man's political identity" (8) resulted in Indian protest movements seeking "to redeem the Indian's masculinity by defeating the British, often fighting against hopeless odds, to free the forme once and for all from the historical memory of their own humiliating defeat . . . This gave a second-order legitimacy to what in the dominant culture of the colony had already become the final differentiæ of manliness" (9).

29. See Stoler 1995, 104-105 on these fears in nineteenth-century Java.

30. And the process could also work the other way. Canny argues that

> In a few baronies in Ulster and Munster those Irish tenants who were taken on by British landlords were introduced to an environment in which British settlers were in a majority, and where this occurred the Irish usually responded to the dominant cultural influence by becoming conversant in English and by acquiring such obvious English habits as pipe smoking. But absorption could go further than this, and some Irish tenants were prevailed upon to attend services of the Established Church and even to alter their names to Anglicized forms (1988, 45).

31. Canny 1976 notes that these Old English lords took on Gaelic titles and practices to rebel but "held back from making common cause with, and providing leadership to, the Gaelic Irish" (150).

32. See Nandy's analysis of Kipling's colonial experience in India.

Notes to Chapter 2

1. For this and other cartographic theory references, I am indebted to Meaghan Duff.

2. On the struggle against racist sports team names, see Churchill, "Let's Spread the 'Fun' Around," 439-443, in *From a Native Son*. On renamings and conflicts over changing toponyms, see Harley 1990 4, 15 and Rundstrom 9.

3. That relativism is thinkable but abhorrent to an early-seventeenth-century colonialist is shown by Samuel Purchas's anxiety over his collection of religions and customs from around the world. Purchas declares in his dedicatory letter that there is "no admittance of Paritie" among the religions he presents, "all Religions on Earth, as here we show, being equally subject to inequalitie, that is, to the equitie of subordinate Order" (3v).

4. On the functions of this poetry, see also Bradshaw 1979.

5. I am indebted to Jorge Cañizares Esguerra for this reference. Seed's brilliant book talks briefly about Spanish naming practices (190). She claims that the English, virtually alone among the colonizers she treats, did not see naming as important in the colonial world; however, as my arguments in this chapter demonstrate, her evidence is incomplete (174, 190). See also Schmidt on English mapping and naming and Sayre's preface for arguments about English and French naming practices, especially their namings of the native groups they encountered.

6. Although the careers of the second group—Hood, Rodney, and Nelson—are separated by almost two hundred years from the first group's careers, they are linked by more than just a hegemonic imperial imagination. In 1588, Drake and Hawkins (probably with money from piracy and slaving) founded the Chest at Chatham, which became Greenwhich Hospital, a charitable foundation for seamen and the site of the Royal Naval College. As bonuses during their careers, both Rodney and Hood were given the governorship of Greenwhich Hospital.

7. The *OED* defines "metes" as boundaries or limits, especially when used in the phrase "metes and bounds."

8. The word "owner" applies of course to the English system of property that is so familiar to most modern readers as to be invisible.

9. Rountree asserts that "Powhatan groups appear to have taken their names either from their eating habits . . . or from the characteristics of the places in which they lived" (1989, 29).

10. See Day for a discussion of the difficulty of determining meaning for Indian place names. Barbour seems to reject Heckewelder's definition for "Kiquotan" categorically, but he provides no explanation and he uses Heckewelder's Algonquian vocabulary as a source in his 1972 article. If we understand naming as a contest and not solely as a piece of the recoverable historical record we can reinterpret naming "offenses" like the one Day cites as an example of Indian "myopia": "An extreme example of this procedure is Pushaw Lake in Maine, which was named for a white settler named Pushaw. . . This did not prevent an Abenaki from deriving it from *passabákw* 'rising lake water'" (26)

11. Turner disputes this possibility (209), but his developmental framework clearly predetermines his conclusions.
12. On spellings of this name in Virginia texts and maps, see Barbour 1979, 49.
13. See also Nobles 13 and 28.
14. Svetlana Alpers notes that English natives were equally suspicious of surveyors employed by their landlords: "because of the nature of land ownership [in England] the project of surveying was greeted with suspicion on the part of tenant farmers" (148). The French Jesuit missionaries descibe the same suspicion on the part of Native North Americans:

> Vn autre prenant la parole, prit la defense du vin & de l'eau de vie. Non, dit-il, ce ne font pas ces bissons qui nous ostent la vie; mais vos écritures: car depuis que vous auez décry nostre païs, nos fleuues, nos terres, & nos bois, nous mourons tous, ce qui n'arriuit pas deuant que vous vinssiez icy. Nous-nous mismes à rire entendands ces causes nouuelles de leurs maladies. Ie leur dy que nous décriuions tout le monde, que nous décriuions nostre païs, celuy des Hurons, des Hiroquuois; bref toute la terre, & cependant qu'on ne mourit point ailleurs, comme on fait en leur païs, qu'il falloit donc que leur mort prouint d'ailleurs; ils s'y accorderent. (*Jesuit Relations*, vol. 9 206)

> Another one, breaking into the conversation, took up the defense of wine and brandy. "No," said he, "it is not these drinks that take away our lives, but your writings; for since you have described our country, our rivers, our lands, and our woods, we are all dying, which did not happen until you came here." We began to laugh upon hearing these new causes of their maladies. I told them that we described the whole world,—that we described our own country, that of the Hurons, of the Hiroquois, in short, the whole earth; and yet they did not die elsewhere as they did in their country. It must be, then, that their deaths arose from other causes. They agreed to this. (207)

15. For a summary of English attitudes toward Indian land claims in the seventeenth century, see Robinson 1-10. His survey reveals that while the English became gradually sensitive to "the Indian right of occupation" (10), their sensitivity was self-interested and lasted exactly as long as they did not feel pressured to expand their settlements.
16. J. B. Harley reminds us that "maps created for one purpose may be used for others" (1987, 3). Thus Smith's visual argument for Powhatan's power was in no way designed to work against English preeminence, but it could be taken up by other European mapmakers to argue against English possession.

17. Captain John Smith continually named Virginia land in a way that points to his complicated place in the English social network. While he, like other Englishmen, named in order to compliment men of rank, he also named territory after the men with whom he traveled. For a fascinating example of these naming practices see his *Generall Historie*, especially II 172.

18. Although Jacob's story's moral ambivalences did not escape notice in English biblical commentary, John Speed's *Genealogies*, a visual mapping of biblical names bound with most English bibles in the early seventeenth century, gives Jacob a privileged position in Christ's lineage. In the *Genealogies*, Jacob's is one of only three pages illustrated. He is depicted as a dying king prophesying greatness. The picture's caption reads: "Iacob in Egypt and vpon his death-bed prophesied to his sonnes their seuerall euents, to Iudah hee giueth the preheminence, and sheweth that his Scepter should continue till Shilo the Christ should come: In him bee blesseth euery Tribe: Read Genesis 49" (A6v).

19. For a discussion of the Adam story in Genesis as another master narrative of colonialism, see Bach 1997. Seed examines how the English used Genesis 1:28 to claim possession of American land (32-35).

20. Captain John Smith offers the following story on the name: "the greatest rumour is, that a Spanish ship called *Bermudas* was there cast away, carrying Hogges to the West-Indies that swam a shore, and there increased: how the Spaniards escaped is uncertaiane: but they say, from that ship those Iles were first called Bermudas, which till then for six thousand yeares had been namelesse" (II 345). Although the name's origin is probably an explorer's and not a ship's, the swine that fed the English in 1609 probably did come from Spanish ships. See also Lefroy edited manuscript (Butler) 9 and 9n.2, 9-10.

21. As Harley notes, "it is taken for granted in a society that the place of the king is more important than the place of a lesser baron . . . Cartography deploys its vocabulary accordingly so that it embodies a systematic social inequality" (1989, 7).

22. The text I am calling Butler's is the British Museum's Sloane manuscript 750, edited by J. H. Lefroy in 1882 for the Hakluyt Society. Lefroy attributes the manuscript to Captain John Smith, but the museum's catalogue attributes it to Nathaniel Butler. Edward Arber apparently thought that Governor Daniel Tucker had written the manuscript, but that attribution seems impossible given the text's vituperative attitude toward Tucker. While the text does support Butler vociferously, I am not fully convinced that it is Butler's text, but I will call the text Butler's as that is the accepted attribution by Bermuda's historiographers.

23. Jourdain's 1610 text ends "The barmudas lyeth in the height of two and thirty degrees and a halfe, of Northerly latitude, Virginia bearing directly from it, west, north west, two hundred and thirty leagues" (24).

24. See also Nobles 25.
25. Smith tells the story (II 363-64); Butler's account, 79-83, is considerably more detailed and condemnatory of Tucker than Smith's.
26. Butler says that Norwood conspired with Tucker to rig the survey in the governor's favor, but that the governor's machinations were foiled by the Company (105 and following). See Ives's discussion of the charge (*Rich Papers* 36).
27. Although all the adventurers upheld the system of rank order, they were themselves a divided and contentious group. The adventurers contained both merchants and noblemen whose interests might and did conflict at times. Butler describes a contest between the lords and the merchants over choosing the next governor of the islands (122-23). There was an ongoing struggle for power between Sir Edwin Sandys and the Earl of Warwick's faction. Some of the other disputes are described by Butler 246-50.
28. The *Rich Papers'* editor, Vernon A. Ives, comments that "Gurnet Rock ... was probably named for its resemblance to a gurnet, a large-headed fish" (141n.2).
29. This word in the text has two dots in the middle of the line between "cock" and "boat."
30. Presumably people who haunt ale houses.
31. Captain John Smith writes of the naming of Nonsuch plantation in Virginia (I 271). He says the name was given because of the exceptional beauty of the place.

Notes to Chapter 3

1. See Altman, Baker (1992, 1997), Baldo, Hopkins, Maley (1997a), and Murphy (1997 and 1999). See Murphy 1997 on Macmorris's shifting name and *Henry V*'s two texts.
2. For an amplification of the argument Hadfield presents in *Shakespeare and Ireland*, see Hadfield's *Literature, Travel, and Colonial Writing in the English Renaissance*, particularly chapter 4.
3. See Hadfield 1998, 263 on *The Island Princess* also. For an analysis of a Renaissance play in the context of the emerging global market, see Howard 1996.
4. See Bartley's chapter 2.
5. Renaissance drama is also replete with incidental references to the emerging Atlantic world that collectively helped to reinforce stereotypes. For example, see Thomas Heywood's *A Woman Killed With Kindness* when Sir Charles prepares to sell his sister Susan to Sir Francis and tells her to forget that he is her brother and instead to imagine him "Some barbarous outlaw, or uncivil kerne" (XIV 5). "Kerne" signifies here as an Irishman as well as as a generic name for ruffian; indeed, I would argue that in 1607 these senses were inseparable.

6. Brown uses Chapman, Jonson, and Marston's 1605 play *Eastward Hoe* as a "prelude" to his magisterial collection of and commentary upon documents from England's colonial efforts in North America (I 29).

7. The explorer/colonist was the literalization of Jonson's figurative stance. In our current historical position we see Jonson as a man between two worlds, and his texts are obsessed with the men who stood between worlds, with the figure of the "New World," with the poet as explorer. Jonson's texts show a fascination with the idea of discovery and exploration and with explorers and adventurers—figures who really could stand between worlds. Drummond tells his readers that Jonson planned to write the story of his journey to Scotland and to call it "discovery." And envisioning his role as a national poet not only as moralizer and judge but also as explorer, he called his commonplace book "discoveries." Unlike his friend John Donne, Jonson did not plan to become a colonist, but he was living among men who were reading and speaking of Virginia and Ireland, and men who were planning to make their worlds there.

8. For essential work on theatrical collaboration in the period, see Masten. His brilliant book shows that collaboration was the norm in the period. My chapter's focus on Jonson's plays might seem to work against that formulation, but as Masten argues, again Jonson is an anomaly pointing to the future of singular authorship. Importantly, *Eastward Hoe* was not Jonson's autonomous work, a point to remember when this chapter groups it with Jonson's later plays.

9. Sigalas also notes the *Utopia* reference (92).

10. For an account of the revisions, see the textual introduction to the play in the Herford, Simpson and Simpson Jonson works, especially pages 498-99.

11. Of course, as we have seen in the introduction and in chapter 2, these groups are not always in agreement. And plays do not necessarily represent them as agreeing with one another. One interesting discussion of the groups' differences occurs in *The Devil is an Asse* when Plutarchus responds to his father's plan to make him into a gentleman:

> I doe not wish to be one, truely Father.
> In a descent, or two, wee come to be
> Iust i'their state, fit to be coozend, like 'hem.
> And I had rather ha'tarryed i'your trade:
> For, since the *Gentry* scorne the Citty so much,
> Me thinkes we should in time, holding together,
> And matching in our owne tribes, as they say,
> Haue got an *Act* of *Common Councell*, for it,
> That we might coozen them out of *rerum natura*.
> (VI:III.i.27-35)

12. For an exploration of this topos and its meaning for Renaissance cultures, see Knapp.

13. See Jonson's poem, "The Vision of Ben Jonson, on the Muses of His Friend M. Drayton" (Hereford and Simpson and Simpson, VIII: 396-98). But see also Jonson's comments about Drayton, I:136 l.153, 137 ll.161-62. All references to Jonson's texts will be to this edition by volume, page number, and line numbers for poems and by volume, act, scene, and lines for plays.

14. Though often Jonson's plays keep these categories separate—i.e., Bartholomew Cokes will never be anything but a gull—at times Jonson alludes to an essential fluidity between cozeners and gulls. For an example see the gull Fitz-dottrel's (almost successful) attempt at cozening at the end of *The Devil is an Asse.*

15. In this regard, we should take seriously *Bartholomew Fair*'s grumpy Stage-keeper's comment that the play doesn't represent the "real" Smithfield. See Haynes's 1992 analysis of beggar plays, especially p. 103.

16. The lotteries became a point of contention between the parties fighting over the Virginia Company's demise. Rich's party, especially, contended that lottery funds had been misused and diverted for private gain. For accusations on both sides, see *Records* III 117, 141, 147, 152-53, 184, 215.

17. See Craven; see also the introductory essay to *Three Proclamations concerning the Lottery for Virginia 1613 - 1621*, Robert Johnson 1966 and 1967 and O'Brien.

18. See also the 1612 Alderman Johnson text discussed in the introduction.

19. Central to this analysis is Paster's marvelous book. Sexuality is an anachronistic word here. For more on the connections between male sexual desire for women and (lower) class status, see Bach 1999.

20. See Greenblatt 1988 and Kupperman.

21. See Campbell 126. See also Mason and Greenblatt 1991.

22. See Hawkins 216.

23. "Discovery" is one of Overdo's favorite words. Overdo's self-identification as a new Columbus shows that "discovery" is active in his discourse in its relatively recent sense. The first usage the *OED* records of "discovery" as "the finding out or bringing to light of that which was previously unknown" is from a 1553 text collected in Hakluyt's *Voyages*, 1589.

24. Although tobacco smoking was depicted constantly as an urban vice, it was also always identified with its colonial source. See Overdo's comment in *Bartholomew Fair*: "that tawney weede tabacco . . . Whose complexion is like the Indians that vents it" (VI: II.vi.21-4).

25. See Greenblatt 1988, 56. *The New Inne*'s Irish Nurse's name is spelled differently throughout the text. I have arbitrarily chosen the first spelling.

26. See Butler for a discussion of the court's values in relationship to Caroline parliamentary politics.

27. See Hadfield 1998, especially chapter 2.

28. On liminality and the carnivalesque quality of the fair, see Haynes 1984, Susan Wells, and McCanles 1977.

29. See Mullaney 1988.

30. Sanders notes that "the Light Heart functions as a quasi-pastoral setting; it is not London and yet not quite the country either, but on the margins of each" (555).

31. *The New Inne* acknowledges this capitalization when Lovel comments that were Lady Frampul's love for him true, "The *Spanish* Monarchy, with both the *Indies*, / Could not buy off the treasure of this kisse" (VI:III.ii.261-62).

32. See, for example, a Company letter to Sir George Yeardley in 1618: "We had also sent out men and ship at this tyme, but that it hath pleased god to keep her wind bound in Ireland" (*Records* II 136).

33. This broadside is also collected in *Three Proclamations concerning the Lottery for Virginia 1613 - 1621*.

34. Feest reads this image; from a detailed look at the apparel and weapons he makes a case for these representations as "our earliest and most direct pictures of the Indians of what is now Virginia" (13). But he concedes the conventionality of their postures. Ruth Wright, 303, believes that these Indians were copied from John White's earlier drawings. On European representations of Indians, see Sturtevant.

35. For additional evidence of this idea, see Oberg 53-54.

36. Captain John Smith reports that on his way to Virginia he stopped in the West Indies, "where, with a loathsome beast like a Crocodil, called a Gwayn, Tortoises, Pellicans, Parrots, and fishes, we daily feasted" (II 137). Feest 11 speculates that the turtles at the feet of the Indians may possibly "be taken as Indian symbols with some religious or sociological significance." Whether or not turtles were important to the Indians in Virginia, I see them in this representation as English markers of Virginia as a land of exotic commodities.

37. See Rackin's essential work on down-stage characters voicing resistance to the progression of elite male history in Shakespeare's history plays (1990). See also Weimann 1978 and 1992.

38. See Haynes 1992 on social divisions in Jonson's audiences. He suggests that "differentiating his audience into more or less fit and unfit, after bringing them all under contract, is a project Jonson pursued through his whole theatrical career" (132). Also see Jean Howard, 1994, on divided audiences.

39. See Morgan.

40. The herald, Pyed-Mantle, draws Pecunia's pedigree which shows "The rich *mynes of Potosi*. / The *Spanish mynes* i' the *West-Indies* . . . The *mynes o'Hungary*, this of *Barbary*" and even, and tellingly, "The *Welsh-myne*" (VI:IV.iv.20-23).

41. See Robertson.

Notes to Chapter 4

1. These inaugural spectacles are commonly called pageants, although in the period they might be called "triumphs," "solemnities," "entertainments," and/or "shewes." Thomas Heywood (or his printer) typically gave his a title and added "expressed in sundry triumphs, pageants and shows." I will use the words "triumphs" and "pageants" interchangeably.

2. The recent collection, *The Politics of the Stuart Court Masque*, for example, contains many excellent readings of masques and masque culture; however, although its editors claim that the masque "was . . . the site of negotiation over England's most pressing problems," none of its essays treats the colonial implications of masques (7). Raman and Siddiqi write about East Indian material in a few pageants. Hall and Siddiqi discuss *The Masque of Blackness*. Skiles Howard's and Andrew Murphy's books treat *The Irish Masque at Court*. See further work on that masque cited later in the chapter. McLeod states that "many a court masque and Lord Mayor's Show celebrated trade through symbols of imperial domination" (21).

3. The "East" is a construction that depends on Europe as the center. Jerry Brotton argues that the idea of the East that we now understand only became active with Mercator's seventeenth-century mapping of the world (169). Thus the East India Company was formed only at the beginning of a new conceptualization of the world that separated Europe from a traditional identity that included constitutive interaction with territory that we now call the Middle East.

4. These entertainments' insistent use of the word "complexion" to denote skin color puts paid to the still persistent idea that that word could not signify in that way when *The Merchant of Venice*'s Portia states "Let all of his complexion choose me so" (TLN 1053). In James Shirley's *The Trivmph of Peace*, one of the Caroline court's black-guard enters, saying "I never saw [a Maske] afore, I am one of the Guard, though of another complexion" (D2v). Middleton's King of the Moores asks, "do's my Complexion draw/ So many Christian Eyes, that never saw/ A King so blacke before?" (*Trivmphs of Truth* 1613, B4v). Error, in that masque, is furious to "see such a deuot humility take hold of that complexion" (Cv).

5. An antimasque is the period of debased revelry before the appearance of the court masquers. See Orgel 1965 for Jonson's role in developing the antimasque and the masque form. Although Orgel's work has been supplemented by such important work as Martin Butler's, *The Jonsonian Masque* and *The Illusion of Power* remain essential on the masque form.

6. For courtiers' ambitions for African colonization in the early seventeenth century, see Brenner 172. Brenner also discusses later merchant plans for settlement in Africa (177-78).

7. See Brenner's discussion of this symbiosis, particularly 54-59. On that symbiosis as expressed in the pageants and masques, see Nancy Wright. The

crown also used merchant money made in Eastern trade to finance its military efforts in Ireland. A 1618 letter reads, "Your Lordship heard before this time that the marchands of middleb. & the East Indies haue vundertaken to furnish the Exchequer with 50000l; of which his majestie hath bin pleased to assigne for Ireland 12000l" (quoted in Herford, Simpson, and Simpson X, 577).

8. See also Lowe's and Suleri's introductions.

9. Cockayne is an example both of the links between the city and the court and of the links between colonial efforts in the East and in the Atlantic world. Brenner argues that Cockayne profited from his close ties to Charles I's court (223); Brenner also notes Cockayne's various colonial investments: he was a Levant Company merchant as well as being prominent in the East India Company (78). In addition, Cockayne's factor, Matthew Craddock, was a leading Virginia tobacco trader.

10. See Moody, Martin and Byrne 204-205.

11. See Young, chapter 1, for a discussion of the institution of Greenwich mean time and the place of England in the postcolonial world.

12. See Brenner 85 for the conflict between grocers and the Spanish Company for the control of the trade in Spanish goods (imported from Spain's colonies). Brenner also notes the involvement of two prominent grocers in Virginia trading (147).

13. It is crucial to note the power dynamic even in this seemingly benign definition. Each of the adjectives here begs the question, "to whom?"

14. Sir Thomas Middleton was a grocer by affiliation and a great Merchant Adventurer. On Middleton, see Brenner 16n.33 and 87n.104. Although Middleton was a grocer, the speech Brenner quotes on 87 shows that he identified with merchants rather than with the retail grocers. It is unlikely that the two Middletons were related.

15. See Siddiqi's reading of this triumph, 152-60.

16. Except where otherwise noted, all quotations from Jonson's masques will be from the Herford, Simpson, and Simpson Complete Works, volume VII, by line number. I refer to this edition in this chapter as the Oxford edition.

17. James M. Smith similarly argues that "Ireland's idealized incorporation was always a fiction demanding the effacement of contemporary realities" (300).

18. See McLeod, chapter 1, for a discussion of Spenser's framing of Ireland in spatial terms.

19. James M. Smith reads these Irish footmen as representing the degenerate Old English who will be replaced by the properly English New English settlers.

20. In the Oxford edition, this entertainment is called *The Entertainment at Highgate*. Dekker is more optimistic about the empire in his 1603 *Magnificent Entertainment*. His character Zeal declares, "And then so rich an Empyre, whose fayre brest, / Contaynes foure Kingdomes by your entrance blest" (I).

21. See Moody et al. 210-15.

22. In chapter 1, see the discussion of the historical and cultural consequences of the trope of Ireland as a licentious woman.

23. See the discussion of Irish bards in chapter 1. Also see Canny, *Kingdom* for a historical account of the bards in Ireland who seem to have conformed more to Spenser's fears than to Jonson's hopes.

24. The word "comic" resounds in the previously scant commentary on this masque. More recently, as James Smith's work indicates, critics have attended to the implications of this comedy. See Blank's chapter 5 for a reading of the dialect in both *The Irish Masque at Court* and *For the Honor of Wales*. Murphy's 1999 book has an interesting and pertinent reading of *The Irish Masque at Court*. Skiles Howard's seminal book on the politics of court dancing also has an important reading on physical movement and colonialism in this masque. See pages 126-28. Also see Lindley; he details a relationship of the masque to the contemporary dispute in the Irish Parliament. Lanier has a short reading of the masque (338-39). See also Jones and Stallybrass.

25. Reaction to the original masque was uniformly negative. Herford, Simpson, and Simpson record the original comments in their commentary on *Pleasure* (X, 575-77).

26. See Marcus, *Politics* 125-27 for a cogent reading of *Wales*.

27. See the entire argument about Wales in Shakespeare's history plays, esp. 169-75.

28. In this masque, as Lindley notes, that moment is also marked by a shift from prose to verse.

29. The 1640 folio prints the final word of this speech as "you." For the textual difficulties associated with the dialect in this masque see Fumerton 1999 (103).

30. This definition and the definition of "jargon" above are taken from the second edition of the unabridged *Random House Dictionary of the English Language*.

31. See also Siddiqi on *Blackness*.

32. Of course this is powerful compliment to James who can, in the masque's logic, make the blind see, but is also a way of making "Indian" riches compatible with honor and virtue. I thank Constance Jordan for indicating this aspect of the masque.

33. See also Gillies 673-75. Parry reads this masque in the context of Prince Henry's Protestant colonial ambitions (103-105).

34. Obviously, as England's empire developed, these groups were differentiated and received differential treatment at the hands of their various oppressors. Their "darknesses" became defined against each other as well as against a singular dominant imaginary whiteness. These entertainments install that whiteness against an only rudimentarily differentiated darkness.

35. This word does not appear in the *OED*. Evans speculates that it is a misprint for ochre-dust. E. A. J. Honigmann, who edited the text included in *A Book of Masques* edited by Hill, replaces the word with orsidue, "a gold-coloured alloy of copper and zinc, rolled into a very thin leaf, used to ornament toys, etc." (176).

36. Campion clearly drew his imagery from the frontispiece of new atlases, which often displayed the sections of the world as women.

37. Obviously, this masque is also collapsing all Europeans together, as well as all Asians and Africans. However, only America is displayed as solely a commodity.

38. See also William Browne's 1614 masque, *Ulysses and Circe*. For the use of the Circe myth in Irish colonial texts, see Carroll 1993 and the introduction to Beacon xxxiii.

39. On the date of this masque see the Oxford editors' commentary (X, 596).

40. Pearl 64 calls the world in the Moon, "an imaginatively satiric version of London." Goldberg 127 notes the resemblance of this world to London.

41. See Newman for a reading of Jonson's critique of the fashion industry.

42. Epicene is the term for a Greek or Latin noun which may denote either sex. These Epicoenes are another version of the conceit of the grammatical concept applied to people that Jonson first employed in his 1609 play of that name. See Garber and Jean Howard 1994b for discussions of the sumptuary laws. There is a vast literature on the boy actor. Interesting recent speculations include Phyllis Rackin's seminal articles in *PMLA*, and Stephen Orgel, "Nobody's Perfect." The most comprehensive materials on sex roles include Henderson, Usher, and McManus and Woodbridge.

43. *A Book of Masques*, edited by Hill, 194-97 provides pictures of masque Indians in feathered costumes. Mason 65n.28 speculates that "the feather skirt as an element in the iconography of the American Indian may have helped to foster the identification of Indians with the Wild Men. The portrayal of the body hair of the latter is often done in a manner suggesting feathers rather than hair. This is evident in the setting of an Indian in a feather skirt side by side with a hairy wild man wearing a bird mask on one of a series of prints from around 1600 illustrating various masked ball costumes." See Sturtevant for a discussion of the origins of the stereotype of the Indian in feathers.

44. For example in Smith's *Generall Historie*, Book III, II, 212, he tells us that though he set his former Indian prisoners Kemps and Tassore free, "so well they liked our companies they did not desire to goe from us."

45. This masque has an extremely intricate textual history, to which this reading only barely refers. See the Oxford editors introduction to the text.

46. James had created Villiers marquess in 1618; he made him the Duke of Buckingham in 1623.

47. For the details of the ballad's later history and the masque's performances, see the Oxford edition's commentary (X, 633-35, 612).

48. See Taussig on the notion of wildness.

49. "Tawney" is a color word that pervades descriptions of Indians by Englishmen. For example, in Raleigh's *Discoverie*, he says that some of the Indians he saw in the West Indies were "well fauored" and "but for their tawnie colour may bee compared to anie of *Europe*" (33).

50. Captain John Smith noted in his *General Historie of Virginia* that the Virginia Indians were generally "of a colour browne when they are of any age" (II 114).

51. See *OED* "ball" 10b for the use of "ball" as ball of soap.

52. In 1633 gypsies were still being used as representative "savages." See Thomas Carew's antimasque of gypsies in his *Coelum Britanicum*.

Notes to Chapter 5

1. See Stallybrass and White, Wayne 1984, Orgel 1991, Loewenstein 1991, and Newton.

2. Loewenstein 1991 claims that Jonson's earlier published masques are bids for authority, especially as they relate to his contest with Daniel for patronage.

3. See Goldberg, esp. 61, for a brilliant exposition of this argument.

4. I do not mean to minimize the differences between the two collections. Both do promote "the English as colonizers." In Hakluyt's texts, the English are preeminently merchants. Hakluyt wants English economic dominion. Purchas's collections promote the activities of missionaries. His collections aim at the Christianization of the world. He wants to extend the influence of the Anglican Church, and to that end he collects evidence of religion. But in the end the projects feed a similar aim.

5. Hakluyt's collection of colonial accounts was first published in 1599. The texts are compiled without much editorial interference. The first edition of *Purchas his Pilgrimage* was issued in 1614. *Purchas* is heavily edited and full of marginal commentary, much of it indicating authorship.

6. Kirkpatrick Sale has recently challenged this assignment of roles. I will have much more to say about this in the rest of this chapter.

7. For an impressive assessment of the place of women across Native North America today and in the past, and an account of their disempowerment through history, see Jaimes and Halsey. Naomi Quinn reviews the anthropological evidence. See also Mathes and Kidwell.

8. Sale argues that what Smith described here "had nothing to do with concubinage; these men and women almost assuredly made up the central council of Wahunsencka's village" (299).

9. See Rountree's creative interpretation of these observations (1998).

10. In addition to those sources noted below, see the collection *Women and Colonization*, edited by Mona Etienne and Eleanor Leacock. The information

in the following paragraphs is largely taken from Jaimes's and Halsey's beautifully researched and written article.

11. This problem is returned to again and again in contemporary Native American literature.

12. Rackin 1990 gives an excellent account of the power that English women did have and the difficulty that English patriarchal ideology had with that reality (191-93). See Ezell's essential book and also Lewalski.

13. For an account of traveler's "sightings" of Amazons, see Mason, especially 100-11.

14. The witch in English literature and culture has attracted a lot of attention in recent years. See Belsey, Stallybrass 1982, Newman, Mason, and Adelman. Adelman's footnotes are an especially rich source of material.

15. Taussig's chapter on "Wildness," 209-220, gives an exceptional history of and commentary on this thrill.

16. See Sale 299-30. He also points to Englishmen's inability to see Powhatan women in their powerful roles. Also see Quitt 235-36 on women's roles in the encounter.

17. Philip L. Barbour's excellent edition of Smith unfortunately explains this away as follows: "That John Smith, who presumably saw something of the Tatars in 1603, relied on a work 350 years old for support and background is neither inept nor unfitting. Time stands still, here and there" (III, 191, n. 2). For a seminal and incisive critique, see Johannes Fabian's *Time and the Other: How Anthropology Makes its Object*.

18. I refer of course to Edward Said's seminal work on the "hegemony of European ideas about the Orient" (7).

19. In *The Subject of Tragedy*, Catherine Belsey offers examples of this limitation in analyses of both paintings and plays.

20. This attitude is reflected in *Measure for Measure*, where we can read all of the Duke's conspiracies as in the service of his attempt to reassign Mariana a place in society: "Why, you are nothing then; neither maid, widow, nor wife!" (TLN 2550-51, V. i. 176).

21. On the effacement of translation in encounter literature, see Cheyfitz; David Murray's book has an important discussion of Indian language in colonial texts.

22. Derrida's essay "Signature Event Context" exposes the fantasy of pure unmediated communication.

23. See Fumerton 1991 for a fascinating discussion of food in Jacobean masquing.

24. For an interesting interpretation of the implications of this fact in the early years of the settlement, see Quitt.

25. Of course, as my introduction indicates, many Englishmen did join the Indians. See especially note 29.

26. Bell offers a good overview of ritual studies and its vicissitudes.

Notes to Epilogue

1. The gift shop inside the Jamestown Settlement museum is much more restrained and less problematic, but it seems to do less business with the public. The Colonial National Historical Park's gift shop has a large section of books, nicely selected to tell the various histories surrounding the site.
2. The National Park is also devoted to the later history of settlement in Jamestown—into what we now call colonial times (the eighteenth century).
3. As Barbara Kirshenblatt-Gimblett notes, "the issue of who is qualified to perform culture is thorny because it reveals the implicit privileging of descent over consent in matters of cultural participation" (431). Kirshenblatt-Gimblett's comment applies to her observations of ethnic festivals, but it touches on the difficulties of having white Americans play Native Americans. The Jamestown Settlement does not require the costumed interpreters in the recreated Powhatan village to be Native American. Such a requirement would perhaps be oppressive, but in this white observer's opinion, the white interpreters in the village in the summer of 1998 were enormously more sensitive to Native American rights issues than the average white person, precisely because they worked with Native American colleagues.
4. See Churchill 1996.
5. Across from Smith's wall in the museum is perhaps its most compelling feature from my perspective: a map that lights up first the Indian settlements up the river, then the early English settlements, and finally the settlement pattern a bit later when the English have seized most of the Indians' land. That map calls into question many of the museum's other exhibits, including its hagiography of Smith.
6. For a debate about Smith's empiricism, see Jehlen and Hulme. Jehlen is responding in part to Bach 1994. See also Kathleen Brown.
7. Tree ring data that has recently been collected from Jamestown island may complicate the story of the colony's difficulties. According to that data, the early years of English settlement at Jamestown were part of the most severe drought in 770 years (Stahle et al.). Perhaps the English were not lazy at all but were actually unable to grow food. The data makes the English pressure on the Indians to supply the colony look even more brutal.
8. This cult oddly continues to this day. See the 1998 films *Elizabeth* and *Shakespeare in Love*.
9. For powerful statements about the current situation and claims for justice see Deloria and Lytle and also see Churchill's 1993 *Struggle for the Land*.
10. Since 1994, the Association for the Preservation of Virginia Antiquities has sponsored major excavations on its part of the island. They have published the results of those excavations in a series of illustrated pamphlets available at the National Historical Park gift shop. The artifactual findings are partic-

ularly interesting. Those excavations may change our understanding of the period, although as yet the results have been interpreted solely through the lens of the English texts, especially Smith's.

11. The Jamestown Settlement museum and the Colonial National Historical Park are currently planning for the 2007 anniversary of the settlement. The Settlement museum will be completely renovated. There are plans to make the outdoor fort site and the Powhatan village more accurate with reference to the recent archaeological excavations.

Bibliography

Act of Attainder of Shane O'Neill 1569. *Irish Statutes* 1786. *Irish History From Contemporary Sources 1509–1610.* Ed. Constantia Maxwell. London: George Allen & Unwin Ltd. 1923: 174-75.

Adelman, Janet. "'Born of woman': Fantasies of Maternal Power in *Macbeth.*" *Cannibals, Witches and Divorce.* Ed. Marjorie Garber. Baltimore: The Johns Hopkins University Press, 1987.

Alpers, Svetlana. *The Art of Describing: Dutch Art in the Seventeenth Century.* Chicago: The University of Chicago Press, 1984

Altman, Joel. "'Vile Participation': The Amplification of Violence in the Theater of *Henry V.*" *Shakespeare Quarterly* 42:1 (1991): 1-32.

Anderson, Benedict. "Exodus." *Critical Inquiry* 20:2 (1994): 314-27.

Andrews, John. *Ireland in Maps.* Dublin: The Dolmen Press, 1961.

Ariel: Special Issue 26:1 (1995).

Ashcroft, Bill, Gareth Griffiths, and Helen Tiffin. *The Empire Writes Back.* London: Routledge, 1989.

Ashcroft, Bill, Gareth Griffiths, and Helen Tiffin, eds. *The Post-Colonial Studies Reader.* London: Routledge, 1995.

Avery, Bruce. "Mapping the Irish Other." *ELH* 57 (Summer 1990): 263-79.

Axtell, James. *The European and the Indian.* New York: Oxford University Press, 1981.

——— *The Invasion Within.* New York: Oxford University Press, 1985.

Bach, Rebecca Ann. "Bearbaiting, Dominion, and Colonialism." *Race, Ethnicity, and Power in the Renaissance.* Ed. Joyce Green MacDonald. Madison: Fairleigh Dickinson University Press, 1997.

——— "The Homosocial Imaginary of *A Woman Killed With Kindness.*" *Textual Practice* 12(3) (1998): 503-24.

——— "Producing the 'New World': The Colonial Stages of Ben Jonson and Captain John Smith." Diss. University of Pennsylvania, 1994.

Baker, David J. *Between Nations: Shakespeare, Spenser, Marvell, and the Question of*

Britain. Stanford: Stanford University Press, 1997.

———— "'Wildehirisshman': Colonialist Representation in Shakespeare's *Henry V.*" *English Literary Renaissance* 22:1 (1992): 37-61.

———— "Where is Ireland in *The Tempest.*" *Shakespeare and Ireland.* Eds. Mark Thornton Burnett and Ramona Way. New York: St. Martin's Press, 1997b. 68-88.

Baldo, Jonathan. "Wars of Memory in *Henry V.*" *Shakespeare Quarterly* 47:2 (1996): 132-59.

Balzano, Wanda. "Irishness—Feminist and Post-Colonial." *The Post-Colonial Question: Common Skies, Divided Horizons.* Ed. Iain Cambers and Lidia Curti. New York: Routledge, 1996: 92-98.

Banerjee, Pompa. "Milton's India and *Paradise Lost.*" *Milton Studies* 37 (1999): 142-165.

Barbour, Philip L. "The Earliest Reconnaissance of the Chesapeake Bay Area: Captain John Smith's Map and Indian Vocabulary." *The Virginia Magazine of History and Biography.* 79:3 (1971): 280-302.

———— "The Earliest Reconnaissance of the Chesapeake Bay Area: Captain John Smith's Map and Indian Vocabulary." Part 2. *The Virginia Magazine of History and Biography.* 80 (1972): 21-51.

———— "On Considering the Feasibility of Establishing Key-Spellings for Indian Place-names in the Index to *The Complete Works of Captain John Smith.*" *Papers of the Twelfth Algonquian Conference.* Ed. William Cowan. Ottawa: Carleton University, 1981: 21-30.

———— "Variant Spelling of Virginia and Maryland Indian Place-names Before 1620." *Papers of the Tenth Algonquian Conference.* Ed. William Cowan. Ottawa: Carleton University, 1979: 43-59.

Barish, Jonas. *The Anti-theatrical Prejudice.* Berkeley: University of California Press, 1981. Barish, Jonas, ed. *Ben Jonson.* Englewood Cliffs: Prentice-Hall, 1963.

Bartels, Emily C. "The Double Vision of the East: Imperialist Self-Construction in Marlowe's *Tamberlaine, Part One.*" *Renaissance Drama* n.s. 23 (1992): 3-24.

———— "*Othello* and Africa: Postcolonialism Reconsidered." *The William and Mary Quarterly* 54 (1997): 45-64.

———— *Spectacles of Strangeness: Imperialism, Alienation, and Marlowe.* Philadelphia: University of Pennsylvania Press, 1993.

Bartley, J. O. *Teague, Shenkin and Sawney: Being an Historical Study of the Earliest Irish, Welsh and Scottish Characters in English Plays.* Cork: Cork University Press, 1954.

Barton, Anne. "*The New Inn* and the Problem of Jonson's Late Style." *ELR* 9 (1979): 395-418.

Bates, Catherine. "The Politics of Spenser's *Amoretti.*" *Criticism* 33:1 (1991):73-89

Beacon, Richard. *Solon His Follie.* Ed. Clare Carroll and Vincent Carey. Binghamton: Medieval and Renaissance Texts and Studies, 1996.

Beare, Philip O'Sullivan. *Historiae Catholicae Iberniae Compendium.* 1621. *Irish History From Contemporary Sources 1509–1610.* Ed. Constantia Maxwell. London: George Allen & Unwin Ltd., 1923, 211.

Bell, Catherine. *Ritual Theory, Ritual Practice.* New York: Oxford University Press, 1992.

Belsey, Catherine. *The Subject of Tragedy.* London and New York: Methuen, 1985.

Bennett, Tony. *The Birth of the Museum.* London: Routledge, 1995.

Bernard, John D. "Spenserian Pastoral and the *Amoretti.*" *ELH* 47:3 (1980): 419–32.

Bevington David, and Peter Holbrook, eds. *The Politics of the Stuart Court Masque.* Cambridge: Cambridge University Press, 1998.

Bieman, Elizabeth. "'Sometimes I . . . mask in myrth lyke to a Comedy.'" *Spenser Studies* 4 (1983): 131–41.

Black, Jeanette D. *The Blathwayt Atlas.* Vol. 2. Commentary. Providence: Brown University Press, 1975.

Blank, Paula. *Broken English: Dialects and the Politics of Language in Renaissance Writings.* London: Routledge, 1996.

Boelhower, William. "Inventing America: a model of cartographic semiosis." *Word & Image* 4:2 (1988): 475–97.

Bottigheimer, Karl S. "Kingdom and Colony: Ireland in the Westward Enterprise 1536 – 1660." *The Westward Enterprise.* Ed. K. R. Andrews, N. P. Canny, and P. E. H. Hair. Detroit: Wayne State University Press, 1979: 45–64.

Bradshaw, Brendan. "Native reaction to the Westward Enterprise: a case-study in Gaelic ideology." *The Westward Enterprise.* Ed. K. R. Andrews, N. P. Canny and P. E. H. Hair. Detroit: Wayne State University Press, 1979: 65–80.

Bradshaw, Brendan. "Edmund Spenser on Justice and Mercy." *The Writer as Witness.* Ed Tom Dunne. Cork: Cork University Press, 1987.

Bradshaw, Brendan, Andrew Hadfield and Willy Maley. *Representing Ireland.* Cambridge: Cambridge University Press, 1993.

Breitenberg, Mark. *Anxious Masculinity in Early Modern England.* Cambridge: Cambridge University Press, 1996.

Brenner, Robert. *Merchants and Revolution: Commercial Change, Political Conflict, and London's Overseas Traders, 1550–1653.* Princeton: Princeton University Press, 1993.

Brotton, Jerry. *Trading Territories: Mapping the Early Modern World.* Ithaca: Cornell University Press, 1997.

Brown, Alexander. *The Genesis of the United States.* 2 vols. New York: Russell & Russell Inc., 1964.

Brown, Kathleen M. *Good Wives, Nasty Wenches, and Anxious Patriarchs: Gender, Race, and Power in Colonial Virginia.* Chapel Hill: The University of North Carolina Press, 1996.

Browne, William. *The Masques of the Inner Temple (Ulysses and Circe). A Book of Masques in Honour of Allardyce Nicoll.* Ed. R. F. Hill. Cambridge: At the University Press, 1967.

"Bryskett, Lodowick." *Dictionary of National Biography.*

Burt, Richard. *Licensed by Authority: Ben Jonson and the Discourses of Censorship.* Ithaca: Cornell University Press, 1993.

Butler, Martin. "Late Jonson." *The Politics of Tragicomedy.* Ed. Gordan McMullan and Jonathan Hope. London: Routledge, 1992: 166-88.

———— "'We are one mans all': Jonson's *The Gipsies Metamorphosed." The Yearbook of English Studies* (1991): 253-73.

By His Maiesties Councell for Virginia. London: F. Kingston for W. Welby, 1613, STC #24833.6.

Cairns, David, and Shaun Richards. *Writing Ireland: Colonialism, Nationalism, and Culture.* Manchester: Manchester University Press, 1988.

Camden, William. *Britannia: or, a Chorographical Description of the Flourishing Kingdoms of England, Scotland, and Ireland* . . . Translated from the Edition Published by the Author in MDCVII. Enlarged by the latest discoveries by Richard Gough. London: Printed for John Stockdale: Piccadilly by S. Gonell, Little Queen St, 1806.

Campion, Thomas. *The description of a Maske: Presented in the banqueting roome at White Hall.* London: Printed by E.A. for Laurence Lisle, 1614.

Campion, Edmund. *History of Ireland.* Ed. Ware 1633. *Irish History From Contemporary Sources 1509-1610.* Ed. Constantia Maxwell. London: George Allen & Unwin Ltd. 1923, 252-53.

Campbell, Mary B. *The Witness and the Other World.* Ithaca: Cornell University Press, 1988.

Canning, Rick G. "'Ignorant, Illiterate Creatures': Gender and Colonial Justification in Swift's *Injured Lady* and *The Answer to the Injured Lady." ELH* 64 (1997): 77-97.

Canny, Nicholas. "Edmund Spenser and the Development of an Anglo-Irish Identity." *The Yearbook of English Studies* 13 (1983): 1-19.

———— *The Elizabethan Conquest of Ireland.* Sussex: The Harvester Press, 1976.

———— *Kingdom and Colony: Ireland in the Atlantic World 1560 - 1800.* Baltimore: The Johns Hopkins University Press, 1988.

———— *The upstart earl: A study of the social and mental world of Richard Boyle, first Earl of Cork 1566 - 1643.* Cambridge: Cambridge University Press, 1982.

Carney, James. *The Irish Bardic Poet.* Dublin: The Dolmen Press, 1967.

Carroll, Clare. "The Construction of Gender and the Cultural and Political Other in *The Faerie Queene* V and *A View of the Present State of Ireland:* The Critics, the Context, and the Case of Radigund." *Criticism* 32:2 (1990):163-92.

———— "Representation of Women in Some Early Modern English Tracts on the Colonization of Ireland." *Albion* 25:3 (1993): 379-94.

Carter, Paul. *The Road to Botany Bay: An Exploration of Landscape and History.* Chicago: The University of Chicago Press, 1987.

Cavanagh, Sheila T. "'The fatal destiny of that land': Elizabethan views of Ireland." *Representing Ireland.* Ed. Brendan Bradshaw, Andrew Hadfield, and Willy Maley. Cambridge: Cambridge University Press, 1993, 116-31.

———— "'Such Was Irena's Countenance': Ireland in Spenser's Prose and Poetry." *Texas Studies in Literature and Language* 28:1 (1986): 24-50.

Chalfont, Fran C. *Ben Jonson's London.* Athens: University of Georgia Press, 1978.

Chamberlain, John. *The Letters of John Chamberlaine*. Ed. Norman Egbert McClure. 2 vols. Philadelphia: The American Philosophical Society, 1939.

Chambers, Iain and Lidia Curti, eds. *The Post-Colonial Question: Common Skies, Divided Horizons*. New York: Routledge, 1996.

Chapman, George. *The Memorable Maske of the Two Honorable Houses or Inns of Court; the Middle Temple, and Lincolns Inne*. London: Printed by F.K. for George Norton, 1613.

———— *Selected Poems*. Ed. Eirian Wain. Manchester: Fyfield Books, 1978.

Cheyfitz, Eric. *The Poetics of Imperialism*. New York: Oxford University Press, 1991.

Churchill, Ward. "Deconstructing the Columbus Myth." *From a Native Son: Selected Essays on Indigenism, 1985–1995*. Boston: South End Press, 1996.

———— *Fantasies of the Master Race*. Monroe: Common Courage Press, 1992.

———— *From a Native Son: Selected Essays on Indigenism, 1985–1995*. Boston: South End Press, 1996.

———— *Struggle for the Land*. Monroe, Maine: Common Courage Press, 1993.

Clarke, G. N. G. "Taking Possession: The Cartouche as Cultural Text in Eighteenth-Century American Maps." *Word & Image* 4:2 (1988): 455–74.

Cobo, Father Bernabe. *History of the Inca Empire*. 1653. Trans. Roland Hamilton. Austin: University of Texas Press, 1979.

Coleman, James. "Edmund Spenser." *Journal of the Cork Historical and Archaelogical Society* 2nd series 3 (1894): 89–100.

———— "The Poet Spenser's Wife." *Journal of the Cork Historical and Archaelogical Society*. 2nd series 1 (1895): 131–33.

Cooper, Frederick, and Ann L. Stoler. "Tensions of Empire." *American Ethnologist* 16:4 (1989): 609–21.

Copland, Patrick. *Virginia's God be Thanked, or A Sermon of Thanksgiving for the Happie successe of the affayres in VIRGINIA this last yeare*. London: Printed by I.D. for William Sheffard and Iohn Bellamie, 1622 (STC #5727).

Coughlan, Patricia, ed. *Spenser and Ireland: An Interdisciplinary Perspective*. University College, Cork: Cork University Press, 1989.

Crashaw, William. *A Sermon Preached in London before the right honorable the Lord Lawarre* . . . London: Printed for William Welby, 1610.

Craven, Wesley Frank. *Dissolution of the Virginia Company*. Gloucester, Mass: P. Smith, 1964.

————. *An Introduction to the History of Bermuda*. Bermuda: The Bermuda Maritime Museum Association, 1990.

————. *The Virginia Company of London, 1606–1624*. The Virginia 350th Anniversary Celebration Corporation, 1957.

Cronon, William. *Changes in the Land*. New York: Hill and Wang, 1983.

Crosby, Alfred W. "Ecological Imperialism: the Overseas Migration of Westrn Europeans as a Biological Phenomenon." *The Texas Quarterly* XXI:1 (1979): 10–22.

A Declaration for the certaine time of drawing the great standing Lottery. London 1615. STC #24833.8

A Declaration of the State of the Colonie and Affaires in VIRGINIA: With the Names of the Adventurors, and Summes adventures in that Action. By His Maiesties Counseil for Virginia. 22 Junij. 1620. London: Printed by T.S. 1620.

Dasenbrook, Reed Way. "The Petrarchan Context of Spenser's *Amoretti.*" *PMLA* 100:1 (1985): 38-50.

Davenant, William. *The Triumphs of the Prince D'Amovr.* London: Printed for Richard Meighen, 1635.

Davies, R. R. "Colonial Wales." *Past and Present* 65 (November 1974): 3-23.

Day, Gordon M. "Indian Place-Names as Ethnohistoric Data." *Actes Du Huitieme Congres Des Algonquinistes.* Ed. William Cowan. Ottawa: Carleton University, 1977: 26-31.

De Grazia, Margreta, and Peter Stallybrass. "The Materiality of the Shakespearean Text." *Shakespeare Quarterly* 44:3 (1993): 255-83.

De la Warr, Thomas West, Baron. *The relation of the Right Honourable the lord De-La-Warre, Lord Gouernour and Captaine Generall of the Colonie, planted in Virginia.* London: Printed by William Hall, for William Welbie, 1611.

Dekker, Thomas. *The Second Part of the Honest Whore. The Dramatic Works of Thomas Dekker.* Vol. 2. Ed. Fredson Bowers. Cambridge: At the University Press, 1964.

────── *London's Tempe or the Feild of Happiness.* October 29, 1629. London, 1629.

────── *The magnificient Entertainment Given to king Iames, Queen Anne his wife, and henry Frederick the Prince, vpon the day of his Maiesties Tryumphant Passage (from the tower) through his Honorourable Citie.* 15 march, 1603. London: T.C. for Tho: Man the yonger, 1604.

Deloria, Vine, Jr., and Clifford M. Lytle. *American Indians, American Justice.* Austin: University of Texas Press, 1983.

Derrida, Jacques. *Limited Inc.* Evanston: Northwestern University Press, 1988.

Dictionary of the Gaelic Language. Compiled and published under the direction of the Highland Society of Scotland in two volumes Willaim Balckwood, Edinburgh, 1828.

Dictionary of the Irish Language [B]. Arranged by Maura Carney and Máirín O Daly. Dublin; the Royal Irish Academy, 1975.

Dollimore, Jonathan. *Radical Tragedy.* Chicago: The University of Chicago Press, 1984.

Donaldson, Ian. *Jonson's Magic Houses: Essays in Interpretation.* Oxford: Clarendon Press, 1997.

Donne, John. *A Sermon upon the Eighth Verse of the First Chapter of the Acts of the Apostles Preached to the Honourable Company of the Virginian Plantation,* 13 November 1622. London: Printed for Thomas Iones, 1624. STC #7051.

Donow, Herbert S. *A Concordance to the Sonnet Sequences of Daniel, Drayton, Shakespeare, Sidney, and Spenser.* Carbondale: Southern Illinois University Press, 1969.

Drayton, Michael. *The Works of Michael Drayton.* Ed. J. William Hebel. 5 vols. Oxford: Basil Blackwell, 1961.

Duanaire Mhéig Uidhir: The Poembook of Cú Chonnacht Mâg Uidhir Lord of Fermanagh

1566–1589. Ed. David Greene. Dublin: Dublin Institute for Advanced Studies, 1972.

Dufrene, Phoebe. "Contemporary Powhatan Art and Culture: Its Link with Tradition and Implications for the Future." *Papers of the Twenty-Second Algonquian Conference.* Ed. William Cowan. Ottawa: Carleton University, 1991: 125-36.

Dunlop, Robert. "The Plantation of Munster 1584–1589." *English Historical Review* 3 (1888): 250-69.

Dunlop, Alexander. "Calendar Symbolism in the 'Amoretti.'" *Notes and Queries* January (1969): 24-26.

Dunne, T. J. "The Gaelic Response to Conquest and Colonisation." *Studia Hibernica* 20 (1980): 7-30.

Dyer, Richard. "Entertainment and Utopia." *The Cultural Studies Reader.* Ed. Simon During. London: Routledge, 1993. 271-83.

────── *White.* London: Routledge, 1997.

Ellis, Steven G. *Tudor Ireland: Crown, Community and the Conflict of Cultures, 1470 - 1603.* London: Longman, 1985.

Ellyson, J. Taylor. *The London Company of Virginia.* New York: 1908.

Encylopædia Britannica. 9th ed. 1878.

Etienne, Mona and Eleanor Leacock, eds. *Women and Colonization.* New York: Praeger Publishers, 1980.

Evans, Herbert Arthur. *English Masques.* London: Blackie & Son, Ltd., 1897.

Evans, Robert C. *Ben Jonson and the Poetics of Patronage.* Lewisburg: Bucknell University Press, 1989.

────── *Jonson and the Contexts of His Time.* Lewisburg: Bucknell University Press, 1994.

Ezell, Margaret J. M. *The Patriarch's Wife: Literary Evidence and the History of the Family.* Chapel Hill: The University of North Carolina Press, 1987.

Fabian, Johannes. *Time and the Other.* New York; Columbia University Press, 1983.

Fanon, Frantz. *Black Skin, White Masks.* New York: Grove Press, Inc., 1967.

────── "National Culture." In Bill Ashcroft, Gareth Griffiths, and Helen Tiffin, eds. *The Post-Colonial Studies Reader.* London: Routledge, 1995.

Feest, Christian F. "The Virginia Indian in Pictures, 1612–1624." *The Smithsonian Journal of History* 2:1 (1967): 1-30.

Ferrar, Nicholas. *Sir Thomas Smith's Misgovernment of the Virginia Company.* Ed. with an introduction by D. R. Ransome. Cambridge: The Roxbourghe Club, 1990.

Fogarty, Anne. "The Colonization of Language: Narrative Strategy in *A View of the Present State of Ireland* and *The Faerie Queene,* Book VI." *Spenser and Ireland: An Interdisciplinary Perspective.* Ed. Patricia Coughlan. University College, Cork: Cork University Press, 1989.

Ford, John. *The Chronicle History of Perkin Warbeck, 'Tis Pity She's a Whore and Other Plays.* Ed. Marion Lomax. Oxford: Oxford University Press, 1995.

Fukuda, Shohachi. "The Numerological Patterning of *Amoretti* and *Epitalamion.*" *Spenser Studies* 9 (1991): 33-48.

Fuller, Mary C. *Voyages in Print: English Travel to America, 1576–1624.* Cambridge: Cambridge University Press, 1995.

Fumerton, Patricia. *Cultural Aesthetics.* Chicago: The University of Chicago Press, 1991.

———— "Homely Accents: Ben Jonson Speaking Low." *Renaissance Culture and the Everyday.* Ed. Patricia Fumerton and Simon Hunt. Philadelphia: University of Pennsylvania Press, 1999.

Garber, Marjorie. *Vested Interests.* New York: Routledge, 1992.

Gibbs, Donna. *Spenser's Amoretti: A Critical Study.* Hants: Scolar Press, 1990.

Gillies, John. "Shakespeare's Virginian Masque." *ELH* 53:4 (1986): 673-707.

———— *Shakespeare and the Geography of Difference.* Cambridge: Cambridge University Press, 1994.

Gilroy, Paul. *The Black Atlantic: Modernity and Double Consciousness.* Cambridge: Harvard University Press, 1993.

Gleach, Frederic W. *Powhatan's World and Colonial Virginia.* Lincoln: University of Nebraska Press, 1997.

Goldberg, Jonathan. *James I and the Politics of Literature.* Stanford: Stanford University Press, 1989.

Gordon, D. J. "Chapman's *Memorable Masque.*" *The Renaissance Imagination: Essays and Lectures by D. J. Gordon.* Ed. Stephen Orgel. Berkeley: University of California Press, 1975.

Greenblatt, Stephen. *Learning to Curse.* New York: Routledge, 1990.

———— *Marvelous Possessions.* Chicago: The University of Chicago Press, 1991.

————. *Renaissance Self-Fashioning.* Chicago: The University of Chicago Press, 1980.

———— *Shakespearean Negotiations.* Berkeley: The University of California Press, 1988.

———— "The Touch of the Real." *Representations* 59 (1997): 14-29.

Greene, David, ed. *Duanaire Mhéig Uidhir: The Poembook of Cú Chonnacht Mág Uihir Lord of Fermanagh 1566–1589.* Dublin: Dublin Institute for Advanced Studies, 1972.

Greene, Roland. "The Colonial Wyatt: Contexts and Openings." *Rethinking the Henrician Era.* Ed. Peter C. Herman. Urbana: University of Illinois Press, 1994. 240-66.

———— "Petrarchism among the Discourses of Imperialism." *America in European Consciousness, 1493–1750.* Ed. Karen Ordahl Kupperman. Chapel Hill: University of North Carolina Press, 1995. 130-65.

Greene, Thomas M. "The Balance of Power in Marvell's 'Horatian Ode.'" *ELH* 60:2 (1993): 379-96.

Hadfield, Andrew. "'Hitherto she ne're could fancy him': Shakespeare's 'British' Plays and the Exclusion of Ireland." *Shakespeare and Ireland.* Eds. Mark Thornton Burnett and Ramona Way. New York: St. Martin's Press, 1997. 47-67.

———— *Spenser's Irish Experience: Wilde Fruit and Salvage Soyl.* Oxford: Clarendon Press, 1997b.

———— *Literature, Travel, and Colonial Writing in the English Renaissance* 1545–1625. Oxford: Clarendon Press, 1998.

Haile, Edward Wright. *Jamestown Narratives: Eyewitness Accounts of the Virginia Colony.* Champlain, Virginia: Roundhouse, 1998.

Hakluyt, Richard. *Virginia richly valued, By the description of the maine land of Florida, her neighbor. Written by a Portugall gentleman of Elvas, emploied in all the action, and translated out of Portugese by Richard Hakluyt.* London: Printed by Felix Kyngston, 609. March of America Facsimilie Series Number 12. Ann Arbor: UMI, 1966.

Hall, Catherine. "Histories, Empires and the Post-Colonial Moment." *The Post-Colonial Question: Common Skies, Divided Horizons.* Ed. Iain Chambers and Lidia Curti. New York: Routledge, 1996. 65-77.

Hall, Kim F. "Guess Who's Coming to Dinner? Colonization and Miscegenation in *The Merchant of Venice.*" *Renaissance Drama* n.s. 23 (1992): 87-111.

———— *Things of Darkness: Economies of Race and Gender in Early Modern England.* Ithaca: Cornell University Press, 1995.

Hall, Stuart. "When Was 'The Post-Colonial'? Thinking at the Limit." *The Post-Colonial Question: Common Skies, Divided Horizons.* Ed. Iain Chambers and Lidia Curti. New York: Routledge, 1996. 242-20.

Hamilton, A. C. *The Spenser Encyclopedia.* Toronto: The University of Toronto Press, 1990.

Hamor, Ralph. *A True Discourse of the Present Estate of Virginia.* London: John Beale for William Welby, 1615. STC 12736 facsimile. Amsterdam: Theatrum Orbis Terrarum Ltd., 1971.

Hanafin, Patrick. "Defying the female: the Irish constitutional text as phallocentric manifesto." *Textual Practice* 11:2 (1997): 249-73.

Handler, Richard, and Eric Gable. *The New History in an Old Museum: Creating the Past at Colonial Williamsburg.* Durham: Duke University Press, 1997.

Harley, J. B. "Cartography, Ethics and Social Theory." *Cartographica* 27:2 (1990): 1-23.

———— "Deconstructing the Map." *Cartographica* 26:2 (1989): 1-20.

———— "The Map and the Development of the History of Cartography." *The History of Cartography, Volume I: Cartography in Prehistoric, Ancient, and Medieval Europe and the Mediterranean.* Ed. J. B. Harley and David Woodward. Chicago: The University of Chicago Press, 1987. 1-42.

———— "Maps, Knowledge, and Power." *The Iconography of Landscape: Essays on the symbolic representation, design and use of past environments.* Ed. Denis Cosgrove and Stephen Daniels. Cambridge: Cambridge University Press, 1988a: 277-312.

———— "Silences and secrecy: the hidden agenda of cartography in early modern Europe." *Imago Mundi* 40 (1988b): 57-76.

Harriot, Thomas. *A Briefe and True Report of the New Found Land of Virginia.* London 1590. Reprinted New York: Dover, 1972.

Hatch, Charles E., Jr. *The First Seventeen Years: Virginia, 1607-1627.* Williamsburg: The Virginia 350th Anniversary Celebration Corporation, 1957.

Hawkins, Harriet. "The Idea of a Theater in Jonson's *The New Inn.*" *Renaissance Drama* 9 (1966): .

Haynes, Jonathan. "Festival and the Dramatic Economy of Jonson's *Bartholomew Fair.*" *ELH* 51:4 (Winter 1984): 645-68.

—— *The Social Relations of Jonson's Theater.* Cambridge: Cambridge University Press, 1992.

Haynes-McCoy, G. A. "The Completion of the Tudor conquest and the advance of the counter-reformation, 1571–1603." *A New History of Ireland.* Ed. T. W. Moody, F. X. Martin, and F. J. Byrne. Oxford: Clarendon Press, 1976.

Hechter, Michael. *Internal Colonialism.* London: Routledge & Kegan Paul, 1975.

Helgerson, Richard. *Self-Crowned Laureates: Spenser, Jonson, Milton and the Literary System.* Berkeley: University of California Press, 1983.

Henderson, Katherine Usher, and Barbara F. McManus. *Half Humankind.* Urbana and Chicago: University of Illinois Press, 1985.

Hendricks, Margo. "Managing the Barbarian: *The Tragedy of Dido, Queen of Carthage.*" *Renaissance Drama* n.s. 23 (1992): 165-88.

Hendricks, Margo, and Patricia Parker, eds. *Women, "Race," and Writing in the Early Modern Period.* London: Routledge, 1994.

Henley, Pauline. *Spenser in Ireland.* Cork: Cork University Press, 1928.

Herr, Cheryl. "The Erotics of Irishness." *Critical Inquiry* 17:1 (1990): 1-34.

Heywood, Thomas. *Londini Emporia, or Londons Mercatura.* London: Nicholas Okes, 1633. *Thomas Heywood's Pageants.* Ed. David Bergeron. New York: Garland, 1986.

—— *Londini Status Pacatus: or London's Peaceable Estate.* London: Iohn Okes, 1639. *Thomas Heywood's Pageants.* Ed. David Bergeron. New York: Garland, 1986.

—— *A Woman Killed With Kindness.* Ed. Brian Scobie. New York: W. W. Norton and Company, 1985.

Highley, Christopher. *Shakespeare, Spenser and the Crisis in Ireland.* Cambridge: Cambridge University Press, 1997.

Hill, R. F. Ed. *A Book of Masques in Honour of Allardyce Nicoll.* Cambridge: At the University Press, 1967.

Hopkins, Lisa. "Neighbourhood in *Henry V.*" *Shakespeare and Ireland.* Eds. Mark Thornton Burnett and Ramona Way. New York: St. Martin's Press, 1997. 9-26.

Howard, Jean E. "Crossdressing, the Theatre, and Gender Struggle in Early Modern England." *Shakespeare Quarterly* 39 (1988): 418-40.

—— "An English Lass amid the Moors: Gender, Race, Sexuality, and National Identity in Heywood's *The Fair Maid of the West.*" *Women, "Race," and Writing in the Early Modern Period.* Ed. Margo Hendricks and Patricia Parker. London: Routledge, 1994. 101-17.

—— "Mastering Difference in *The Dutch Courtesan.*" *Shakespeare Studies* 24 (1996): 105-17.

—— *The Stage and Social Struggle in Early Modern England.* London: Routledge, 1994.

Howard, Skiles. *The Politics of Courtly Dancing in Early Modern England*. Amherst: University of Massachusetts Press, 1998

Hudson, Kenneth. "How Misleading Does an Ethnographical Museum Have to Be." *Exhibiting Cultures*. Ed. Ivan Karp and Steven D. Lavine. Washington: Smithsonian Institution Press, 1991: 457-64.

Hulme, Peter. *Colonial Encounters: Europe and the Native Caribbean 1492–1797*. London: Methuen, 1986.

——— "Critical Response 1: Making No Bones: A Response to Myra Jehlen." *Critical Inquiry* 20:1 (1993): 179-86.

——— "Polytropic man: Tropes of Sexuality and Mobility in Early Colonial Discourse." *Europe and its Others*. Vol. 2 of *Proceedings of the Essex Conference on the Sociology of Literature*, July 1984. Ed. Francis Barker, Peter Hulme, Margaret Iversen, and Diana Loxley. Colchester: University of Essex, 1985.

Hunter, G. K. "'Unity' and Numbers in Spenser's *Amoretti*." *Yearbook of English Studies* 5 (1975): 39-45.

Ireland Statistical Abstract 1995. Dublin: Stationary Office, 1996.

Irish History From Contemporary Sources 1509–1610. Ed. Constantia Maxwell. London: George Allen & Unwin Ltd., 1923.

Jaimes, M. Annette, with Theresa Halsey. "American Indian Women: At the Center of Indigenous Resistance in North America." *The State of Native America*. Ed. M. Annette Jaimes. Boston: South End Press, 1992.

Jehlen, Myra. "Critical Response II: Response to Peter Hulme." *Critical Inquiry* 20:1 (1993): 187-91.

——— "History Before the Fact; Or, Captain John Smith's Unifinished Symphony." *Critical Inquiry* 19:4 (1993):677-92.

Jenkins, Raymond. "Spenser and the Clerkship in Munster." *PMLA* 47 (1932):109-21.

——— "Spenser: The Uncertain Years 1584–1589." *PMLA* (1938): 350-62.

——— "Spenser with Lord Grey in Ireland." *PMLA* 52 (1937): 338-53.

Johnson, Robert. *The New Life of Virginea: Declaring the Former Svccesse and Present estate of that plantation, being the second part of* Noua Britannia. Published by the authoritie of his Maiesties Counsell of Virginia. London: Imprinted by Felix Kyngston for William Welby, 1612.

——— *Nova Britannia*. London: Printed for Samuel Macham, 1609. Facsimile. New York: Printed for J. Sabin, 1867.

Johnson, Robert C. "The Lotteries of the Virginia Company 1612 - 1621." *The Virginia Magazine of History and Biography* 74 (July 1966): 259-92.

——— "The 'Running Lotteries' of the Virgiania Company." *VMHB* (July 1967): 156-65.

Johnson, William C. "Gender Fashioning and the Dynamics of Mutuality in Spenser's *Amoretti*." *English Studies* 74:6 (1993): 503-19.

——— *Spenser's Amoretti*. Lewisburg: Bucknell University Press, 1990.

Jones, Ann Rosalind, and Peter Stallybrass. "Dismantling Irena: The Sexualizing of

Ireland in Early Modern England." *Nationalisms and Sexualities.* Ed. Andrew Parker, Mary Russo, Doris Summer, and Patricia Yeager. New York: Routledge, 1992.

Jones, Inigo, and William Davenant. *The Temple of Love A Masque.* London: Printed for Thomas Walkey, 1634.

Jonson, Ben. *The Complete Masques.* Ed. Stephen Orgel. New Haven: Yale University Press, 1969.

——— *The Workes of Beniamin Jonson.* London: Printed by William Stansby, 1616.

——— *The Workes of Benjamin Jonson. The Second Volume.* London: Printed for Richard Meighen, 1640.

——— *Ben Jonson* [Works]. Ed. C. H. Herford and Percy and Evelyn Simpson. 11 vols. London: Oxford at the Clarendon Press, 1925-1952.

Jourdain, Silvester. *A Discovery of the Barmvdas, otherwise called the Ile of Divels.* London 1610.

——— *A Plaine Description of the Barmvdas, now called the Sommer Ilands* . London: Printed by W. Stansby, for W. Welby, 1613.

Joyce, P. W. *The Origin and History of Irish Names of Places.* Dublin: M. H. Gill & Son, 1883.

Kanneh, Kadiatu. "Feminism and the Colonial Body." *The Post-Colonial Studies Reader.* Ed. Bill Ashcroft, Gareth Griffiths, and Helen Tiffin. London: Routledge, 1995. 346-48.

Kaske, Carol V. "Spenser's *Amoretti* and *Epithalamion* of 1595: Structure, Genre, and Numerology." *ELR* 8 (1978): 271-95.

Keane, Patrick J. *Yeats, Joyce, Ireland, and the Myth of the Devouring Female.* Columbia: University of Missouri Press, 1988.

Kiberd, Declan. *Inventing Ireland.* Cambridge: Harvard University Press, 1995.

Kidwell, Clara Sue. "The Power of Women in Three American Indian Societies." *The Journal of Ethnic Studies.* Vol. 6, No. 3 (Fall 1978): 113-21.

Kincaid, Jamaica. "Alien Soil." *The Best American Essays 1994.* Ed. Tracy Kidder. Boston: Houghton Mifflin Company, 1994.

——— *A Small Place.* New York: Penguin Books, 1988.

Kirshenblatt-Gimblett, Barbara. "Objects of Ethnography." *Exhibiting Cultures.* Ed. Ivan Karp and Steven D. Lavine. Washington, D.C.: Smithsonian Institution Press, 1991: 386-443.

Klein, Lisa M. "'Let us love, dear love, lyke as we ought': Protestant Marriage and the Revision of Petrarchan Loving in Spenser's *Amoretti.*" *Spenser Studies* X (1992): 109-37.

Knapp, Jeffrey. *An Empire Nowhere.* Berkeley: The University of California Press, 1992.

Knott, Eleanor, ed. *The Bardic Poems of Tadhg Dall Ó Huiginn (1550–1591).* Vols. XXII and XXIII of Irish Texts Society. Lúndain: Simpkin, Marshall, Hamilton, Kent & Co, Ltd., 1922, 1926.

Knott, Eleanor. *Irish Classical Poetry. Early Irish Literature*. Intro. James Carney. New York: Barnes & Noble, Inc., 1966.

Kupperman, Karen Ordahl. *Settling with the Indians: The Meeting of English and Indian Cultures in America, 1580–1640*. Totowa, N.J.: Rowman and Littlefield, 1975.

Lane, Ralph. Letter to Sir Philip Sidney. August 12, 1585. *Archaeologia Americana* IV (1860): 17-18.

Lanier, Douglas. "Fertile Visions: Jacobean Revels and the Erotics of Occasion." *SEL* 39:2 (1999): 327-356.

Le Jeune, Paul. "Relation de ce qui c'est passé en la Nouvelle france, en l'année 1636." In Kébec, August 28, 1636. *The Jesuit Relations and Allied Documents*. Ed. Reuben Gold Thwaites. Vol. 9. Cleveland: The Burrows Brothers Company, 1897.

Lefroy, J. H., ed. [Butler] *The Historye of the Bermudas or Summer Islands*. Hakluyt Society, 1st ser., LXV. London, 1882.

―――― *Memorials of the Discovery and Early Settlement of the Bermudas or Somers Islands 1515–1685*. 2 vols. London: Longmans, Green, and Co., 1877.

Lemly, John. "'Make odde discoveries!' Disguises, Masques, and Jonsonian Romance," *Comedy from Shakespeare to Sheridan*. Ed. A. R. Braunuller and J.C. Bulman. Newark: University of Delaware Press, 1986.

Lever, J. W. *The Elizabethan Love Sonnet*. London: Metheun & Co., 1956.

Lewalski, Barbara Kiefer. *Writing Women in Jacobean England*. Cambridge: Harvard University Press, 1993.

Lewis, Clifford M., and Albert J. Loomie. *The Spanish Jesuit Mission in Virginia 1570–1572*. Chapel Hill: The University of North Carolina Press, 1953.

Lewis, G. Malcolm. "The Indigenous Maps and Mapping of North American Indians." *The Map Collector* 9 (1979): 25-32.

Lim, Walter S. H. *The Arts of Empire: The Poetics of Colonialism From Raleigh to Milton*. Newark: University of Delaware Press, 1998.

Limon, Jerzy. *The Masque of Stuart Culture*. Newark: University of Delaware Press, 1990.

Lindley, David. "Embarrassing Ben: The Masques for Frances Howard." *ELR* 16 (1986): 343-59.

Loewenstein, Joseph. "A Note on the Structure of Spenser's *Amoretti*: Viper Thoughts." *Spenser Studies* 8 (1990): 311-23.

―――― "Printing and 'The Multitudinous Presse': The Contentious Texts of Jonson's Masques." *Ben Jonson's 1616 Folio*. Ed. Jennifer Brady and W. H. Herendeen. Newark: University of Delaware Press, 1991.

Londonderry and the London Companies 1609–1629. Being a Survey and Other Documents Submitted to King Charles I by Sir Thomas Phillips. Belfast: His Majesty's Stationary Office, 1928.

Lowe, Lisa. *Critical Terrains: French and British Orientalisms*. Ithaca: Cornell University Press, 1991.

Lupton, Julia Reinhard. "Home-Making in Ireland: Virgil's Ecologue I and Book VI of *The Faerie Queene.*" *Spenser Studies* VIII (1987): 119–45.

MacArthur, Janet H. *Critical Contexts of Sidney's* Astrophil and Stella *and Spenser's* Amoretti. Victoria: University of Victoria, 1989.

MacCarthy-Morrogh, Michael. *The Munster Plantation: English Migration to Southern Ireland 1583–1641.* Oxford: the Clarendon Press, 1986.

Mackenthun, Gesa. *Metaphors of Dispossession: American Beginnings and the Translation of Empire, 1492–1637.* Norman: University of Oklahoma Press, 1997.

Maley, Willy. "Shakespeare, Holinshed and Ireland: Resources and Con-texts." *Shakespeare and Ireland.* Eds. Mark Thornton Burnett and Ramona Way. New York: St. Martin's Press, 1997. 213–34.

———— *Salvaging Spenser: Colonialism, Culture, Identity.* London: Macmillan, 1997b.

Marcus, Leah S. *The Politics of Mirth.* Chicago: University of Chicago Press, 1986.

———— *Puzzling Shakespeare: Local Reading and its Discontents.* Berkeley: University of California Press, 1988.

Marotti, Arthur. "'Love is not love': Elizabethan Sonnet Sequences and the Social Order." *ELH* 49:2 (1982): 396–428.

Marvell, Andrew. *Miscellaneous Poems by Andrew Marvell, Esq.* London: Printed for Robert Boulter, at the Turks-Head in Cornhill, 1681.

———— *The Poems and Letters of Andrew Marvell.* Ed. H. M. Margoliouth. 3rd ed. in 2 vols. Oxford: At the Clarendon Press, 1971.

Maske of Flowers. London: Printed by N.O. for Robert Wilson, 1614.

Mason, Peter. *Deconstructing America.* London and New York: Routledge, 1990.

Masten, Jeffrey. *Textual Intercourse.* Cambridge: Cambridge University Press, 1997.

Mathes, Valerie Shirer. "A New Look at the Role of Women in Indian Society." *American Indian Quarterly.* Vol. 2 No. 2 (Summer 1975): 131–39.

Maus, Katherine Eisaman. *Ben Jonson and the Roman Frame of Mind.* Princeton: Princeton University Press, 1984.

McCanles, Michael. "Festival in Jonsonian Comedy." *Renaissance Drama* 8 (1977): 203–19.

———— *Jonsonian Discriminations: The Humanist Poet and the Praise of True Nobility.* Toronto: University of Toronto Press, 1992.

McCary, Ben C. *Indians in Seventeenth-Century Virginia.* Williamsburg: The Virginia 350th Anniversary Celebration Corporation, 1957.

McKeon, Michael. "Pastoralism, Pluralism, Imperialism, Scientism: Andrew Marvell and the Problem of Mediation." *Yearbook of English Studies* 13 (1983): 46–67.

McLeod, Bruce. *The Geography of Empire in English Literature.* Cambridge: Cambridge University Press, 1999.

McMullan, Gordon. *The Politics of Unease in the Plays of John Fletcher.* Amherst: The University of Massachusetts Press, 1994.

Memmi, Albert. *The Colonizer and the Colonized.* Boston: Beacon Press, 1965.

Mercator, Gerard. *Gerardi Mercatoris Altas or A Geographicke description of the Regions,*

Countries and Kingdomes of the World. Trans. Henry Hexham. Amsterdam: Henry Hondi, 1636.

Middleton, Thomas. *Honorable Entertainments Composed for the Seruice of this Noble Cittie.* London: G.E. 1621. Reprinted for the Malone Society. London: Oxford University Press, 1953.

———— *The Inner Temple Masque or Masque of Heros.* London: Printed for John Browne, 1619.

———— *The Triumphs of Honor and Vertue.* October 29, 1622. London: Nicholas Okes, 1622.

———— *The Trivmphs of Loue and Antiquity.* An Honourable Solemnitie performed through the Citie, at the confirmation and establishment of the Right Honourable Sir William Cockayn, Knight. October 29, 1619. London: Nicholas Okes, 1619.

———— *The Triumphs of Truth.* October 29, 1613. London: Nicholas Okes, 1613

Miles, Rosalind. *Ben Jonson: His Craft and Art.* Routledge: London, 1990.

Miller, Shannon. *Invested with Meaning: The Raleigh Circle in the New World.* Philadelphia: University of Pennsylvania Press, 1998.

Montrose, Louis Adrian. "'The perfecte paterne of a Poete': The Poetics of Courtship in *The Shepheardes Calendar.*" *Texas Studies in Literature and Language* 21:1 (1979): 34–67.

———— "Spenser's domestic domain: poetry, property, and the Early Modern subject." *Subject and object in Renaissance culture.* Ed. Margreta DeGrazia, Maureen Quilligan, and Peter Stallybrass. Cambridge: Cambridge University Press, 1996: 83–130.

———— "The Work of Gender in the Discourse of Discovery." *Representations* 33 (1991):1–41.

Morgan, Edmund S. *American Slavery American Freedom.* New York: W. W. Norton & Company, Inc., 1975.

Morton, Thomas. *New England Canaan.* Amsterdam, 1937. *Tracts and Other Papers.* Ed. Peter Force. 2: 1–128.

Moryson, Fynes. *An Itinerary Containing His Ten Yeeres Travell through the Twelve Dominions of Germany, Bohmerland, Sweiterland, Netherland, Denmarke, Poland, Italy, Turky, France, England, Scotland & Ireland.* 1617. Reprinted Glasgow: James MacLehose and Sons. 4 vols. New York: The Macmillan Company, 1907.

Mullaney, Steven. "The New World on Display: European Pageantry and the Ritual Incorporation of the Americas." *New World of Wonders.* Ed. Rachell Doggett. Washington, D.C.: The Folger Shakespeare Library, 1992.

———— *The Place of the Stage.* Chicago: The University of Chicago Press, 1988.

Munday, Anthony. *Camp-Bell or the Ironmongers Faire Feild.* October 29, 1609. *Pageants and Entertainments of Anthony Munday.* Ed. David M. Bergeron. New York: Garland, 1985.

———— *Chruso-thriambos: The Triumphes of Golde.* October 29, 1611. London: William Iaggard, 1611.

———— *Chrysanaleia: The Golden Fishing.* October 29, 1616. London: George Purstowe, 1616.

———— *London's Love, To the Royal Prince Henrie Meeting Him on the River of Thames, at his returne from Richmonde.* Thursday, last of May 1610. London: Printed by Edward Allde for Nathaniell Forsbrooke, 1610.

———— *The Tivmphes of re-vnited Britania.* October 29, 1605. London: W. Jaggard. *Pageants and Entertainments of Anthony Munday.* Ed. David M. Bergeron. New York: Garland, 1985.

Murphy, Andrew. "Gold Lace and a Frozen Snake: Donne, Wotton and the Nine Years War." *Irish Studies Review* 8 (1994): 9-11.

———— "'Tish ill done': *Henry the Fift* and the Politics of Editing." *Shakespeare and Ireland.* Eds. Mark Thornton Burnett and Ramona Way. New York: St. Martin's Press, 1997: 213-34.

———— *But the Irish Sea Betwixt Us: Ireland, Colonialism, and Renaissance Literature.* Lexington: The University Press of Kentucky, 1999.

Murray, David. *Forked Tongues.* Bloomington: Indiana University Press, 1991.

Nandy, Ashis. *The Intimate Enemy: Loss and Recovery of Self under Colonialism.* Delhi: Oxford University Press, 1983.

Neely, Carol Thomas. "The Structure of English Renaissance Sonnet Sequences." *ELH* 45:3 (1978): 359-89.

Nelson, Ann Thrift. "Woman in Groups: Women's Ritual Sodalities in Native North America." *The Western Canadian Journal of Anthropology.* Vol. VI, No. 3 (1976): 29-67.

Newman, Karen. *Fashioning Femininity.* Chicago: The University of Chicago Press, 1991.

Newton, Richard C.. "Jonson and the Reinvention of the Book." In *Classic and Cavalier.* Ed. Claude J. Summers and Ted-Larry Pebworth. Pittsburgh: University of Pittsburgh Press, 1982.

Nobles, Gregory H. "Straight Lines and Stability: Mapping the Political Order of the Anglo- American Frontier." *The Journal of American History* 80:1 (1993): 9-35.

Norbrook, David. *Poetry and Politics in the English Renaissance.* London: Routledge and Kegan Paul, 1984.

Oberg, Michael Leroy. *Dominion and Civility: English Imperialism and Native America, 1585-1685.* Ithaca: Cornell University Press, 1999.

O'Brien, Terence H. "The London Livery Companies and the Virginia Company." *VMHB,* 137-55.

Orgel, Stephen. *The Illusion of Power.* Berkeley: University of California Press, 1975.

———— *Impersonations; The performance of gender in Shakespeare's England.* Cambridge: Cambridge University Press, 1996.

———— *The Jonsonian Masque.* New York: Columbia University Press, 1981.

———— "Nobody's Perfect: Or Why Did the English Stage Take Boys for Women." *South Atlantic Quarterly* 88 (Winter 1989): 7-29.

—— "To Make Boards Speak." *Renaissance Drama* n.s. 1 (1968): 121-52.

—— "What is a Text?" In *Staging the Renaissance.* Ed. David Scott Kastan and Peter Stallybrass. New York: Routledge, 1991.

Original Writings & Correspondence of the Two Richard Hakluyts. Ed. E.G. R. Taylor. 2 vols. London: The Hakluyt Society, 1935.

Palmer, Daryl W. "Ben Jonson as Other: Recent Trends in the Criticism of Jonson's Drama." *Research Opportunities in Renaissance Drama* 31 (1992): 1-11.

Parry, Graham. "The Politics of the Jacobean Masque." *Theatre and Government Under the Stuarts.* Ed. J. R. Mulryne and Margaret Shewring. Cambridge: Cambridge University Press, 1993. 87-117.

Paster, Gail Kern. *The Body Embarrassed: Drama and the Disciplines of Shame in Early Modern England.* Ithaca: Cornell University Press, 1993.

Pearl, Sara. "'Sounding to present occasions': Jonson's masques of 1620-5." *The Court Masque.* Ed. David Lindley. Manchester: Manchester University Press, 1984.

Percy, George. *Observations gathered out of a Discourse of the Plantaion of the Southerne Colonie of Virginia by the English, 1606.* *Purchas His Pilgrimes.* in Five Bookes. London: Printed by William Stansby for Henrie Fetherstone, 1625. Vol. 4, 1685-1690.

Phillips, Sir Thomas. *Londonderry and the London Companies 1609 - 1629. Being a Survey by Sir Thomas Phillips.* Belfast, 1928.

Pratt, Mary Louise. *Imperial Eyes: Travel Writing and Transculturation.* London: Routledge, 1992.

Prescott, Anne Lake. "The Thirsty Deer and the Lord of Life: Some Contexts for *Amoretti* 67-70." *Spenser Studies* VI (1985): 33-76.

A Publication by the Counsell of Virginea, touching the plantation there. London: T. Haueland for William Welby, 1610, STC #24831.7.

Purchas, Samuel. *Purchas his Pilgrimage, or Relations of the World and the Religions Observed in Al Ages and Places discouered, from the Creation unto this Present.* London: Printed by William Stansby for Henry Fethestone, 1617.

Quinn, David Beers. "The Munster Plantation: Problems and Opportunities." *Journal of the Cork Historical and Archaelogical Society* LXXI (1966): 19-40.

—— "Bermuda in the Age of Exploration and Early Settlement." *Bermuda Journal of Archaeology and Maritime History* 1 (1989): 1-23.

Quinn, Naomi. "Anthropological Studies on Women's Status." *Annual Review of Anthropology.* (1977): 181-225.

Quitslund, Jon A. "Spenser's *Amoretti* VIII and Platonic Commentaries on Petrarch." *Journal of the Warburg and Courtauld Institutes* 36 (1973): 256-76.

Quitt, Martin H. "Trade and Acculturation at Jamestown, 1607 - 1609: The Limits of Understanding." *The William and Mary Quarterly.* 3rd. series 52:2 (1995): 227-58.

Rackin, Phyllis. "Androgyny, Mimesis, and the Marriage of the Boy Heroine on the English Renaissance Stage." *PMLA* 102 (1987): 29-41.

———— *Stages of History: Shakespeare's English Chronicles.* Ithaca: Cornell University Press, 1990.

Raleigh, Sir Walter. *The Discovery of the Large, Rich, and Beautiful Empire of Guiana.* Reprinted from the edition of 1596. Ed. Sir Rober H. Schomburgk. London: Hakluyt Society, 1848.

Raman, Shankar. "Imaginary Islands: Staging the East." *Renaissance Drama* n.s. XXVI (1995): 131-61.

Randall, Dale B. J. *Jonson's Gypsies Unmasked.* Durham, N.C.: Duke University Press, 1975.

The Records of the Virginia Company of London. Susan Myra Kingsbury Ed. 4 vols. Washington: United States Government Printing Office, 1933.

Repertory of the Inrolments on the Patent Rolls of Chancery in Ireland. Ed. J. C. Erck. *Irish History From Contemporary Sources 1509–1610.* Ed. Constantia Maxwell. London: George Allen & Unwin Ltd. 1923: 252-53.

Rich, Barnaby. *A New Description of Ireland.* (1610). *Irish History From Contemporary Sources 1509 - 1610.* Ed. Constantia Maxwell. London: George Allen & Unwin Ltd., 1923: 340-41.

The Rich Papers: Letters From Bermuda 1615–1646. Ed. Vernon A. Ives. Toronto: The University of Toronto Press, 1984.

Roach, Joseph. *Cities of the Dead.* New York: Columbia University Press, 1996.

Robertson, Karen. "Pocahontas at the Masque." *Signs* 21:3 (1996): 551-83.

Robinson, W. Stitt, Jr. *Mother Earth: Land Grants in Virginia, 1607–1699.* Jamestown: The Virginia 350th Anniversary Celebration Corporation, 1957.

Rountree, Helen C. *Pocahontas's People: The Powhatan Indians of Virginia Through Four Centuries.* Norman: University of Oklahoma Press, 1990.

———— *The Powhatan Indians of Virginia: Their Traditional Culture.* Normon: University of Oklahoma Press, 1989.

———— "Powhatan Indian Women: The People Captain John Smith Barely Saw." *Ethnohistory* 45:1 (1998): 1-29.

"Royal Proclamation Against the Earl of Tyrone." (1595) *Irish History From Contemporary Sources 1509–1610.* Ed. Constantia Maxwell. London: George Allen & Unwin, Ltd., 1923.

Rundstrom, Robert A. "Mapping, Postmodernism, Indigenous People and the Changing Direction of North American Cartography." *Cartographica* 28:2 (1991): 1-12.

Said, Edward W. *Culture and Imperialism.* New York: Knopf, 1992.

———— *Orientalism.* New York: Vintage Books, 1978.

Sale, Kirkpatrick. *The Conquest of Paradise.* New York: Penguin Books, 1990.

Sanders, Julie. "'The Day's Sports Devised in the Inn': Jonson's *The New Inn* and Theatrical Politics." *Modern Language Review* 91:3 (1996): 545-60.

Sandys, George. *Ovid's Metamorphosis Englished by George Sandys.* The Eighth Edition. London: Printed for A. Roper, 1690.

Saxey, William to the Earl of Essex, October 19, 1599. *Irish History From Contem-*

porary Sources 1509–1610. Ed. Constantia Maxwell. London: George Allen & Unwin, Ltd., 1923: 191.

Sayre, Gordon M. *Les Sauvages Américains: Representations of Native Americans in French and English Colonial Literature*. Chapel Hill: The University of North Carolina Press, 1997.

Scanlon, Thomas. *Colonial Writing and the New World 1583–1671*. Cambridge: Cambridge University Press, 1999.

Schmidt, Benjamin. "Mapping an Empire: Cartographic and Colonial Rivalry in Seventeenth-Century Dutch and English North America." *The William and Mary Quarterly.* 3rd ser. LIV:3 (1997): 549-78.

Seed, Patricia. *Ceremonies of Possession in Europe's Conquest of The New World 1492-1640*. Cambridge: Cambridge University Press, 1995.

Shakespeare, William. *The Complete Works of Shakespeare*. Ed. David Bevington. New York: HarperCollins, 1992.

————— *The First Folio of Shakespeare*. Prepared by Charlton Hinman. New York: W. W. Norton & Company, 1968.

————— *The tragoedy of Othello, The Moore of Venice*. London: Printed by N.O., 1622. Reproduced in *Shakespeare's Plays in Quarto*. Ed. Michael J. B. Allen and Kenneth Muir. Berkeley: University of California Press, 1981.

Shirley, James, and Inigo Jones. *The Triumph of Peace*. February 3, 1633. London: Printed by Iohn Norton for William Cooke, 1633.

Siddiqi, Yumna. "Dark Incontinents: The Discourses of Race and Gender in Three Renaissance Masques." *Renaissance Drama* n.s. XXIII (1992): 139-63.

Sidney, Sir Philip. *An Apology for Poetry or The Defense of Poesy.* Ed. Geoffrey Shepherd. London: Thomas Nelson and Sons, Ltd., 1965.

Sigalas, Joseph. "Sailing against the Tide: Resistance to Pre-Colonial Constructs and Euphoria in *Eastward Hoe!*" *Renaissance Papers* (1994): 85-94.

Sinfield, Alan. *Literature in Protestant England 1560–1660*. London: Croom Helm, 1983.

Slights, William W. E. *Ben Jonson and the Art of Secrecy*. Toronto: University of Toronto Press, 1994.

Smith, Abbot Emerson. *Colonists in Bondage: White Servitude and Convict Labor in America 1607–1776*. Chapel Hill: The University of North Carolina Press, 1947.

Smith, James M. "Effaced History: Facing the Colonial Contexts of Ben Jonson's *Irish Masque at Court*." *ELH* (1998): 297-321.

Smith, Captain John. *The Complete Works of Captain John Smith*. Ed. Philip Barbour. 3 vols. Chapel Hill: The University of North Carolina Press, 1986.

Speed, John. *The Genealogies Recorded in the Sacred Scriptures, According to euery Family and Tribe. With the Line of Ovr Sauiour Iesvs Christ . . .* [F. Kingston] 1638.

————— *A Prospect of the Most Famovs Parts of the World . . . Together With all the Prouinces, Counties and Shires, contained in that large Theater of great Brittaines Empire.* London: Printed by John Dawson for George Humble. *Theatrum Orbis Terrarvm.*

Series of Atlases in Facsimile. Third Series, Vol. VI. Intro. R. A. Skelton. Amsterdam: Theatrum Orbis Terrarvm Ltd., 1966.

Spelman, Henry. *Relation of Virginia*. London: Printed for Jas. F. Hunnewell, at the Chiswick Press, 1872.

Spenser, Edmund. *The Faerie Queene*. Ed. Thomas P. Roche, Jr. New York: Penguin Books, 1978.

———— *A View of the Present State of Ireland*. Ed. W. L. Renwick. London: Scholartis Press, 1934.

———— *The Yale Edition of the Shorter Poems of Edmund Spenser*. Ed. William A. Oram, Einar Bjorvand, Ronald Bond, Thomas H. Cain, Alexander Dunlop, and Richard Schell. New Haven: Yale University Press, 1989.

Spurr, David. *The Rhetoric of Empire*. Durham: Duke University Press, 1993.

Stahle, David W., Malcolm K. Cleveland, Dennis B. Blanton, Matthew D. Therrell, and David A. Gay. "The Lost Colony and Jamestown Droughts." *Science* April 24: 280 (1998): 564-67.

Stallybrass, Peter. "Macbeth and Witchcraft." *Focus on Macbeth*. Ed. J. R. Brown. London: Routledge, 1982.

Stallybrass, Peter, and Allon White. *The Politics and Poetics of Transgression*. Ithaca: Cornell University Press, 1986.

Stillman, Robert E. "Spenserian Autonomy and the Trial of New Historicism: Book Six of *The Faerie Queene*." *ELR* 22:3 (1992): 299-314.

Stoler, Ann Laura. *Race and the Education of Desire: Foucault's History of Sexuality and the Colonial Order of Things*. Durham: Duke University Press, 1995.

———— "Rethinking Colonial Categories: European Communities and the Boundaries of Rule." *Comparative Studies in Society and History* 31 (1989): 134-61.

Strachey, William. *For the Colony in Virginea Britannia Lawes Divine, Morall and Martiall, &.* London: Printed for Walter Burre, 1612.

———— *The Historie of Travell Into Virginia Britania*. London: The Hakluyt Society, 1953.

———— *A true reportory of the wracke, and redemption of Sir Thomas Gates Knight; vpon, and from the Ilands of the Bermudas: his comming to Virginia, and the estate of that Colonie then, and after, vnder the gouernment of the Lord La Warre, Iuly 15, 1610*. Purchas His Pilgrimes. London: Printed by William Stansby for Henrie Fethestone, 1625. Vol. 4, 1734-56.

Sturtevant, William C. "The Sources for European Imagery of Native Americans." *New World of Wonders*. Ed. Rachel Dogget. Washington: The Folger Shakespeare Library, 1992: 25-33.

Suleri, Sara. *The Rhetoric of English India*. Chicago: The University of Chicago Press, 1992.

Sweeney, John Gordon III. *Jonson and the Psychology of Public Theater*. Princeton: Princeton University Press, 1985.

Symonds, William. *Virginia A Sermon Preached at White-Chappel*. April 25, 1609.

Published for the Benefit and Vse of the Colony. London: Printed by I. Windet, for Eleazar Edgar, and William Welby, 1609.

Taussig, Michael. *Shamanism, Colonialism, and the Wild Man.* Chicago: University of Chicago Press, 1987.

Thomas, Nicholas. *Colonialism's Culture: Anthropology, Travel and Government.* Princeton: Princeton University Press, 1994.

Thompson, Charlotte. "Love in an Orderly Universe: A Unification of Spenser's *Amoretti,* "Anacreontics," and *Epithalamion.*" *Viator* 16 (1985): 277-335.

Three Proclamations concerning the Lottery for Virginia 1613–1621. Providence: The John Carter Brown Library, 1907.

Townshend, Aurelian. *Albion's Triumph.* London: Printed by Aug: Matthewes for Robert Allet, 1631.

Townshend, Aurelian, and Inigo Jones. *Tempe Restored.* London: Printed by A.M. for Robert Allet and George Baker, 1631.

Turnbull, David. *Maps Are Territories: Science Is an Atlas.* Chicago: The University of Chicago Press, 1993.

Turner, E. Randolph. "Socio-Political Organization within the Powhatan Chiefdom and the Effects of European Contact, A.D. 1607–1646." *Cultures in Contact: The Impact of European Contacts on Native American Cultural Institutions A.D. 1000–1800.* Ed. William W. Fitzhugh. Washington: Smithsonian Institution Press, 1985. 193-224.

U.S. Department of Commerce. *1990 Census of Population: Social and Economic Characteristics of Virginia.* September 1993.

Verner, Coolie. "The First Maps of Virginia, 1590–1673." *The Virginia Magazine of History and Biography* 58:1 (1950): 3-15.

Villeponteaux, Mary A. "'With her own will beguyld'": The Captive Lady in Spenser's *Amoretti.*" *Explorations in Renaissance Culture* 14 (1988): 29-39.

Virginia Company. *A declaration of the state of the colonie and affaires in Virginia.* 1620 STC #24841.2.

Vizenor, Gerald. *Manifest Manners: Postindian Warriors of Survivance.* Hanover: Wesleyan University Press, 1994.

Waller, Gary. *Edmund Spenser: A Literary Life.* New York: St. Martin's Press, 1994.

Waterhouse, Edward. *A Declaration of the State of the Colony and Affaires in VIRGINIA. With A Relation of the Barbarous Massacre in the time of peace and League, treacherously executed by the Native Infidels vpon the English, the 22 of March last.* London: G.Eld for Robert Mylbourne, 1622.

Wayne, Don E. "Drama and Society in the Age of Jonson: An Alternative View." *Renaissance Drama.* 13 (1982): 103-29.

———— *Penshurst.* Madison: University of Wisconsin Press, 1984.

Webster, John. *The Duchess of Malfi.* London: A&C Black, 1993.

Webster, John. *Monuments of Honor.* London: Nicholas Okes, 1624. Reprinted in R.T.D. Sayle. *Lord Mayor's Pageants of the Merchant Taylor's Company in the 15th, 16th & 17th Centuries.* London: The Eastern Press, Ltd., 1931.

Weimann, Robert. "Representation and Performance: The Uses of Authority in Shakespeare's Theater." *PMLA* 107:3 (1992): 497-510.

———— *Shakespeare and the Popular Tradition in the Theater: Studies in the Social Dimension of Dramatic Form and Function.* Baltimore: Johns Hopkins University Press, 1978.

Wells, Susan. "Jacobean City Comedy and the Ideology of the City." *ELH* 48 (1981): 37-60.

Wells, Robin Headlam. "Poetic Decorum in Spenser's *Amoretti.*" *Cahiers Elisabethains* 25 (April 1984): 9-21.

———— "*Semper Eadem:* Spenser's 'Legend of Constancie.'" *Modern Language Review* 73 (1978): 250-55.

White, John. *The American Drawings of John White.* Ed. Paul Hulton and David Beers Quinn. Chapel Hill: The University of North Carolina Press, 1964.

Williams, Raymond. *Keywords.* New York: Oxford University Press, 1976.

Womack, Peter. *Ben Jonson.* London: Basil Blackwell, 1986.

Woodbridge, Linda. *Women and the English Renaissance.* Urbana and Chicago: University of Illinois Press, 1984.

Woren, Blair. "Andrew Marvell, Oliver Cromwell, and the Horatian Ode." *Politics of Discourse: The Literature and History of Seventeenth-Century England.* Ed. Kevin Sharpe and Steven Zwicker. Berkeley: University of California Press, 1987.

Wright, Nancy E. "'Rival traditions': civic and courtly ceremonies in Jacobean London." *The Politics of the Stuart Court Masque.* Ed. David Bevington and Peter Holbrook. Cambridge: Cambridge University Press, 1998.

Wright, Ruth C. "Descriptive Notes to the Illustrations." *Illustrated English Social History.* Ed. George M. Trevelyan. Hammondsworth, England: Longmans, 1964: II, 303.

Young, Robert J. C. *Colonial Desire: Hybridity in Theory, Culture and Race.* London: Routledge, 1995.

Index

Adelman, Janet, 255 n.14
Africans, 29, 159, 160–1, 174, 176–7, 222,
    231, 239 n.43, 250 n.6
Algonquian, 14, 80–4, 184, 201, 233 n.7,
    243 n.10
allegory, colonial, 145–7, 184
Alpers, Svetlana, 244 n.14
Altman, Joel, 246 n.1
Amazons, 2, 49–51, 54, 63–4, 181, 205,
    207–8, 210, 255 n.13
Anderson, Benedict, 4
Andrews, John, 85
Andrews, Lynn, 214
Anglican Church, Anglicans, 14, 16, 252
    n.32, 254 n.4
    see also Christianity; conversion
Anne, Queen, 163, 166, 174, 175, 198, 206
Ansley, Joan, 46
"antics," 187, 195
Antigua, 23, 26, 28, 30, 77–8, 238 n.33
antimasques, 150, 165–71, 173, 174, 178,
    179, 181–2, 183, 186, 195, 206–8, 209,
    250 n.5, 254 n.53
"antiselves," 124, 126, 131, 137, 138, 140,
    141, 146, 151, 155, 161, 177, 181, 185
Arber, Edward, 245 n.22
Archer, John Michael, 239 n.40
Argall, Samuel, 21, 183
Arundell, Peter, 7, 9, 22
Avery, Bruce, 165
Axtell, James, 237 n.29

Bach, Rebecca Ann, 123, 245 n.19, 248
    n.19, 256 n.6
Bacon, Francis, "On Masques and
    Triumphs," 195
Bagenal, Marshal, 43
Baker, David J., 40, 113, 246 n.1
Baldo, Jonathan, 246 n.1
Balzano, Wanda, 242 n.28
Banerjee, Pompa, 239 n.40
Barbados, 164
Barbour, Philip, 80, 89, 91, 196, 243 n.10,
    244 n.12, 255 n.17
"bards," bardic poetry, 32, 52, 59–61, 70–3,
    167–8, 252 n.23
Bartels, Emily C., 239 n.39, n.40
Bartley, J.O., 246 n.4
Barton, Anne, 130
Bates, Catherine, 40, 47, 240 n.8
Beare, Philip O'Sullivan, 43–44
Bell, Catherine, 216, 256 n.26
Belsey, Catherine, 255 n.14, n.19
Bennett, Tony, 221
Bermuda Company, 32, 98, 114
Bermuda(s), 3, 4, 5, 12–13, 29, 32, 42–3,
    93–112, 114, 157, 186, 239 n.43,
    n.44
    area of London, 120, 136
Bermudez, Juan, 93
Bernard, John D., 240 n.5
Black, Jeanette, 93, 99
Blackfriars theater, 125, 135, 143, 146

blackness, 11, 29–31, 33, 45, 150, 155, 158–63, 174–6, 180, 250 n.4
  see also whiteness
Blank, Paula, 252 n.24
Blount, Charles, Lord Mountjoy, 71
Boelhower, William, 101
Boyle, Elizabeth, 39, 45–7, 240 n.14
Boyle, Richard, 39, 45–7, 241 n.21
Bradshaw, Brendan, 233 n.2, 243 n.4
Breitenberg, Mark, 239 n.4
Breminham, Lord, 167
Brenner, Robert, *Merchants and Revolution,* 151, 153, 157, 164, 239 n.43, 250 n.6, n.7, 251 n.9, n.12, n.14
Brent, Nathaniel, 173
Brotton, Jerry, 250 n.3
Brown, Alexander, 233 n.3, 247 n.6
Brown, Kathleen, 256 n.6
Browne, William, *Ulysses and Circe,* 253 n.39
Bryskett, Lodowick, 38, 46–7
Butler, Martin, 184–5, 248 n.26, 250 n.5
Butler, Nathaniel, 96–7, 102–11, 117, 118, 239 n.41, 245 n.20, n.22, 246 n.25, n.26, n.27, 248 n.26

Cairns, David, 242 n.27
Camden, William, 58
  Britannia, 72–3, 241 n.23
Campbell, James, 161
Campbell, Mary B., 129, 248 n.21
Campbell, Thomas, 160
Campion, Edmund, 77
Campion, Thomas, 164, 175, 179, 253 n.37
Canning, Rick C., 62
Canny, Nicholas, 240 n.14, 241 n.21, 242 n.26, n.30, n.31, 252 n.23
Capp, William, 236 n.20
Carew, Thomas, *Coelum Britanicum,* 254 n.53
Carleton, Sir Dudley, 96, 97, 168, 173, 237 n.26
Carney, James, 60
Carroll, Clare, 41, 51, 233 n.2, 253 n.39
Carter, Paul, 68, 75, 85, 86
cartography, 82, 245 n.21
  see also maps, mapping
Castaneda, Carlos, 214
categorization, 3–7, 9, 23, 72, 75, 111, 124, 174

Catholicism, Catholics, 14, 16, 28–9, 65, 73–6, 95, 169
Chamberlaine, John, 96–7, 237 n.26
Chapman, George, 117, 118, 177
  Eastward Hoe, 116–20, 124–7, 128, 137, 141, 143, 145
  The Memorable Maske, 175, 177
  "A Relation of the Second Voyage to Guiana," 41
Charles I, 46, 153, 189–90, 251 n.9
  as Prince of Wales, 170, 183, 210
Cheyfitz, Eric, 173, 237 n.31, 255 n.21
Chickahominy, 4, 19, 233 n.7, 235 n.12
children
  English, 4, 14, 18, 19, 20
  Indian, 15, 16, 18, 19, 20, 28
Christianity, 15, 18–19, 22, 30, 32, 53, 55, 75, 91–2, 117, 123, 138, 147–8, 150, 159–60, 176, 204
  see also Anglican Church; conversion
Churchill, Ward, 28, 34, 69, 214, 231, 236 n.16, 243 n.2, 256 n. 4, n. 9
Clarke, G.N.G., 87, 101
Cobo, Father Bernabe, 95
  History of the Inca Empire, 74–8
Cockayne, Sir William, 153, 156, 157, 251 n.9
Cohen, Walter, 239 n.40
colonial desire, 10, 37, 93, 118, 119, 152, 185
colonial fantasy, 82, 108, 110, 116, 118, 165
Colonial National Historical Park, 222, 225, 226, 230–2, 256 n.1, 257 n.11
colonizer/colonized, 4–6, 7, 10, 14–15, 16, 23, 40, 146–7, 152, 204, 234 n.9, 235 n.14
Columbus, Christopher, 23, 128, 131, 135, 223, 248 n.23
commerce, 83, 86, 97, 141, 150–1, 156–7, 159–60, 250 n.2, 250–1 n.7, 251 n.12
Constable, Sir William, 164
conversion, 15, 18–19, 91–3, 117, 122, 138, 140–1, 146, 150, 157–60, 162, 204, 217, 254 n.4
Cook, James, 68, 75
coodinated oppressions, 6, 30–1, 50, 62, 69, 149–50

Copland, Patrick, 116
  Virginia's God be Thanked, 140–3
Cornwalleis, Sir William, 166, 186
Coughlan, Patricia, 41
court masques, *see* masques
Crashaw, William, 116, 138, 160
Craven, Wesley Frank, 237 n.25, 248 n.17
Crofton, John, 46
Cromwell, Oliver, 65
Cronon, William, 25, 26, 77, 80, 238 n.35
cultural genocide, 34, 91, 213–14, 219, 236 n.16
cultural relativism, 71–3, 123, 228, 243 n.3

Dale, Sir Thomas, 13–14, 21, 87, 202
Daniel, Samuel, 195, 254 n.2
  Delia, 52, 240 n.16
Dasenbrook, Reed Way, 40, 47
Davenant, William
  The Temple of Love, 179
  The Triumphs of the Prince D'Amovr, 179–80
Davies, R.R., 170, 171, 234
Day, Gordon M., 243 n.10
De Bry, 178
Declaration of the state of the colonie and affaires in Virginia, 25, 115, 122
  for 1620, 120, 122, 164
  for 1622, 91, 133
DeGrazia, Margreta, 83, 108
Dekker, Thomas
  London's Tempe or the Feild of Happines, 161
  Magnificent Entertainment, 251 n.20
  The Second Part of the Honest Whore, 114
Deloria, Vine, 256 n.9
Derrida, Jacques, 255 n.22
Desmond rebellion, 3–4, 44, 71, 75
de Soto, Fernando, 5–6, 141
  see also Hakluyt, Richard
de Wilton, Lord Grey, 31, 47, 57–9, 71
Dickenson, Jane, 8–9, 14–15
Doncaster, Viscount, 164
Donne, John, 27, 116, 141–3, 151, 162, 247 n.7
Donow, Herbert S., 51
Drake, Francis, 78, 131, 135, 137, 243 n.6

Drayton, Michael, 30, 116, 248 n.13
  Idea, 52
  Poem to Sandys, 121–4, 141
Duff, Meaghan, 242 n.1
Dunlop, Robert, 240 n.12
Dunne, T.J., 62
Dyer, Richard, 34, 174, 239 n.45

East India Company, 143, 150, 152, 250 n.3, 251 n.9
East Indies, 140–1, 146, 149–51, 159–61, 163, 175, 179, 239 n.40, 250 n.2, 251 n.7
Edward I, 170
Elias, Norbert, 123
Elizabeth (film), 256 n.8
Elizabeth I, 3, 24, 40, 41, 44, 46, 52–3, 55, 73–4, 78, 163, 205, 228, 256 n.8
Elizabeth (daughter of Edward, Duke of Buckingham), 184–5, 210–11
Elizabeth (daughter of James I), 84, 119, 175–6, 210
Ellis, Steven G., 241 n.24
embodiment, 41, 87
  see also feminization
England, imperial identity of, 1–5, 9, 21, 31, 33, 57, 67, 115, 126, 149–52, 185–7, 230–2
escheatage, 4, 31, 44–5, 46, 71, 153, 163, 167, 240 n.13
Esguerra, Jorge Ca izares, 243 n.5
Etienne, Mona, 255 n.10
Evans, Herbert, 253 n.36
Ezell, Margaret J.M., 205, 210, 255 n.12

Fabian, Johannes, 255 n.17
"fame," 168
Fanon, Frantz, 6, 43
Feest, Christian F., 249 n.34, n.36
feminization, 41–3, 53
  of Guyana, 41, 117
  of Indians, 126
  of Ireland, 32, 41, 59–63, 68, 240 n.9, 242 n.28
  of Virginia, 115–16
  of Wales, 169
Ferrar, John, 9, 14
Ferrar, Nicholas, 14, 18

Ffrethorne, Richard, 17–18
firearms, 16–17, 85, 232
Fletcher, John, 113
   The Island Princess, 114, 246 n.3
   The Sea Voyage, 114
Florida, "Floridians," 5, 99, 177
Fogarty, Anne, 59
food, 17–18, 21, 25, 94–5, 105–7, 203,
   214–16, 255 n.23, 256 n.7
Ford, John, *The Chronicle History of Perkin
   Warbeck,* 114
Frances, Duchess of Richmond and Lenox,
   210
Freeman, Ralph, 157
Fuller, Mary, 28, 233 n.5, 240 n.16
Fumerton, Patricia, 252 n. 29, 255 n.23

Gable, Eric, 224
Gaelic, 4, 6, 17, 30, 32, 44, 59–61, 64–5,
   70–4, 77, 237 n.29, 242 n.26, n.28,
   n.31
Garber, Marjorie, 253 n.43
Garway, Henry, 157
Gates, Sir Thomas, 5, 83, 93, 94, 98, 105
Genesis, book of, 92, 245 n.19
gentility, 1–3, 145
Gibbs, Donna, 47, 55
Gillies, John, 252 n.34
Gilroy, Paul, 29
Gleach, Frederic W., 235 n. 13
Gln Dwr (Owen Glendower), 171
Goldberg, Jonathan, 253 n.41, 254 n.3
Gramsci, Antonio, 104
Greenblatt, Stephen, 26, 39, 41, 238 n.38
   Renaissance Self-Fashioning, 40
Greene, Roland, 31, 239 n.39
Gunpowder Plot, 16
Guy, Arthur, 91
Guyana, 41–2, 117, 177, 240 n.10, 241 n.20
gypsies, 31, 124, 132, 134, 143, 148, 184–9,
   254 n.53

Hadfield, Andrew, 41, 113, 233 n.2, 237
   n.24, 246 n.2, n.3, 248 n.27
Hakluyt, Richard, the younger, 5–6, 7, 28,
   42, 61, 181, 196, 197, 210, 237 n.24,
   254 n.4, n.5
   Principal Navigations, 129

*Virginia richly valued, By the description of
   the maine land of Florida, her neighbor,* 5,
   141
Voyages, 248 n.23
Hall, Catherine, 235 n.14
Hall, Kim, 30, 41, 160–1, 174, 233 n.5
   Things of Darkness, 27
Hall, Stuart, 234 n.10
Halsey, Theresa, 204, 254 n.7, 255 n.10
Hamor, Ralph, 13–14, 21, 236 n.18
   *A True Discourse of the Present Estate of
   Virginia,* 202
Hanafin, Patrick, 62
Handler, Richard, 224
Harjo, Joy, 28
Harley, J.B., 69, 85, 89–91, 105, 112, 243
   n.2, 244 n.16, 245 n.21
Harriot, Thomas, 140
   *A Briefe and True Report of the New Found
   Land of Virginia,* 138
Hawkins, Harriet, 248 n.22
Hawkins, Sir John, 78, 243 n.6
Haynes, Jonathan, 135, 248 n.15, 249 n.28,
   n.38
Haynes-McCoy, G.A., 241 n.24
Hechter, Michael, 25
Heckewelder, John, 83, 243 n.10
Helgerson, Richard, 38, 115
Henderson, Katherine Usher, 253 n.43
Hendricks, Margo, 239 n.45
Henley, Pauline, 64, 240 n.6
Henry, Prince of Wales (son of James I), 81,
   163, 170, 252 n.34
Herbert, Sir Gerard, 173
   *Historye of the Bermudaes or Summer
   Islands,* 1, 29
Herford, C.H., 133, 247 n.10, 248 n.13,
   250–1 n.7, 251 n.16, 252 n.25
Herr, Cheryl, 242 n.28
Heywood, Thomas, 250 n.1
   *Londini Status Pacatus: or Londons Peaceable
   Estate,* 157–8
   A Woman Killed with Kindness, 246 n.5
Highley, Christopher, 233 n.2, 235 n.15,
   237 n.29, 239 n.39
Holland, Hugh, *Pancharis,* 124
Holliday, Sir Leonard, 155
Honigmann, E.A.J., 253 n.36

Hood, Samuel, 78, 243 n.6
Hope theater, 125, 135
Hopkins, Lisa, 246 n.1
Houghton, Sir Richard, 164
Howard, Frances, 177
Howard, Jean E., 246 n.3, 249 n.38, 253 n.43
Howard, Skiles, 206, 250 n.2, 252 n.24
Hudson, Kenneth, 229–30
Hulme, Peter, 9, 10, 107, 183, 236 n.20, 238 n.34, 240 n.10, 256 n.6
Hunter, G.K., 240 n.5
hybridity, 4–6, 23

India, 180, 221
 see also East Indies
Indians, 3–25, 28–9, 33–5, 69–70, 79–93, 107, 122–3, 182–4
 early modern representations of , 121, 126, 138–40, 151, 160–1, 175–80, 185
 in the Jamestown Settlement museum, 221–32
 John Smith and, 191–219
 "undifferentiated," 6, 31, 33, 86, 141–3, 150–2, 175–80, 185, 210, 221–2
Ireland, 3–4, 6, 20, 67–9, 115, 149, 153–6, 163–4, 168–9, 224
 see also Irish, representations of; Spenser, Edmund
Ireland island, Bermuda, 110–12
Irish, representations of, 114, 131–2, 133–4, 150, 165–8, 171–3
Ives, Vernon A., 246 n.26, n.28

Jacob and Esau, 92, 245 n.18
Jaimes, M. Annette, 204, 254 n.7
James, M. Annette, 204, 254 n.7
James I, 14, 26, 80–2, 87, 89, 91, 98, 122, 124, 146, 164–70, 175–6, 184–6, 189, 196, 200, 228, 252 n.32, 254 n.47
 Demonologie, 207
James City, 19, 81, 82
James River, 87, 91
James town (Jamestown), 79–80
Jamestown Settlement museum, 34, 221–31
Jehlen, Myra, 238 n.38, 256 n.6
Jenkins, Raymond, 240 n.6
Jobson, Francis, 85

Johnson, Robert, 10, 12, 14–15, 17
 Nova Brittania, 16
Jones, Ann Rosalind, 240 n.9, 252 n.24
Jones, Inigo, 175, 193, 195–6
 Albion's Triumph, 180
 The Memorable Maske, 175
 Tempe Restord, 179
 The Temple of Love, 179
Jonson, Ben, 29, 33–4, 115–16, 145, 146, 151, 155, 92, 93–6
 Bartholomew Fair, 119–41, 195
 Devil is an Asse, The, 127, 247 n.11, 248 n.14
 Eastward Hoe, 116–20, 124–7, 128, 137, 141, 143, 145
 The Entertainment at Highgate, 166, 251 n.20
 Epicoene, 145, 253 n.43
 "An Epistle to Sir Edward Sacvile, now Earl of Dorset," 136
 For the Honor of Wales, 31, 150, 169–73
 Gypsies Metamorphos'd, 31, 184–90
 Hymenaei, 164
 Irish Masque at Court, 1, 31, 41, 150, 164–73, 188, 198–9
 Love Freed From Ignorance and Folly, 198–9
 Masque of Beauty, 174–5
 Masque of Blackness, 30–1, 174–5, 250 n.2
 Masque of Queenes, 206–8
 The New Inne, 120, 125, 127–35, 136–8, 140, 141, 143–4
 News From the New World Dicovered in the Moon, 181–4
 "Ode Allegoric," 124
 Pleasure Reconciled to Virtue, 169
 Prince Henry's Barriers, 163–5, 167
 The Staple of News, 144–8, 182, 191
 The Vision of Delight, 183
Jordan, Constance, 252 n.33
Josephy, Alvin M., Jr., 192
Jourdain, Silvester, 105–6
 A Discovery of the Barmvdas, 97, 245 n.23

Kanneh, Kadiatu, 43, 45
Keane, Patrick J., 242 n.28
Kiberd, Declan, 62
Kicoughtan (Kiccotan), 82–3
Kidwell, Clara Sue, 254 n.7

Kincaid, Jamaica, 23, 24, 26, 28, 30, 69–70, 77–9, 238 n.33
Kipling, Rudyard, 242 n.32
Kirshenblatt-Gimblett, Barbara, 256 n.3
Knapp, Jeffrey, *An Empire Nowhere*, 3, 9, 238 n.37
Kupperman, Karen Ordahl, 236 n.21, 248 n.20

Lane, Ralph, 7–8, 233 n.5
Lanier, Douglas, 252 n.24
Leacock, Eleanor, 255 n.10
Lefroy, J.H., 237 n.32, 245 n.20, n.22
Leman, John, 161
Lemly, Jon, 144
Lever, J.W., 47, 51, 55
Lewalski, Barbara Kiefer, 255 n.12
Limon, Jerzy, 177
Lindley, David, 252 n.24, n.28
livery companies, 24, 33, 149, 152–3
Lluellen, 171
Loewenstein, Joseph, 195–6, 254 n.1, n.2
London stage, 33, 112, 113–48
   see also Lord Mayor's pageants; theater
Lord Mayor's pageants, 31, 33, 141, 149–65, 185, 210, 250 n. 1, 2, 7, 252 n.32
lotteries, 33, 96, 115, 124–5, 127, 138–40, 141, 155, 248 n.16, 17, 249 n.33
Lowe, Lisa, 251 n.8
Lytle, Clifford, 256 n.9

Mac an Bhaird, Fearghal Óg, 60
MacArthur, Janet H., 240 n.5
MacCarthy-Morrogh, Michael, 44, 55, 85, 240 n.13
Mackenthun, Gesa, 237 n.23, 238 n.34, 240 n.11
Mág Uidhir, (Maguire) Cú Chonnacht, Lord of Fermanagh, 60, 72
Maguire, Hugh, (son of Cú Chonnacht), 43–4, 60–1
Maley, Willy, 233 n.2, 246 n.1
Mandeville, John, 115–16, 128–32, 135, 136
   Mandeville's Travels, 129, 242 n.25
maps, mapping, 3–4, 13, 32, 69, 76, 84–93, 94, 99–112, 145, 153, 154
   see also naming
Marcus, Leah, 233 n.4, 252 n.26

Marotti, Arthur, 47, 240 n.6
Marston, John, 116–18
   Eastward Hoe, 116–20, 124–7, 128, 137, 141, 143, 145
Martin, John, 236 n.16
Martyr, Peter, *Decades of the New World*, 42, 234 n.9
Martz, Louis, 47
Maske of Flowers, 177–8
Mason, Peter, 248 n.21, 253 n.44, 255 n.13, n.14
masques, 165–90
   see also Jonson, Ben; Smith, John
"massacre," 1622 *see* 1622 rebellion
Masten, Jeffrey, 247 n.8
Mathes, Valerie Shirer, 254 n.7
McCanles, Michael, 249 n.28
McClintock, Anne, 239 n.45
McLeod, Bruce, 233 n.6, 250 n.3, 251 n.18
McManus, Barbara, 253 n.43
McMullan, Gordon, 113–14
McSheehy, Rory, 55
Means, Russell, 214
Memmi, Albert, 204
Mercator, Gerard, 89, 250 n.3
merchants, 10, 79–80, 125, 149–64, 239 n.43, 246 n.27, 251 n.7, n.14, 254 n.4
metropole, 3–14, 30, 33, 94, 120–2, 124–7, 134–5
"mews," 63
Middleton, Thomas (grocer), 158, 251 n.14
Middleton, Thomas
   The Inner Temple Masque or Masque of Heros, 174
   The Triumphs of Loue and Antiquity, 155–6
   The Triumphs of Honor and Vertue, 156–7, 159–60, 161–3
   The Triumphs of Truth, 158–9, 250 n.4
Miller, Shannon, 239 n.3, 240 n.7
Milton, John, 51
Monacan, 4, 233 n.7
Montrose, Louis Adrian, 38, 41, 239 n.39, 240 n.10
Moore, Richard, 96–7, 103–4, 107–8, 110, 111–12
More, Sir Thomas, *Utopia*, 116–17, 118, 247 n.9
Morgan, Edmund S., 237 n.29, 249 n.39

Morton, Thomas, *New English Canaan*, 42
Moryson, Fynes, 68–69, 110, 123, 168–9, 241 n.22
   An Itinerary Containing His Ten yeeres Travell, 64, 70–74
Mullaney, Steven, 249 n.29
Munday, Anthony
   Camp-bell, 160
   Chruso-thriambos, 161
   Chrysanaleia, 161
   London's Love, to the Royal Prince Henrie, 170
   The Trvumphs of re-vnited Britania, 155, 156
Munster, 3–4, 28, 38–9, 44–7, 64, 75–6, 85–6, 241 n.22, 242 n.30
Murphy, Andrew, 27, 152, 246 n.1, 250 n.2, 252 n.24
Murray, David, 237 n.31, 255 n.21

naming
   in Bermuda, 93–7, 110–12
   in Ireland, 70–4, 76–7, 78, 85
   Kincaid on, 77–8, 79
   in Virginia, 76, 79–84, 85, 86–93
   see also maps, mapping
Namontack, 145–6
Nandy, Ashis, 16, 71–2, 242 n.28, n.32
Nansemond, 4, 233 n.7
Naunton, Sir Robert, 20
nation, English, 8, 30, 31, 67, 157, 170
Needham, Sir Robert, 43
Nelson, Ann Thrift, 211
Nelson, Horatio, 77–8, 243 n.6
New England (US), 25–6, 42, 61, 80, 99, 111, 157–8, 191, 211, 238 n. 35
New English, in Ireland, 4, 6, 21–2, 39, 45, 61, 70–1, 74, 112, 251 n.19
New Historicists, 26, 38
Newman, Karen, 253 n.42, 255 n.14
Newport, Captain, 93, 145, 201, 237 n.30
Newton, Richard, 254 n.1
Nobles, Gregory H., 244 n.13, 246 n.24
Norbrook, David, 41
Norburne, Nicholas, 91
Norris, Thomas, 55
Norton, Robert, 199
Norwood, Richard, 13, 32, 99–106, 110–12, 153, 246 n.26

"A Note of the Shipping, Men, and Provisions, sent and Provided for Virginia," 234 n.8
Nuce, Sir William, 9

Oberg, Michael Leroy, 249 n.35
O'Brien, Terence H., 248 n.17
Ó Cléirigh, An Giolla Riabhach, 60
Ó Huiginn, Tadhg Dall, 60, 241 n.25
Ó hEoghusa, Eochaidh, 60–1, 241–2 n.25
Old English, 4, 6, 28, 30, 44, 48, 63–5, 67, 74, 167, 242 n.31, 251 n.19
O'Neill, Hugh (second earl of Tyrone), 43, 55, 75–7
O'Neill, Shane (first earl of Tyrone), 55, 68, 76–7, 78
Opechancanough, 19–23, 235 n.12
Orgel, Stephen, 172, 181
   The Illusion of Power, 189–90, 198–9, 239 n.4, 250 n.5, 253 n.43, 254 n.1
O'Sullivan, Philip, 43–4
otherness, 3, 40–1, 43, 50, 63–4, 115, 120, 123, 131, 134–8, 148–52, 161, 169, 174–5, 181
   see also antiselves
Ovid, *Metamorphoses*, 1, 4, 121, 239 n.39

pageants, *see* Lord Mayor's pageants
Pagett, Lord, 103
Paine, Henry, 107–9, 227
Parker, Patricia, 239 n.45
Parry, Graham, 252 n.34
Paster, Gail Kern, 248 n.19
Patawomeke, 7–9, 233 n.7
Pearl, Sara, 253 n.41
Pemberton, Sir James, 161
Percy, George, 22, 75, 82, 89
Perquin, Francis, 140
Peru, 74–5
Petrarch, Petrarchism, 31, 37, 40, 47–8, 51, 53, 59
Phettiplace, William, 196
Philip III, King of Spain, 27, 189
Phillips, Sir Thomas, 153–55, 165
pirates, 13, 15, 129, 136, 228, 243 n.6
Pocahontas, 18, 21, 22, 34, 144–8, 173, 183, 197–201, 206, 208, 222, 228, 236 n.18
Pocahontas (film), 225

Pope, Peter, 140–43
Pory, John, 10–11, 22
postcolonial studies, 24, 28, 32, 68–9, 116, 152, 225, 235 n.14
Potts, John, 8, 9
Powell, William, 81, 140
Powhatan (chief), 19, 20, 21, 22, 89, 91, 145, 173, 178, 197–203, 209, 234–5 n.12
Powhatan Indians, 4, 16, 21, 22, 34, 80–1, 83, 86, 89, 182–3, 191–3, 201–18, 222–32
Powhatan (river), 89, 91
Pratt, Mary Louise, 116, 130, 146–7
Prescott, Anne Lake, 54–5, 241 n.19
Pring, Martin, 141
Proby, Peter, 156–7, 163
    A Publication by the Counsell of Virginea, touching the plantation there, 126
Purchas, Samuel
    Purchas his Pilgrimage, 91, 98, 129, 178, 196–7, 210, 234 n.9, 243 n.3, 254 n.4, n.5

"Queen of Appamatuck," 203, 205
Quinn, David Beers, 44, 85
Quinn, Naomi, 254 n.7
Quitt, Martin H., 255 n.16. n.24

racial difference, 174
Rackin, Phyllis, 169, 249 n.37, 253 n.43, 255 n.12
Raleigh, Sir Walter, 7, 42, 45, 46, 116, 119, 177, 228
    Discoverie of the Large Rich and Bewtiful Empyre of Guiana, 117, 240 n.10, 254 n.50
Raman, Shankar, 250 n.2
Rambuss, Richard, 240 n.7, n.15
Randall, Dale B.J., *Jonson's Gypsies Unmasked*, 184–5, 187
Rapp, Rayna, 204
rebellion, 14–15, 170–1
    in Ireland, 43–6, 57, 71, 75–7, 85, 166–8, 241 n.22
    1622, 9, 17–18, 19, 23, 83, 86, 92, 133, 184
Reynolds, Sir Carey, 164

Rich, Barnaby, 71, 74
Rich, Sir Nathaniel, 12–15, 102
Rich, Robert, 103, 109
Rich, Robert (second Earl of Warwick), 109
Richards, Shaun, 242 n.27
Roach, Joseph, 238 n.36
Roanoke colony, 7–8, 233 n.5, 235 n.12
Robertson, Karen, 249 n.41
Robinson, W. Still, Jr., 81–2, 244 n.15
Rodney, George Brydges, 77–8, 243 n.6
Rolfe, John, 18–19, 178, 230
Rountree, Helen, 80–1, 233 n.7, 235 n.12, 243 n.9, 255 n.9
Rubruqis, William, *Iterarium*, 209–10
Rundstrom, Robert A., 243 n.2

Said, Edward, 255 n.18
Sale, Kirkpatrick, *The Conquest of Paradise*, 183, 254 n.6, n.8, 255 n.16
Sanders, Julie, 134, 249 n.30
Sandys, Sir Edwin, 9, 14, 17–20, 25, 81, 102, 239 n.44, 246 n.27
Sandys, George, 1, 9, 15, 121–3, 239 n.39
Savage, Thomas, 22
"savages," savagery, 3–4, 9–10, 33, 47–8, 84, 115, 120–4, 141–4, 155–6, 208
Saxey, William, 44–5
Sayre, Gordon M., 234 n. 11, 237 n.23, 243 n.5
Scanlon, Thomas, 28
Schmidt, Benjamin, 243 n.5
Seaventure, 93, 97
Seed, Patricia, 243 n.5, 245 n.19
sermons, 26
    see also Copland, Patrick; Crashaw, William, Donne, John
Shakespeare, William, 1, 2–3, 14, 83, 233 n.4
    Plays: *1 Henry IV,* 169
        2 Henry IV, 129
        Henry V, 11, 94, 113, 246 n.1
        King Lear, 209
        Macbeth, 206
        Measure for Measure, 255 n.20
        The Merchant of Venice, 250 n.4
        Othello, 9

The Tempest, 120, 141
Two Gentlemen of Verona, 1, 14–15, 19–20, 23, 27, 29
sonnets, 152
Shakespeare in Love (film), 256 n.8
Shirley, James, *The Triumph of Peace*, 250 n.4
Siddiqi, Yumna, 250 n.2, 251 n.15, 252 n.31
Sidney, Sir Philip, 2, 7, 27, 71, 233 n.5
Astrophel and Stella, 41, 52, 240 n.16
Sigalas, Joseph, 126, 247 n.9
Simpson, Evelyn, 133, 247 n.10, 248 n.13, 250–1 n.7, 251 n.16, 252 n.25
Simpson, Percy, 133, 247 n.10, 248 n.13, 250–1 n.7, 251 n.16, 252 n.25
1622 Powhatan rebellion, 9, 17–18, 19, 23, 83, 86, 92, 133, 184
slave trade, 29, 78, 157–8, 222, 239 n.43, 239 n.44
Smith, James M., 251 n.17, 19, 252 n.24
Smith, John, 15, 22, 34, 83, 94–5, 127, 145, 173, 182, 195, 201–19, 221, 224, 226–8, 230–2
Generall Historie of Virginia, New-England, and the Summer Isles, 22, 34, 98, 191–9, 208–16
Map of Virginia, 89, 195, 212
A True Relation of such occurences and accidents of note, as hath hapned in Virginia, 203
The True Travels, Adventures, and Observations of Captaine John Smith, 209, 210, 211
A Virginia Maske, 34, 191–219
Smith, Sir Thomas, 10, 14, 17, 153
Smithfield, 120, 125, 130–6, 199–200, 248 n.15
Sommer Islands, *see* Bermudas
Sommer Island Company, 101–2, 153
Sommers, Sir George, 5, 12–13, 93–9, 105–9, 111
sonnet sequences, 51–3, 240 n.16
Spain, Spanish, 4, 5, 13–14, 27, 31, 57, 69, 74–6, 86, 93, 95–7, 101, 104, 111, 132–3, 185–6, 221, 234 n.9, 234–5 n.12, 238 n.37, 245 n.20, 249 n.31, 249 n.40, 251 n.12

Speed, John, 99, 104–5, 245 n.18
Spelman, Henry, 7, 9, 22–3, 187, 235 n.12
Spenser, Edmund, 2–3, 14, 15, 23, 67–9, 70–1, 73–4, 114, 151, 152, 165–8, 179, 187, 205
Amoretti, 27, 31–2, 37–65, 151
Colin Clouts Come Home Againe, 45, 46–7, 54
Epithalamion, 32, 41, 45, 46, 62–4, 65
The Faerie Queene, 2–3, 31, 37–40, 47–8, 49–52, 53–4, 57, 59, 62, 63–4, 179
A View of the Present State of Ireland, 31–2, 40, 48–9, 52–3, 56–61, 63–4, 67
see also Ireland
Spenser, Edmund (grandson), 65
Spenser, Hugolin, 65
Spenser, Peregrine, 65
Spenser, Sylvanus, 65
Spenser, William, 65
Spurr, David, 67–8
Stafford, Sir T., 63
stage, London, 33, 112, 113–48
see also Lord Mayor's pageants; theater
Stallybrass, Peter, 83, 108, 240 n.9, 252 n.24, 254 n.1, 255 n.14
Stoler, Ann Laura, 5, 10, 242 n.29
Strachey, William, 5, 42, 82–3, 89–99, 105–9, 123, 203, 212–13, 215, 234 n.9, 236 n.22, 237 n.29, 239 n.41
Lawes Divine, Morall and Martiall, 21, 30
Sturtevant, William C., 249 n.34, 253 n.44
Suleri, Sara, 251 n.8
Summers, Sir George, *see* Sommers, Sir George
sumptuary laws, 10, 11, 253 n.43
Swift, Jonathan, *Injured Lady*, 62
Symonds, William, 42–3

Taussig, Michael, 254 n.49, 255 n.13
theater, English, 114, 118, 121, 124–48, 188–9, 207, 216
see also London stage
Thomas, Nicholas, 151
Thompson, Charlotte, 240 n.8
Tilden, Freeman, 230–1, 232
Tindall, Robert, "Draughte of Virginia," 87–9, 103

tobacco, 8–9, 12, 23, 104, 120, 126, 131, 136, 157, 161, 177–8, 180, 184, 228, 238 n.37, 248 n. 24, 251 n.9
Todkill, Anas, 196
toponyms, 69, 74–5, 80, 83, 91, 98–9, 102, 105–6, 110, 243 n.2
    see also maps; naming
Townshend, Aurelian
    Albion's Triumph, 180
    Tempe Restord, 179
translation, 7, 19, 22–3, 60–1, 72, 121–2, 212–13, 217
treasure, colonial, 108, 132, 145–8, 150, 161–2, 180–1
triumphs, *see* Lord Mayor's pageants
Tucker, Daniel, 101–4, 109, 111, 245 n.22, 246 n.25, 246 n.26
Turnball, David, 104
Turner, E. Randolph, 244 n.11
Tyrone, 77
    Earl of, *see* Maguire, Hugh

Ulster, 24, 64, 68, 153–5, 163, 165–7, 242 n.30
Rebellion, 43, 55, 75–6, 166
undertakers, 38, 44–5, 64, 241 n.22

Verner, Coolie, 87
Villeponteaux, Mary A., 47–8
Villiers, George, Duke of Buckingham, 184, 254 n.7
Virginia Company, 5, 6, 9–12, 14–21, 25, 27–9, 42, 80–4, 86–92, 97, 99, 119–22, 124–6, 138–43, 144, 145, 146, 152–3, 164–5, 178–9, 183, 184, 189, 191, 200–1
    lotteries, 33, 96, 115, 124–5, 127, 138–40, 141, 155, 248 n.16, 17, 249 n.33
Virginia Council (London), 9, 11, 22, 24–5, 30, 82, 83, 87, 123

Virginia Generall Assembly, 19–22, 82, 83
Vizenor, Gerald, 116, 124, 143

Wales, Welsh, 3, 25, 33, 169–71, 252 n.27
    see also Jonson, Ben, *For the Honour of Wales*
Waller, Gary, 53, 240 n.15
Warwick, Earl of, 14, 109, 246 n.27
Waterhouse, Edward, 86, 91–2, 133, 236 n.20
Wayne, Don E., 135, 195
Webster, John
    The Duchess of Malfi, 114
    Monuments of Honor, 160–1
Weimann, Robert, 249 n.37
Wells, Robin Headlam, 40, 240 n.8
Wells, Susan, 249 n.28
West, Thomas, Lord De La Warre, 24
White, Allon, 254 n.1
White, John, 140, 249 n.34
whiteness, 29–31, 33, 149–51, 159–62, 174–80, 185–9, 192, 214, 226, 231–2, 252–3 n.3, 253 n.5
Whittaker, Jabez, 25
Wiffin, Richard, 196
Williams, Raymond, *Keywords*, 172
Williamsburg, Virginia, 221, 224, 230
witches, 169, 181, 192, 206–10, 217, 255 n.14
Woodbridge, Linda, 253 n.43
Wright, Nancy, 250–1 n.7
Wright, Ruth, 249 n.34
Wyatt, Francis, 8–9, 84–5
Wynston, Thomas, 236 n.20

Yeardley, George (Yardly ch. 2), 19, 22, 79–80, 249 n.32
Young, Robert J.C., 251 n.11

Zuñiga, Don Pedro de, 27
Žižek, Slavov, 34

Printed in the United States
By Bookmasters